D0765521

AN INTRODUCTION TO CRYSTAL CHEMISTRY

AN INTRODUCTION TO CRYSTAL CHEMISTRY

BY

R. C. EVANS

Fellow of St Catharine's College and Lecturer in the Department of Mineralogy and Petrology University of Cambridge

SECOND EDITION

CAMBRIDGE
AT THE UNIVERSITY PRESS
1964

PUBLISHED BY
THE SYNDICS OF THE CAMBRIDGE UNIVERSITY PRESS

Bentley House, 200 Euston Road, London N.W. 1
American Branch: 32 East 57th Street, New York 22, N.Y.
West African Office: P.O. Box 33, Ibadan, Nigeria

First edition	1939
Reprinted with corrections	1946
	1948
	1952
Second edition	1964

Printed in Great Britain at the University Printing House, Cambridge
(Brooke Crutchley, University Printer)

To
my wife

CONTENTS

PART II
SYSTEMATIC CRYSTAL CHEMISTRY

SOME IMPORTANT FIGURES

SOME IMPORTANT TABLES

PREFACE

In the years which have elapsed since the first edition of this work was published in 1939 X-ray diffraction has become firmly established as the most powerful tool at our disposal for the study of the solid state. Twenty-five years ago enough structures were known for the general principles of crystal architecture to be understood; to-day the number of known structures has been multiplied many times and these principles have taken their rightful place as an integral part of modern structural chemistry. Even so, it is only at an advanced level that the integration is as yet complete, and it is in any case not only the chemist to whom the study of solids is of interest: the physicist, the metallurgist, the mineralogist and many others must equally be concerned with the relationship between properties of matter and atomic arrangement. For this reason there still remains a need for a presentation of the fundamental principles of crystal chemistry at an elementary level, with the minimum of descriptive detail, and it is this need which the present work seeks to satisfy. Although the new edition has been completely rewritten, and although the structures described have been reinterpreted in terms of more modern ideas of chemical bonding, an attempt has nevertheless been made to avoid undue expansion and to preserve the essentially elementary character of the original. Only those structures have been described which seem best suited to illustrate the principles here advanced, and these principles have been presented in a manner which demands little specialized crystallographic knowledge on the part of the reader.

The value of diagrams as a means of illustrating structural principles needs no emphasis. Much care has therefore been devoted to the preparation of the many figures, all of which have been specially drawn, and it is a pleasure to express my thanks to the artists of the Cambridge University Press for the skill with which they have executed this work. References to original literature have been omitted as no longer appropriate now that the needs of the more advanced student are so admirably met by other works, some of which are cited in the bibliography.

The greater part of the task of preparing this edition was undertaken during a period of leave of absence from Cambridge University generously granted to me by the General Board of the University, the

Faculty Board of Geography and Geology, and the Governing Body of St Catharine's College. Part of the time was spent at the National Bureau of Standards in Washington, and it is a pleasure to express my thanks to Dr I. C. Schoonover, Mr H. S. Peiser and their many colleagues for all that they did to make my visit a most stimulating and memorable experience. I am equally indebted to Professor H. G. F. Winkler for his hospitality and for the facilities he placed at my disposal during the months which I spent at his laboratory in the University of Marburg.

No reader of this book can fail to recognize how much I owe to others who have written in this field. To the authors of the works quoted in the bibliography, and not least to the selfless editors of the reference works, I express my warm thanks. I am also indebted to many friends (too numerous to mention individually) whose counsels I sought and whose opinions and advice, so generously given, have largely determined the scope and character of what I have written.

Finally, but above all, I am indebted to my wife, but for whose encouragement this book might never have been written. From the day we first discussed the project (at sea off the coast of California) I have been sustained by her interest and support and by her acceptance, with characteristic cheerfulness, of the sacrifices involved. As a token of what I owe to her inspiration this work is dedicated to her with love.

R. C. E.

CAMBRIDGE
September 1963

PART I

GENERAL PRINCIPLES OF CRYSTAL ARCHITECTURE

1 INTRODUCTION

EARLY HISTORY

1.01. Crystals have been known to man from the earliest times, for, as we now realize, the rocks of the Earth's crust, the sands of the desert and, indeed, almost all forms of solid matter are crystalline in character. The Egyptians who operated turquoise mines in the Sinai peninsula, possibly as early as the sixth millenium before Christ, must have been aware of the beauty and geometrical perfection of many naturally occurring minerals, and Theophrastus, in his treatise *On Stones* published in the fourth century B.C., describes the angular shape and regular form of crystals of garnet. Rock crystal or quartz from India was studied by the geographer Strabo (b. 64 B.C.), who, impressed by its resemblance to ice, applied to it the term *κρύσταλλος* from which our 'crystal' is derived, and the same mineral was later described by the elder Pliny (A.D. 23–79) in his *Natural History*. References to diamond, sapphire and many other gem stones are numerous throughout the Bible, and by the first century of the Christian era jewels were so prized that we find contemporary Egyptian writers giving recipes for the preparation of artificial stones.

Artificially prepared crystals were formed in many of the processes of alchemy, and Libavius in 1597 was the first to recognize that the geometrical habit of these crystals was often characteristic of the salts concerned. In the next century crystals were described by Boyle, Leeuwenhoek and others, but it was Steno, distinguished also as a physiologist and later as Bishop of Titiopolis, who made the most important contribution by remarking in 1669 that quartz crystals, whatever their origin or habit, always preserved the same characteristic interfacial angles. This observation was extended to other crystals by Guglielmini during the period 1688–1705, and was further extended by de l'Isle in 1772. A few years later, the invention of the contact goniometer by Carangeot in 1780 provided the means for a more precise study of crystal form than had previously been possible, and in the years immediately following this invention a great mass of crystallographic data appeared. As early as 1783 de l'Isle published four volumes incorporating his observations on over four hundred mineral specimens, observation which added precision to his earlier studies and enabled him to formulate the law of constancy of interfacial angle. A year later, in

1784, Haüy enunciated the laws governing crystal symmetry, and paved the way for his later discovery of the law of rational indices, which, in 1801, he substantiated by a comprehensive survey of the mineral kingdom. Thus by the first year of the nineteenth century the fundamental laws of morphological crystallography had been established.

1.02. The researches of de l'Isle and Haüy may well be said to have laid the foundations of modern crystallography, but it was not until the invention of the reflecting goniometer by Wollaston in 1809 that further progress was possible. The contact goniometer had served to provide the data for early researches on crystal symmetry, but an instrument of an altogether different order of accuracy was required before crystallography could claim the title of an exact science. The application of the reflecting goniometer to crystallographic research resulted in the rapid accumulation of a great wealth of exact experimental data not only on naturally occurring minerals but also on artificial crystals of chemical importance, and it was these data which provided the material for Mitscherlich's discovery of isomorphism in 1819 and of polymorphism in 1822. The work of Mitscherlich at once directed attention to the chemical significance of crystal form and habit, and from 1820 onwards crystallographic research for nearly a hundred years was concerned primarily with the relationship of crystalline form to chemical constitution. Here we can mention only a few of the more important contributions. Pre-eminent stands the work of Pasteur on the enantiomorphism of tartaric acid, of the greatest significance in the later developments of chemical theory. Other researches were concerned with the physical and crystallographic properties of substances chemically closely related, and as early as 1840 Kopp observed that the tendency to form mixed crystals increased with increasing similarity in molecular volume. Later in the century Hiortdahl and Groth showed that in a series of organic compounds systematic substitution brought about a progressive change in crystal form, while important observations on the physical properties of substances chemically closely related are associated with the names Retgers, Liebisch, Gossner, Barker and Tutton. The great volume of crystallographic data on both inorganic and organic compounds which accumulated during the period was tabulated in Groth's monumental *Chemische Krystallographie*. This work, published in the years 1906–19, records morphological, optical and other properties of over 7000 crystalline substances.

1.03. Although the work of the second half of the nineteenth century provided many data which have been of the greatest value to subsequent investigators, it cannot be said to have led to developments in any way comparable in importance with the discoveries of Mitscherlich and Pasteur in the first half of the century. In part this was due to the fact that most of the substances investigated were of far too great chemical complexity, but more particularly was it an inevitable consequence of the fact that external crystal form, closely prescribed by rigid rules of symmetry, can necessarily give but a very limited expression of the internal constitution and structure of a crystal. It was recognized that the regularity of external form must have its origin in some more deep-seated regularity of internal arrangement, and that this regularity of internal arrangement must in its turn determine other properties of crystals. But at the time no experimental means were available for the study of internal structure, and in the absence of experimental verification any conclusions could be nothing more than speculations.

This is not to say that speculations on the internal architecture of crystals had not taken place. In 1611 Kepler suggested that the characteristic hexagonal symmetry of snowflakes was due to a regular packing of the constituent particles. This idea was also developed by Descartes in 1637 and by Hooke, who in 1665 remarked in his *Micrographica* that all the crystal forms which he had studied could be simulated by building piles of 'bullets or globules'. A century later, Bergman in 1770 and Haüy in 1784, as a consequence of studies on cleavage, envisaged calcite crystals, of whatever habit, as built up by the packing together of 'constituent molecules' in the form of minute rhombohedral units. In this way Haüy was able to account for the law of rational indices, to which we have already referred above; but such a viewpoint was soon shown to be an essentially artificial one by the consideration that many crystals have no well developed cleavages, while others have cleavage forms (such as the octahedron) which cannot be assembled to fill space. It was not until the present century that Barlow and Pope (1906–7) gave precision to the ideas of crystal structure by assuming the constituent units to be *atoms* and by regarding a crystal as an essentially geometrical entity formed by the packing of spherical atoms, each of a definite and characteristic size. These speculations, as we shall see later, were extraordinarily close to the truth; but still the means were lacking for their experimental confirmation.

THE TWENTIETH CENTURY

1.04. In 1912 Friedrich, Knipping and Laue discovered the diffraction of X-rays by a crystal, thus *proving* for the first time the regular and periodic arrangement of the atoms in a crystal structure. At first the interest of these experiments lay in the light which they threw on the nature of X-rays, but it soon became clear that they were of even greater importance in providing a means not only of establishing the regularity of atomic arrangement in a crystal but also of actually determining just exactly what that arrangement was. Furthermore, as subsequent X-ray studies have shown, this regularity of internal arrangement is a characteristic not only of substances which occur as well-formed crystals but also of metals and many other solids seemingly amorphous in external form. In fact, with very few exceptions, *all* solids are crystalline in this sense, so that on the discovery of X-ray diffraction crystallography may be said to have become the science of the solid state.

If the importance of a discovery is to be measured by the consequences to which it has given rise, that of Friedrich, Knipping and Laue must be ranked as one of the most important in the whole history of science, for it has provided a means of investigating the solid state of a power altogether transcending any previously available. Prior to the development of X-ray methods, the solid state was the least tractable of all the states of matter, and the internal structure of a solid could be deduced only by argument from its physical properties or from its chemical properties in the liquid or gaseous form. X-ray analysis has removed the determination of crystal structure and molecular configuration from the sphere of speculation to that of measurement, and it is not difficult to see that the consequences of such an advance must be of the greatest significance in all branches of chemical theory.

The year following the discovery of X-ray diffraction saw the publication of the first crystal analysis, that of sodium chloride by W. H. and W. L. Bragg, and within a short time many other simple structures had been elucidated. Concurrently, the theory of the diffraction of radiation by a three-dimensional grating was developed, and improved apparatus and new experimental techniques were introduced. By 1926 the number of known structures was sufficiently large for Goldschmidt to be able to formulate the general principles governing the structural architecture of the crystals of simple inorganic compounds: he showed that such structures were the result of the packing together of spherical atoms (or more commonly ions), each of a constant size characteristic of the element concerned, in a manner determined by geometrical considerations, thus confirming the speculations of Barlow and Pope of twenty years before. It was not, however, only inorganic compounds which were studied. The crystal structures of metallic elements and alloys and of many organic compounds were also elucidated, and by the beginning of the Second World War, when the first edition of this book was published, the general principles underlying the structural characteristics of all known types of solid matter had become clear.

Recent progress

1.05. In the years following the war advances in the science of X-ray crystallography have been both numerous and rapid. Experimental equipment has been improved, important developments have taken place in the theory of structure determination, many more workers have interested themselves in this field, several learned journals devoted exclusively to structural studies have been founded, and many more structures, often of great complexity, have been determined. It is true to say, however, that with few exceptions the numerous new structures elucidated have not added notably to the number of known structural types: for the most part they have been further examples of the basic types already well established. But this is not to belittle the work of the modern structural crystallographer. By determining more and more structures with ever higher precision he has demonstrated that in X-ray crystal analysis we have by far the most powerful and precise tool yet available for the study of the solid state. Nevertheless, this tool, powerful though it is, is but one of many available for this study, and the X-ray crystallographer, if he is interested in anything more than the technique of the X-ray method, must also be something of a chemist, a physicist, a metallurgist or a mineralogist. The chemist, to single out but one field of study, is interested in solving his problems by any means at his disposal, and in this connexion the very phrase 'crystal chemistry' is something of an anachronism, implying, as it does, a division of interest where none should exist. We do not speak of 'gas chemistry' or 'liquid chemistry', being quite prepared to absorb the results of studies of these phases into our conspectus of chemistry as a whole. The same should be true of the study of crystal structures; but historically this has not been the case. In the past, the results of X-ray studies have been assimilated but slowly into chemical thought, perhaps because relatively few chemists have enjoyed any crystallographic training, and it is only in the post-war years that theories of valency and of many other properties of the solid state have been developed from a structural basis. The extent of these recent developments is sufficient evidence of the importance of the foundations on which they are based, and all modern accounts of these developments draw heavily, as indeed they must, on the results of structural studies. Nevertheless, it is still true to say that at a more elementary level the integration of structural and other studies is as yet far from complete, for all too often the results of the X-ray analysis of crystals, if discussed at all, are presented in isolation as a discrete topic without any proper consideration of their impact on chemistry as a whole.

The present book is an attempt to rectify this omission. Ideally, as we have just argued, we should present a coherent account of the whole field of chemistry in which the results of structural studies appeared in their rightful place among those of the many other means of determining chemical constitution. This, however, would be a formidable task, and, indeed, an unnecessary one when there already exist so many works on chemistry, valency theory and other aspects of the solid state. We shall therefore presuppose the reader to have some knowledge of general chemical principles, and we shall confine ourselves to a discussion of those properties of solids, directly related to crystal structure, which are not normally considered in detail in chemical works.

2 INTERATOMIC BINDING FORCES AND ATOMIC STRUCTURE

INTRODUCTION

2.01. We have already remarked that the foundations of crystal chemistry may be regarded as having been laid by Goldschmidt in 1926, and that by 1939 the general structural features of all the then known types of solid had been determined. A review of these solids revealed that their crystal structures were primarily a reflexion of the nature of the interatomic forces acting between the constituent atoms: if the nature of these forces was known the structural arrangement could be predicted, while, conversely, if the structural arrangement could be determined the nature of the interatomic forces could be assessed.

Speculations on the nature and origin of the interatomic forces responsible for the coherence and stability of the chemical molecule did not, of course, await the development of X-ray crystal structure analysis, for they date back almost to the foundations of modern chemistry. The electrical character of these forces was suggested tentatively by Desaguliers as early as 1742, and the idea was developed systematically in 1819 by Berzelius, whose views had an important influence on chemistry until the ionic dissociation theory of Arrhenius was introduced later in the century. The publication of Bohr's theory of atomic structure in 1913 provided a physical picture in terms of which interatomic forces could be interpreted, and there grew up the conception of two basically distinct types of force, the ionic, on the one hand, to account for the formation of compounds (such as sodium chloride) containing atoms of essentially different character, and the covalent, on the other hand, to account for the formation of molecules (such as those of chlorine) composed of similar or identical atoms. Theories of these two types of bond were developed almost simultaneously in 1916 by Kossel and by Lewis, respectively, the ionic bond being interpreted as an electrostatic attraction between oppositely charged ions and the covalent bond being attributed to a mutual sharing of electrons between the atoms concerned.

2.02. The development of X-ray crystal structure analysis immediately threw valuable new light on the question of the nature of interatomic forces, for it provided the first means of determining experimentally the atomic configuration of molecules whose form had long been the subject of speculations based on chemical valency theories. One of the earliest, and certainly one of the most important, consequences which followed the first structure analysis was a realization of the fact that no essential distinction exists between the 'chemical' forces responsible for binding together the atoms of the chemical molecule and the 'physical' forces responsible for the coherence of the solid as a whole, and that, in fact, in the great majority of simple compounds the molecule as such has no existence in the solid state. The structural importance of this discovery will be discussed later, but from the point of view of valency theory it was equally important since it opened the way to a physical approach to the interpretation of the chemical bond: physical as well as chemical properties had to be explained in terms of the bond, and, conversely, physical as well as chemical properties threw light on its nature.

This identification of the chemical and the physical bond at once led to the association with the chemist's ionic and covalent bonds of two other types of binding force not previously regarded as chemical in nature at all, namely, the metallic bond responsible for the cohesion of a metal, and the very much weaker van der Waals or residual bond responsible for the crystallization of the inert gases at very low temperatures. These metallic and van der Waals forces, however, did not lend themselves as readily as did the ionic and covalent to any simple explanation in terms of the Bohr theory, and it is only in recent years that the development of quantum mechanics has enabled a qualitative and even quantitative description of these bonds to be given. At the same time quantum theory has furnished a more exact description of the properties of the ionic and covalent bonds, which were previously so successfully described qualitatively in terms of older ideas, so that it is now possible to give a satisfactory theoretical explanation of many of the physical and chemical properties of simple structures.

Interatomic binding forces

2.03. We may summarize the above discussion by saying that a review of the crystal structures of different types of solid emphasizes the dominating role of the interatomic forces in determining the structural arrangement, and that these forces may be conveniently divided into four distinct types:

(1) The *ionic*, *electrovalent*, *heteropolar* or *polar* bond, the force of electrostatic attraction which operates between oppositely charged ions.

(2) The *covalent* or *homopolar* bond, the normal chemical valency bond such as that, say, between the two atoms in the chlorine molecule, or between the carbon and chlorine atoms in the molecule of carbon tetrachloride.

(3) The *metallic* bond, the interatomic force responsible for the cohesion of metal systems in the solid state.

(4) The *van der Waals* or *residual* bond, a weak force of interatomic attraction operating between all atoms and ions in all solids.

Let it be said at once that it is now clear that the distinction between these four bond types is by no means absolute, and that in many crystals the bonds possess an intermediate character, displaying something of the properties of two or more types. We shall return to this point at length later, but initially we shall find it convenient to discuss these bond types in isolation and to illustrate their properties by considering some simple structures in which only bonds possessing little or no intermediate character are found.

2.04. Another important consideration to bear in mind is that in many structures two or more different types of bond may operate simultaneously between different atoms. Thus the molecules in the crystal structures of almost all organic compounds are bound within themselves by strong covalent or 'chemical' bonds whereas they are bound one to another only by much weaker forces, usually of the van der Waals type. Structures in which discrete molecules can be clearly recognized are said to be *molecular*; although they are most common among organic compounds we shall in fact find later that there are many inorganic compounds, and even elements, whose structures are molecular in this sense.

The great majority of inorganic compounds, however, are *non-molecular* in that no molecule can be discerned in the crystal structure. In the structure of sodium chloride, for example, no one sodium atom is associated with one chlorine atom to form a molecule NaCl, while similarly in the structure of calcium carbonate there is no association of calcium atoms and carbonate groups into $CaCO_3$ molecules. Most of these non-molecular structures owe their coherence to ionic or metallic bonds, and when we come to consider the nature of these bonds in detail we shall understand why they do not normally give rise to molecular association.

Because of the broad distinction, just discussed, between the structures of organic compounds on the one hand and inorganic on the other it is easy to understand that the impact of X-ray structural studies has been very different in these two fields of chemistry. In organic chemistry X-ray methods have added precision to the determination of molecular configuration, but usually it has been a configuration already well established on general grounds by other means. In inorganic chemistry, however, X-ray investigations have opened up a field hitherto inaccessible to experimental study, and the results have had a correspondingly profound influence on chemical thought. It is therefore understandable that the greater part of this book should be devoted to a discussion of the structures of inorganic substances.

2.05. All the types of interatomic binding force to which we have referred above are primarily electronic in nature, and the differences between them arise from differences in the electronic structures of the particles concerned. In order that we may be able to understand the origin of these forces and to predict the types of force likely to operate in any particular structure it is therefore necessary that we should have a clear picture of the extranuclear electron distribution in the atoms of the elements, and of the way in which this distribution changes as we pass from one element to another in the Periodic Table. The rest of this chapter is accordingly devoted to a discussion of this topic and will serve as an introduction to the remaining four chapters of Part I, in which the different types of interatomic binding force are considered individually.

THE RUTHERFORD–BOHR ATOM

2.06. It is convenient to start our discussion of atomic structure from the Rutherford–Bohr theory of the atom, as originally advanced by Rutherford to account for the scattering of α particles by matter and as developed by Bohr to explain the emission spectra of the elements. On this theory the atom is pictured as consisting of a nucleus and an extranuclear structure of electrons. The nucleus has a diameter of the order of 10^{-13} cm and is very small compared with the atom as a whole, the diameter of which is about 10^{-8} cm. In spite of its small size, however, the nucleus contributes practically the whole mass of the atom. It carries a positive charge which (measured in terms of the charge of the electron as unit) is equal to the atomic number Z. Surrounding the nucleus, and conferring electrical neutrality on the atom as a whole, is a set of Z

electrons arranged in a series of 'orbits' or 'shells' and moving in these orbits rather as planets round the sun. The orbits, however, differ from those of a planet in that only certain particular orbits are possible. Thus in the case of the hydrogen atom the single extranuclear electron can move only in an orbit of radius r given by

$$r = n^2h^2/4\pi^2me^2,$$

where m is the mass of the electron, e is its charge, h is Planck's constant and n is the *principal quantum number*. This quantum number can assume only the integral values $n = 1, 2, 3, \ldots$, and the radii of the permitted orbits are correspondingly restricted. For $n = 1$, r is 0·532 Å, for $n = 2$, r is 2·128 Å, and so on.* The orbits corresponding to $n = 1, 2, 3, \ldots, 7$ are sometimes described as the $K, L, M, \ldots, Q$ shells, respectively.

On the classical theory an electron travelling in a circular orbit should radiate continuously, thus losing energy and finally falling into the nucleus. Bohr assumed, however, that an electron does not radiate under these conditions, but only when it passes from one orbit to another of lower quantum number. The energy so released, representing the difference in energy of the orbits concerned, appears as monochromatic radiation and constitutes one of the characteristic lines in the emission spectrum of the element; in this way the spectrum of hydrogen was interpreted with remarkable success, although the spectra of other elements could not be so satisfactorily explained.

The most stable state of an atom is that in which all the electrons occupy orbits of the lowest permitted energy, and this is termed the *ground state*. If, however, sufficient energy is supplied an atom may be promoted to an *excited state* in which some of the electrons occupy orbits of higher energy, or in which one or more electrons are removed from the atom altogether; in this latter case the energy required is the *ionization energy*. The excitation energy of an atom is often large compared with that commonly involved in chemical reactions, in which event the atom will occur in the ground state in the crystal structures of its compounds. In many cases, however, the excitation energy is not large, and then we must countenance the possibility that the electronic structure of an atom in a compound will be different from the ground state described below.

* 1 Å = 10^{-8} cm.

Atomic structure

2.07. The atom of hydrogen, of atomic number $Z = 1$, in its ground state has a single electron in the K shell of principal quantum number $n = 1$; the atom of helium ($Z = 2$) has two electrons in this orbit. This, however, is the maximum number of electrons that can be accommodated in the first shell, for it is an essential feature of the Bohr theory that an orbit of principal quantum number n can accept not more than $2n^2$ electrons. Thus the maximum permissible number of electrons in each shell is as follows:

Shell	K	L	M	N	O	P	Q
n	1	2	3	4	5	6	7
Number of electrons	2	8	18	32	50	72	98

In lithium ($Z = 3$) two electrons can be placed in the K shell, but the third must enter the L shell of higher energy. The difference in energy between these shells is immediately reflected in the ionization energies of the lithium atom, 5·36 eV to remove the L electron, a further 75·3 eV to remove the second electron, and a further 121·8 eV to remove the last electron.* Thus the ion Li^+ may be expected to be of common occurrence, whereas the ions Li^{2+} and Li^{3+} will be highly unstable and will readily acquire electrons to revert to a state of lower ionization.

The L shell can accommodate a total of 8 electrons, and as further electrons are added to form the sequence of elements beryllium, boron, carbon, etc., these electrons take their place in the second shell until finally, at the end of the second period, neon ($Z = 10$) is reached, with both the first and second shells fully occupied and with an electronic structure which can be symbolized as (2, 8). The addition of further electrons to form the sequence of elements sodium, magnesium, etc., of the third period requires the formation of a new shell, and in sodium ($Z = 11$) a single electron occupies the M shell of principal quantum number 3; again the ionization energies reflect the difference in energy between this electron and those more tightly bound in the L and K shells.

The third period terminates at argon ($Z = 18$) with the configuration (2, 8, 8), but here a new phenomenon occurs: although the M shell is not yet fully occupied, the further electrons in potassium ($Z = 19$) and calcium ($Z = 20$) enter the N shell. Only after two electrons have taken their place in this shell is the development of the M shell resumed to give the transition elements scandium, titanium, etc.; it is finally com-

* 1 electron volt = $1{\cdot}60 \times 10^{-12}$ erg = 23·06 kcal/mole.

plete in copper ($Z = 29$) with the configuration (2, 8, 18, 1). Similar considerations apply to the later periods. We shall not, however, develop this discussion further at this point for, as we shall shortly see, the application of wave mechanics to the problem of atomic structure demands a revision of the Bohr picture and necessitates a somewhat different description of the electronic structure of the elements, especially in the case of those of higher atomic number.

APPLICATION OF WAVE MECHANICS

2.08. The fundamental conception of wave mechanics is the wave-like nature of the electron, as a result of which we can no longer picture the electron as a physical particle moving in an orbit of definite geometrical form; instead, we can speak only of a probability distribution of electron density. Thus in the case of the single electron in the ground state of the hydrogen atom the circular orbit of the Bohr theory is replaced by a *spherical* probability distribution in which there is a finite probability of finding the electron at indefinitely large distances from the nucleus, but in which the most probable distance proves to be exactly equal to the radius of the Bohr orbit. If we wish to visualize the orbit as having a finite extent we may consider the volume of space in which there is, say, a 95 per cent probability of finding the electron. In this case the orbit becomes a sphere of limited radius. The orbits representing the excited states of the hydrogen atom are spheres of correspondingly larger radii.

Quantum numbers

Principal and subsidiary quantum numbers

2.09. Certain features of the Bohr theory are still appropriate to the wave-mechanical picture: in particular, the orbits can still be grouped in shells characterized by a principal quantum number n, and the maximum permissible number of electrons in any shell is still $2n^2$. Not all the electrons in one shell, however, are identical, for those in a shell of principal quantum number n are distributed over n *sub-shells* characterized by an *azimuthal* or *subsidiary quantum number* l, which can assume any of the values 0, 1, ..., $(n-1)$. Electrons in sub-shells with $l = 0, 1, 2, 3$ are commonly termed s, p, d, f electrons, respectively, and the state of an electron in respect of its principal and subsidiary quantum numbers is usually symbolized by a figure representing n followed by a letter representing l. Thus $1s$ is an electron in the K shell with $l = 0$

(the only permissible value for this shell) while $3d$ is an electron in the M shell with $l = 2$. If it is necessary to refer to the number of electrons in any particular sub-shell this may be indicated by a superscript following the symbol for the sub-shell. For example, $1s^1$ represents a single electron in the s sub-shell of the K shell, $3d^5$ represents five electrons in the d sub-shell of the M shell, and so on.

Electrons in the same sub-shell have very nearly the same energy, but the energies of those in different sub-shells of the same shell, say the $3p$ and $3d$ electrons, are appreciably different. Moreover, the electrons in different sub-shells differ also in that the probability distributions by which they are represented are different in shape, as we shall shortly see.

Magnetic quantum number

2.10. The electrons in each sub-shell are distributed in *atomic orbitals*, the number of which is determined by the *magnetic quantum number m*. For a give value of l, this quantum number can assume the $(2l+1)$ possible values $-l, \ldots, 0, \ldots, +l$, so that the number of orbitals associated with each sub-shell is as follows:

Sub-shell	*s*	*p*	*d*	*f*
l	0	1	2	3
Number of orbitals	1	3	5	7

We shall not normally have occasion to refer to the numerical value of m but it is, nevertheless, essential to keep in mind the number of orbitals in each sub-shell to which it gives rise.

Spin quantum number

2.11. Finally, each orbital can accommodate only two electrons, and this only if they have oppositely directed spins characterized by the values $+\frac{1}{2}$ and $-\frac{1}{2}$ for s, the *spin quantum number*. Since any orbital is uniquely described by the values of n, l and m, any electron is uniquely described by the value of n, l, m and s, and no two electrons in an atom can have the same values for all four quantum numbers. This is the formal expression of Pauli's *exclusion principle*. Again, we shall not normally have occasion to refer to the numerical value of s; it will be sufficient to remember that if two electrons occupy the same atomic orbital they must have opposite spins.

2.12. We may summarize the above discussion by remarking that the state of an electron in the extranuclear structure of an atom can be described by four quantum numbers, not all of which can be the same for any two electrons:

(1) The *principal quantum number*, n. This determines the shell. A shell of principal quantum number n can accommodate $2n^2$ electrons.

(2) The *subsidiary quantum number*, l. This determines the sub-shell within a given shell. For a shell of principal quantum number n there are n sub-shells corresponding to values of l from 0 to $(n-1)$.

(3) The *magnetic quantum number*, m. This determines the atomic orbital within a given sub-shell. For a sub-shell of subsidiary quantum number l there are $(2l+1)$ orbitals.

(4) The *spin quantum number*, s. This determines the sense of the electron spin within a given orbital. For each orbital there are only two possible values, $+\frac{1}{2}$ or $-\frac{1}{2}$.

These conclusions are presented in an alternative form in table 2.01. For the K, L, M and N shells all the associated sub-shells and orbitals are tabulated, and also the maximum permissible number of electrons in each sub-shell and shell. For the O, P and Q shells only certain of the permissible sub-shells are included since the others, of higher subsidiary

Table 2.01. *Electron distribution as a function of principal and subsidiary quantum numbers*

Shell	Sub-shell	Number of orbitals $(=2l+1)$	Maximum number of electrons in sub-shell $(=2(2l+1))$	Maximum number of electrons in shell $(=2n^2)$
K $(n=1)$	$1s$ $(l=0)$	1	2	2
L $(n=2)$	$2s$ $(l=0)$	1	2	8
	$2p$ $(l=1)$	3	6	
M $(n=3)$	$3s$ $(l=0)$	1	2	18
	$3p$ $(l=1)$	3	6	
	$3d$ $(l=2)$	5	10	
N $(n=4)$	$4s$ $(l=0)$	1	2	32
	$4p$ $(l=1)$	3	6	
	$4d$ $(l=2)$	5	10	
	$4f$ $(l=3)$	7	14	
O $(n=5)$	$5s$ $(l=0)$	1	2	(50)
	$5p$ $(l=1)$	3	6	
	$5d$ $(l=2)$	5	10	
	$5f$ $(l=3)$	7	14	
P $(n=6)$	$6s$ $(l=0)$	1	2	(72)
	$6p$ $(l=1)$	3	6	
	$6d$ $(l=2)$	5	10	
Q $(n=7)$	$7s$ $(l=0)$	1	2	(98)

quantum numbers, are not found in the ground state of any known atom up to $Z = 102$; in other words, no atom is known in which the *O*, *P* or *Q* shell is fully occupied. We shall return to this point later.

The shapes of orbitals

2.13. We have already explained that the single orbital of the *s* sub-shell, of subsidiary quantum number $l = 0$, corresponds to a distribution of electron density which is spherical in the sense that there is the same probability of finding an electron in any direction in space (fig. 2.01 *a*).

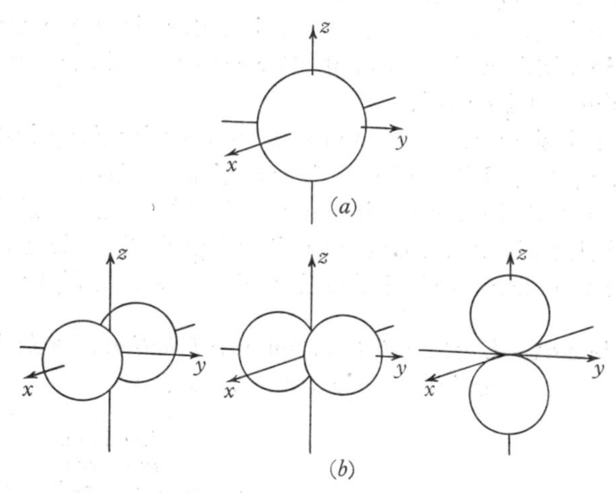

Fig. 2.01. The shapes of (*a*) the *s* and (*b*) the *p* atomic orbitals.

This, however, is not true of the *p*, *d* and *f* sub-shells. Thus each of the orbitals of the *p* sub-shell ($l = 1$) is a dumb-bell shaped figure, and these figures for the three possible orbitals ($m = -1$, $m = 0$, $m = 1$) are arranged with their axes mutually at right angles (fig. 2.01 *b*); if we wish to distinguish between them we may designate the orbitals p_x, p_y and p_z, respectively. As with the *s* orbitals, the figure is not intended to imply that the electrons are strictly confined to the volume shown, for there is always a finite probability of finding an electron indefinitely remote from the nucleus. The length of any radius vector in the figure is, however, a measure of the probability of finding an electron in the direction of that vector, so that for any one of the *p* orbitals the electron is more likely to be found in the direction of the axis than in any other direction. The probability of the electron being found in the nodal plane perpendicular to the axis is zero.

The five d and seven f orbitals are also unsymmetrical figures. Their exact form is a matter of less immediate concern to us than the fact that, in common with the p orbitals, they represent a *directed* distribution of electron density. As we shall shortly see, this spatial direction of the orbitals plays a very important part when we come to consider the inter-atomic binding forces between atoms, and we must remember that it is determined by the subsidiary quantum number l.

The energy levels of orbitals

2.14. We have now to consider the distribution of the electrons between the various possible atomic orbitals in the ground state of the atoms of the elements. Subject to the requirements of the exclusion principle, that no two electrons shall have the same values for all four quantum numbers, the electrons will naturally occupy the states of lowest potential energy. We must therefore consider first the relative energy levels of the different atomic orbitals. These are not independent of atomic number, but vary with Z in a manner represented qualitatively in fig. 2.02.

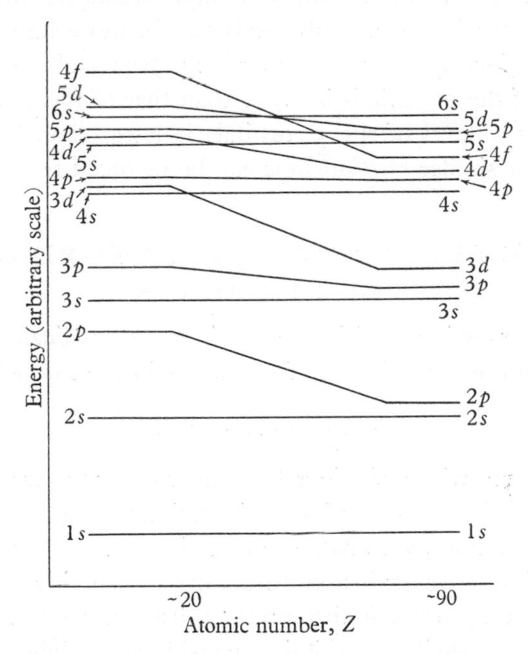

Fig. 2.02. Schematic representation of the variation of the energy levels of atomic orbitals as a function of atomic number.

Certain general points emerge from a study of this figure:

(1) For a given principal and subsidiary quantum number the energy is the same for all the possible atomic orbitals. Thus the energy of the three *p* orbitals is the same for any one value of *n*; similarly the energy of the five *d* orbitals or of the seven *f* orbitals is the same. (This is implicit in the figure only inasmuch as the different orbitals with the same principal and subsidiary quantum numbers are represented by a single line.)

(2) For a given subsidiary quantum number, the energy for any value of *Z* increases with increasing principal quantum number. Thus the energy increases in the sequence of orbitals 1*s*, 2*s*, 3*s*, 4*s*, 5*s*, 6*s*, 7*s*; similarly it increases in the sequences 2*p*, 3*p*, 4*p*, 5*p*, 6*p*; 3*d*, 4*d*, 5*d*, 6*d*; 4*f*, 5*f*.

(3) For a given principal quantum number the energy for any value of *Z* increases with increasing subsidiary quantum number. Thus the energy increases in the sequences of orbitals 2*s*, 2*p*; 3*s*, 3*p*, 3*d*; 4*s*, 4*p*, 4*d*, 4*f*, etc. Moreover, the energy differences between sub-shells of the same principal quantum number may be quite large, and they vary considerably as between elements of low and high atomic number. Thus for certain atomic numbers some sub-shells may be more stable than others of lower principal quantum number, e.g. below about $Z = 20$ the energy level of the 4*s* orbitals is lower than that of the 3*d* orbitals, and again the energy level of the 5*s* orbitals is lower than that of the 4*d* orbitals. Other similar relationships can be deduced from the figure.

THE ELECTRONIC STRUCTURE OF THE ELEMENTS

2.15. We are now in a position to consider systematically the electron distribution in the ground state of the atoms in the Periodic Table.

The first two periods

2.16. Hydrogen, with a single extranuclear electron, has this electron in the *s* orbital of the *K* shell; its configuration may therefore be represented as $1s^1$. Helium, with two electrons, has both in the *s* orbital, but of course with opposite spins, and thus has the configuration $1s^2$. The *K* shell is now full, so that in lithium the third electron must occupy the *s* orbital of the *L* shell, an orbital of considerably higher energy; this element thus has the configuration $1s^2$, $2s^1$. Beryllium, with four electrons, has the configuration $1s^2$, $2s^2$ and the s orbitals of both K and L shells are now fully occupied, in each case, of course, by two electrons of opposite spin. Examination of fig. 2.02 shows that in boron the fifth electron must occupy one of the three $2p$ orbitals, say $2p_x$, to give the configuration $1s^2$, $2s^2 2p_x^1$. The question now arises as to whether a sixth electron, in the carbon atom, will enter the same $2p$ orbital, i.e. p_x, or one of the two as yet unoccupied orbitals $2p_y$ and $2p_z$. The answer to this question is given by Hund's *rule of maximum multiplicity*, namely that as many of the orbitals as possible are occupied by a single electron before any pairing takes place. Thus in carbon the two $2p$ electrons will occupy different orbitals to give the configuration $1s^2$, $2s^2 2p_x^1 2p_y^1$. Similarly, in nitrogen the three $2p$ electrons will each occupy one of the three $2p$ orbitals to give the configuration $1s^2$, $2s^2 2p_x^1 2p_y^1 2p_z^1$. In oxygen there are four $2p$ electrons and it will now be necessary for two of them to occupy the same orbital, say $2p_x$, to give $1s^2$, $2s^2 2p_x^2 2p_y^1 2p_z^1$ as the configuration of this atom. In fluorine, with five $2p$ electrons, two orbitals must be filled to give the configuration $1s^2$, $2s^2 2p_x^2 2p_y^2 2p_z^1$, and in neon, with the configuration $1s^2$, $2s^2 2p_x^2 2p_y^2 2p_z^2$, all the $2p$ orbitals are filled and the K and L shells are fully occupied.

The configurations of the first ten elements in the Periodic Table are summarized in table 2.02. In the left-hand part of the table the configurations are represented schematically, a single arrow representing a single electron and a pair of arrows representing two electrons of opposite spin occupying a single orbital. In the right-hand part of the table the configuration is represented more concisely, and if the rule enunciated in the preceding paragraph is kept in mind it will be seen

Table 2.02. *The electronic configurations of the elements of the first two periods*

Element	Z	K	L				
		$1s$	$2s$	$2p_x$	$2p_y$	$2p_z$	
H	1	↑					$1s^1$
He	2	↑↓					$1s^2$
Li	3	↑↓	↑				$1s^2$, $2s^1$
Be	4	↑↓	↑↓				$1s^2$, $2s^2$
B	5	↑↓	↑↓	↑			$1s^2$, $2s^2 2p^1$
C	6	↑↓	↑↓	↑	↑		$1s^2$, $2s^2 2p^2$
N	7	↑↓	↑↓	↑	↑	↑	$1s^2$, $2s^2 2p^3$
O	8	↑↓	↑↓	↑↓	↑	↑	$1s^2$, $2s^2 2p^4$
F	9	↑↓	↑↓	↑↓	↑↓	↑	$1s^2$, $2s^2 2p^5$
Ne	10	↑↓	↑↓	↑↓	↑↓	↑↓	$1s^2$, $2s^2 2p^6$

Table 2.03. *The electronic configurations of the ground states of the elements*

The configurations for some of the lanthanide and actinide elements are uncertain

Period	Element	Atomic number	K	L		M			N				O				P			Q
			1s	2s	2p	3s	3p	3d	4s	4p	4d	4f	5s	5p	5d	5f	6s	6p	6d	7s
1	H	1	1																	
	He	2	2																	
2	Li	3	2	1																
	Be	4	2	2																
	B	5	2	2	1															
	C	6	2	2	2															
	N	7	2	2	3															
	O	8	2	2	4															
	F	9	2	2	5															
	Ne	10	2	2	6															
3	Na	11	2	2	6	1														
	Mg	12	2	2	6	2														
	Al	13	2	2	6	2	1													
	Si	14	2	2	6	2	2													
	P	15	2	2	6	2	3													
	S	16	2	2	6	2	4													
	Cl	17	2	2	6	2	5													
	A	18	2	2	6	2	6													
4	K	19	2	2	6	2	6		1											
	Ca	20	2	2	6	2	6		2											
	Sc	22*	2	2	6	2	6	1	2											
	Ti	21*	2	2	6	2	6	2	2											
	V	23*	2	2	6	2	6	3	2											
	Cr	24*	2	2	6	2	6	5	1											
	Mn	25*	2	2	6	2	6	5	2											
	Fe	26*	2	2	6	2	6	6	2											
	Co	27*	2	2	6	2	6	7	2											
	Ni	28*	2	2	6	2	6	8	2											
	Cu	29*	2	2	6	2	6	10	1											
	Zn	30*	2	2	6	2	6	10	2											
	Ga	31	2	2	6	2	6	10	2	1										
	Ge	32	2	2	6	2	6	10	2	2										
	As	33	2	2	6	2	6	10	2	3										
	Se	34	2	2	6	2	6	10	2	4										
	Br	35	2	2	6	2	6	10	2	5										
	Kr	36	2	2	6	2	6	10	2	6										
5	Rb	37	2	2	6	2	6	10	2	6			1							
	Sr	38	2	2	6	2	6	10	2	6			2							
	Y	39*	2	2	6	2	6	10	2	6	1		2							
	Zr	40*	2	2	6	2	6	10	2	6	2		2							
	Nb	41*	2	2	6	2	6	10	2	6	4		1							
	Mo	42*	2	2	6	2	6	10	2	6	5		1							
	Tc	43*	2	2	6	2	6	10	2	6	6		1							
	Ru	44*	2	2	6	2	6	10	2	6	7		1							
	Rh	45*	2	2	6	2	6	10	2	6	8		1							
	Pd	46*	2	2	6	2	6	10	2	6	10									
	Ag	47*	2	2	6	2	6	10	2	6	10		1							
	Cd	48*	2	2	6	2	6	10	2	6	10		2							
	In	49	2	2	6	2	6	10	2	6	10		2	1						
	Sn	50	2	2	6	2	6	10	2	6	10		2	2						
	Sb	51	2	2	6	2	6	10	2	6	10		2	3						
	Te	52	2	2	6	2	6	10	2	6	10		2	4						
	I	53	2	2	6	2	6	10	2	6	10		2	5						
	Xe	54	2	2	6	2	6	10	2	6	10		2	6						

Period	Element	Atomic number	*K*	*L*		*M*			*N*				*O*				*P*			*Q*
			1*s*	2*s*	2*p*	3*s*	3*p*	3*d*	4*s*	4*p*	4*d*	4*f*	5*s*	5*p*	5*d*	5*f*	6*s*	6*p*	6*d*	7*s*
6	Cs	55	2	2	6	2	6	10	2	6	10		2	6			1			
	Ba	56	2	2	6	2	6	10	2	6	10		2	6			2			
	La	57*	2	2	6	2	6	10	2	6	10		2	6	1		2			
	Ce	58†	2	2	6	2	6	10	2	6	10	2	2	6			2			
	Pr	59†	2	2	6	2	6	10	2	6	10	3	2	6			2			
	Nd	60†	2	2	6	2	6	10	2	6	10	4	2	6			2			
	Pm	61†	2	2	6	2	6	10	2	6	10	5	2	6			2			
	Sm	62†	2	2	6	2	6	10	2	6	10	6	2	6			2			
	Eu	63†	2	2	6	2	6	10	2	6	10	7	2	6			2			
	Gd	64†	2	2	6	2	6	10	2	6	10	7	2	6	1		2			
	Tb	65†	2	2	6	2	6	10	2	6	10	9	2	6			2			
	Dy	66†	2	2	6	2	6	10	2	6	10	10	2	6			2			
	Ho	67†	2	2	6	2	6	10	2	6	10	11	2	6			2			
	Er	68†	2	2	6	2	6	10	2	6	10	12	2	6			2			
	Tm	69†	2	2	6	2	6	10	2	6	10	13	2	6			2			
	Yb	70†	2	2	6	2	6	10	2	6	10	14	2	6			2			
	Lu	71†	2	2	6	2	6	10	2	6	10	14	2	6	1		2			
	Hf	72*	2	2	6	2	6	10	2	6	10	14	2	6	2		2			
	Ta	73*	2	2	6	2	6	10	2	6	10	14	2	6	3		2			
	W	74*	2	2	6	2	6	10	2	6	10	14	2	6	4		2			
	Re	75*	2	2	6	2	6	10	2	6	10	14	2	6	5		2			
	Os	76*	2	2	6	2	6	10	2	6	10	14	2	6	6		2			
	Ir	77*	2	2	6	2	6	10	2	6	10	14	2	6	9					
	Pt	78*	2	2	6	2	6	10	2	6	10	14	2	6	9		1			
	Au	79*	2	2	6	2	6	10	2	6	10	14	2	6	10		1			
	Hg	80*	2	2	6	2	6	10	2	6	10	14	2	6	10		2			
	Tl	81	2	2	6	2	6	10	2	6	10	14	2	6	10		2	1		
	Pb	82	2	2	6	2	6	10	2	6	10	14	2	6	10		2	2		
	Bi	83	2	2	6	2	6	10	2	6	10	14	2	6	10		2	3		
	Po	84	2	2	6	2	6	10	2	6	10	14	2	6	10		2	4		
	At	85	2	2	6	2	6	10	2	6	10	14	2	6	10		2	5		
	Rn	86	2	2	6	2	6	10	2	6	10	14	2	6	10		2	6		
7	Fr	87	2	2	6	2	6	10	2	6	10	14	2	6	10		2	6		1
	Ra	88	2	2	6	2	6	10	2	6	10	14	2	6	10		2	6		2
	Ac	89*	2	2	6	2	6	10	2	6	10	14	2	6	10		2	6	1	2
	Th	90†	2	2	6	2	6	10	2	6	10	14	2	6	10	1	2	6	1	2
	Pa	91†	2	2	6	2	6	10	2	6	10	14	2	6	10	2	2	6	1	2
	U	92†	2	2	6	2	6	10	2	6	10	14	2	6	10	3	2	6	1	2
	Np	93†	2	2	6	2	6	10	2	6	10	14	2	6	10	4	2	6	1	2
	Pu	94†	2	2	6	2	6	10	2	6	10	14	2	6	10	5	2	6	1	2
	Am	95†	2	2	6	2	6	10	2	6	10	14	2	6	10	7	2	6		2
	Cm	96†	2	2	6	2	6	10	2	6	10	14	2	6	10	7	2	6	1	2
	Bk	97†	2	2	6	2	6	10	2	6	10	14	2	6	10	8	2	6	1	2
	Cf	98†	2	2	6	2	6	10	2	6	10	14	2	6	10	10	2	6		2
	E	99†	2	2	6	2	6	10	2	6	10	14	2	6	10	11	2	6		2
	Fm	100†	2	2	6	2	6	10	2	6	10	14	2	6	10	12	2	6		2
	Mv	101†	2	2	6	2	6	10	2	6	10	14	2	6	10	13	2	6		2
	No	102†	2	2	6	2	6	10	2	6	10	14	2	6	10	14	2	6		2

* Transition elements.

† Lanthanide and actinide elements.

that it is not necessary to distinguish the three $2p$ orbitals from one another; we know, for example, that the four electrons represented as $2p^4$ in oxygen must comprise two sharing one orbital and one in each of the remaining two orbitals.

The third and fourth periods

2.17. The K and L shells being now fully occupied, with two and eight electrons, respectively, the eleventh electron in sodium ($Z = 11$) must enter the M shell to give the configuration $1s^2$, $2s^2 2p^6$, $3s^1$. For the remaining atoms in the third period, magnesium ($Z = 12$) to argon ($Z = 18$), the electrons occupy the $3s$ and $3p$ orbitals in a manner precisely analogous to that discussed above for the second period, thus giving the configurations set out for these elements in table 2.03. There is, however, this important difference: whereas with neon the L shell is filled, with argon the M shell is not fully occupied because there is still room for ten further electrons in the five $3d$ orbitals. It might therefore be expected that in potassium ($Z = 19$) the extra electron would enter one of these orbitals. This, however, is not the case, for, as can be seen from fig. 2.02, the $4s$ orbital of the N shell is energetically more favourable than the $3d$ orbital of the M shell at this value of the atomic number; accordingly, potassium has the configuration $1s^2$, $2s^2 2p^6$, $3s^2 3p^6$, $4s^1$. Similarly, calcium ($Z = 20$) has the configuration $1s^2$, $2s^2 2p^6$, $3s^2 3p^6$, $4s^2$.

After calcium the $3d$ orbital is more stable than the $4s$, and in the elements from scandium ($Z = 21$) to zinc ($Z = 30$) we find electrons entering the $3d$ orbital of the M shell in preference to the N shell (see table 2.03). Only after zinc is the development of the N shell resumed by the occupation of the $4p$ orbitals to give the sequence of elements gallium ($Z = 31$) to krypton ($Z = 36$). The elements from scandium to zinc are termed the elements of the *first transition series*; if they are imagined to be removed from the Periodic Table the remaining elements from potassium to krypton would form a fourth period exactly analogous to the third period from sodium to argon.

The fifth period

2.18. In krypton the s and p orbitals of the N shell are fully occupied but the d and f orbitals are vacant. The elements next following, from rubidium ($Z = 37$) to xenon ($Z = 54$), form a series analogous to that from potassium to krypton. In rubidium and strontium ($Z = 38$) electrons enter the $5s$ orbital, then follows a *second transition series* from yttrium ($Z = 39$) to cadmium ($Z = 48$), and finally from indium ($Z = 49$) to xenon the development of the O shell is resumed. This second series, however, differs from the first in that the end of the series does not represent the saturation of the penultimate shell: in zinc the M shell is full, but in cadmium the $4f$ orbitals of the N shell are still entirely vacant. We shall see the relevance of this point in the next paragraph.

The sixth period

2.19. In caesium ($Z = 55$) one electron occupies the $6s$ orbital of the P shell, and in barium ($Z = 56$) there are two electrons in this orbital. Thereafter the development of the P shell is interrupted, and lanthanum ($Z = 57$) initiates the *third transition series* with one electron in the $5d$ orbital of the O shell. The development of this series, however, proceeds no further at this stage, and in the elements from cerium ($Z = 58$) to lutecium ($Z = 71$) electrons are entering the hitherto vacant $4f$ orbitals of the N shell. These elements constitute the *rare earths* or elements of the *lanthanide series*, and the fact that the differentiating electrons are so deep in the electronic structure is responsible for the close similarity of their chemical properties.

After lutecium the development of the $5d$ orbital is resumed, and the elements hafnium ($Z = 72$) to mercury ($Z = 80$) complete the third transition series initiated at lanthanum. In mercury, as in cadmium, the f orbitals of the penultimate shell remain vacant. Finally, the development of the P shell is resumed by the occupation of the $6p$ orbitals to give the sequence of elements from thallium ($Z = 81$) to radon ($Z = 86$).

The seventh period

2.20. In francium ($Z = 87$) and radium ($Z = 88$) electrons enter the $7s$ orbital of the Q shell. Thereafter, the development of this shell is suspended, and actinium ($Z = 89$) initiates the *fourth transition series*, with one electron in the $6d$ orbital of the P shell. As with the third transition series, however, development proceeds no further, and in all the remaining elements so far known (up to $Z = 102$) electrons are entering the hitherto unoccupied $5f$ orbitals of the O shell. These elements therefore form a series analogous to the rare earths; they are generally termed elements of the *actinide* or *5f series*.

2.21. We have discussed the electronic structure of the atoms at some length because, as we shall see, it plays a dominant part in determining their chemical and physical properties and the way in which they can enter into chemical combination with one another. It is important to remember, however, that the configurations described are those of the ground state of the isolated neutral atom. Often the energy differences between different orbitals are small compared with the energies involved in chemical reactions so that when atoms are associated together alternative electronic configurations may have to be considered; in metal systems this point is of particular importance. Many atoms, too, are readily ionized, either by losing electrons to form positive ions (cations) or by acquiring electrons to form negative ions (anions), and sometimes several alternative states of ionization are possible. In all such cases the electronic configurations will be correspondingly altered.

THE CLASSIFICATION OF THE ELEMENTS

2.22. With these considerations in mind it is convenient to discuss the classification of the elements in terms of the form of Periodic Table presented in table 2.04. This presentation is designed to emphasize the electronic structures of the various elements and enables these elements to be classified under four headings:

Type I. Atoms in which the outermost shell consists of a complete octet of electrons in the s and p orbitals with no other electrons in this shell. These atoms comprise the inert gases. (Helium must be included here; it has a pair of electrons in the $1s$ orbital, but of course no p orbitals are possible in the K shell.) The chemical inactivity of these elements emphasizes the extreme stability of the octet configuration, and this stability is a basic conception in valency theory. These elements have very high ionization potentials, and do not ordinarily occur in the ionized state. In the nature of things, elements of Type I are of very limited importance in the study of crystal structures.

Type II. Atoms in which the differentiating electrons occupy the shell of highest energy. These atoms, together with those of Type I, are designated as 'representative elements' in table 2.04. The series in which they occur are each characterized by a rapid transition from metallic properties on the left to non-metallic properties on the right. The more metallic elements readily lose the electrons in the outermost shell to form positive ions, e.g. potassium readily forms the ion K^+ with the configuration $1s^2$, $2s^2 2p^6$, $3s^2 3p^6$ and calcium forms the ion Ca^{2+} with the same configuration. The more non-metallic elements, on the

Table 2.04. *The periodic classification of the elements*

Period	Types I and II								Type III										Type IV													
	The representative elements								The transition elements										The lanthanide and actinide elements													
1	H 1							He 2																								
2	Li 3	Be 4	B 5	C 6	N 7	O 8	F 9	Ne 10																								
3	Na 11	Mg 12	Al 13	Si 14	P 15	S 16	Cl 17	A 18																								
4	K 19	Ca 20	Ga 31	Ge 32	As 33	Se 34	Br 35	Kr 36	Sc 21	Ti 22	V 23	Cr 24	Mn 25	Fe 26	Co 27	Ni 28	Cu 29	Zn 30														
5	Rb 37	Sr 38	In 49	Sn 50	Sb 51	Te 52	I 53	Xe 54	Y 39	Zr 40	Nb 41	Mo 42	Tc 43	Ru 44	Rh 45	Pd 46	Ag 47	Cd 48														
6	Cs 55	Ba 56	Tl 81	Pb 82	Bi 83	Po 84	At 85	Rn 86	La 57	Hf 72	Ta 73	W 74	Re 75	Os 76	Ir 77	Pt 78	Au 79	Hg 80	Ce 58	Pr 59	Nd 60	Pm 61	Sm 62	Eu 63	Gd 64	Tb 65	Dy 66	Ho 67	Er 68	Tm 69	Yb 70	Lu 71
7	Fr 87	Ra 88							Ac 89										Th 90	Pa 91	U 92	Np 93	Pu 94	Am 95	Cm 96	Bk 97	Cf 98	E 99	Fm 100	Mv 101	No 102	

other hand, readily form negative ions by acquiring electrons, e.g. chlorine forms the ion Cl^-, again with the same configuration as K^+ and Ca^{2+}. These ions are usually colourless and diamagnetic. This is due to the fact that the electrons are all paired in filled orbitals, as, for example, in the three ions just mentioned:

	1s	2s	2p	3s	3p
Cl^-, K^+, Ca^{2+}	↑↓	↑↓	↑↓ ↑↓ ↑↓	↑↓	↑↓ ↑↓ ↑↓

Elements of Type II include many of the most important in inorganic chemistry, and almost all of those of common occurrence in organic compounds. We shall have occasion to discuss the properties of these elements at length.

Type III. Atoms in which the differentiating electrons occupy the shell of second highest energy. These are the elements of the transition series described above. They are all metals and, like the metallic elements of Type II, they can readily form positive ions. They differ from the metal of Type II, however, in that several degrees of ionization are often possible. Thus iron can lose not only the two electrons in the $4s$ orbital to form the ion Fe^{2+} ($1s^2$, $2s^2 2p^6$, $3s^2 3p^6 3d^6$), but also in addition one of the electrons in the $3d$ orbitals to form the more stable ion Fe^{3+} ($1s^2$, $2s^2 2p^6$, $3s^2 3p^6 3d^5$). By contrast, calcium (Type II) readily forms the ion Ca^{2+}, but further ionization, which would involve the destruction of the $3s^2 3p^6$ octet, does not occur. The ions of Type III elements are often coloured and paramagnetic. This is due to the fact that in these ions incomplete orbitals containing unpaired electrons are found. Thus the ions Fe^{2+} and Fe^{3+} just mentioned contain four and five unpaired electrons, respectively, in the $3d$ orbitals:

	1s	2s	2p	3s	3p	3d
Fe^{2+}	↑↓	↑↓	↑↓ ↑↓ ↑↓	↑↓	↑↓ ↑↓ ↑↓	↑↓ ↑ ↑ ↑ ↑
Fe^{3+}	↑↓	↑↓	↑↓ ↑↓ ↑↓	↑↓	↑↓ ↑↓ ↑↓	↑ ↑ ↑ ↑ ↑

Elements of Type III include many of great importance in metallurgical technology.

Type IV. Atoms in which the differentiating electrons occupy the shell of third highest energy. These are the elements of the lanthanide and actinide series already discussed. They are all metals, and, on account of the fact that the differentiating electrons are so deeply buried in the electronic structure, they show great similarity in chemical properties. In common with the elements of Type III, they readily form ions, and these ions are commonly coloured and paramagnetic.

Table 2.05. *The periodic classification of the elements*

The lanthanide and actinide elements are omitted. For these elements see table 2.04

Period	Type II		Type III										Type II					Type I
	1*A*	2*A*	3*A*	4*A*	5*A*	6*A*	7*A*	8			1*B*	2*B*	3*B*	4*B*	5*B*	6*B*	7*B*	0
1	H 1																	He 2
2	Li 3	Be 4											B 5	C 6	N 7	O 8	F 9	Ne 10
3	Na 11	Mg 12											Al 13	Si 14	P 15	S 16	Cl 17	A 18
4	K 19	Ca 20	Sc 21	Ti 22	V 23	Cr 24	Mn 25	Fe 26	Co 27	Ni 28	Cu 29	Zn 30	Ga 31	Ge 32	As 33	Se 34	Br 35	Kr 36
5	Rb 37	Sr 38	Y 39	Zr 40	Nb 41	Mo 42	Tc 43	Ru 44	Rh 45	Pd 46	Ag 47	Cd 48	In 49	Sn 50	Sb 51	Te 52	I 53	Xe 54
6	Cs 55	Ba 56	La 57	Hf 72	Ta 73	W 74	Re 75	Os 76	Ir 77	Pt 78	Au 79	Hg 80	Tl 81	Pb 82	Bi 83	Po 84	At 85	Rn 86
7	Fr 87	Ra 88	Ac 89															

Elements of Type IV are of rare occurrence and we shall have little occasion to discuss crystal structures in which they are found.

An alternative and in some respects more convenient form of the Periodic Table, from which the lanthanide and actinide elements have been omitted, is given in table 2.05.

2.23. Having now considered the electronic structures of the elements in relation to their positions in the Periodic Table, we are ready to discuss the nature of the various types of interatomic binding force and of the structures to which they give rise. This we do in the next four chapters.

3 THE IONIC BOND AND SOME IONIC STRUCTURES

INTRODUCTION

The formation of ions

3.01. The ionic bond is physically the simplest of the four types of interatomic binding force, and arises from the electrostatic attraction between oppositely charged ions. We must therefore first consider the conditions under which ions of the various elements are formed.

The lack of chemical activity of the inert gases and their high ionization potentials reveal the stability of the electronic configurations of these elements. The same configurations can be achieved by Type II elements not too far removed from the inert gases by the loss or gain of one or more electrons. Thus sodium, by losing a single electron, becomes the positively charged cation Na^+, with the neon configuration $1s^2$, $2s^2 2p^6$, and the readiness with which this electron is lost is revealed by the low ionization potential of 5·15 V. Similarly, the magnesium atom can acquire the neon configuration by the loss of two electrons to form the ion Mg^{2+}. Here, however, the tendency to ionize is less marked than in sodium, for now two electrons must be removed and these electrons are more strongly bound owing to the influence of the greater nuclear charge. This is reflected in the high ionization potential of 22·6 V for the formation of the Mg^{2+} ion.

The tendency to form positive ions among the Type II elements is, therefore, most marked in the first two or three groups of the Periodic Table. A few elements in the fourth group form A^{4+} ions, but such highly charged ions are rare and are found chiefly among elements of high atomic number, e.g. Sn^{4+} and Pb^{4+}. These particular ions do not, of course, possess an inert-gas configuration, owing to the interposition of the transition metals between them and the corresponding inert gases, but they do have a structure in which the *s*, *p*, and *d* sub-shells are fully occupied, as can be seen from table 2.03.

Another characteristic of Type II elements of high atomic number is a tendency to display two states of ionization or to occur in a state of ionization two units lower than might be expected, owing to the 'lone pair' effect. Thus tin, thallium and lead form not only the ions Sn^{4+},

Tl^{3+} and Pb^{4+} but also the ions Sn^{2+}, Tl^{+} and Pb^{2+}. Similarly, antimony and bismuth form the ions Sb^{3+} and Bi^{3+}. In all of these examples the pair of $5s^2$ or $6s^2$ electrons in the outermost shell constitute the lone pair.

Certain of the Type II elements can also form negative ions. Thus fluorine and oxygen give rise to the ions F^- and O^{2-}, both with the neon configuration, by acquiring one and two electrons, respectively. This tendency is in practice confined to the Type II elements in the seventh and sixth groups, i.e. the halogens and chalcogens.

3.02. Elements of Types III and IV are all metals, and do not form negative ions. They do, however, readily form positive ions, often with two or more different states of ionization. In these cases the ions cannot, of course, achieve the configuration of an inert gas, and it is accordingly not self-evident in what states of ionization the elements will occur and which of several possible states will be the most stable. We shall revert to this question when we come to discuss crystal structures containing some of these elements.

The ionic bond

3.03. The force of electrostatic attraction between oppositely charged ions constitutes the ionic or electrovalent bond. It must not be assumed that any given structure is necessarily ionic just because it consists of particles some of which can exist as cations and others as anions, for we shall find that in many such cases the binding is in fact of a different kind. Nevertheless, there are numerous structures in which the binding forces are predominantly ionic in character, and it is convenient to introduce our discussion of the structures of ionic crystals by considering the simplest of such examples, namely, the alkali halides and some other compounds of the general composition AX.

SOME SIMPLE IONIC STRUCTURES

The sodium chloride structure

3.04. The structure of sodium chloride was the first to be determined by X-ray diffraction and was elucidated by W. H. & W. L. Bragg in 1913. The arrangement of the ions in the cubic unit cell is shown in fig. 3.01.*

* Readers unfamiliar with crystal-structure diagrams are advised first to read appendix 2.

One feature of this structure is immediately apparent: there is no trace of any molecule NaCl. Instead, each sodium ion is symmetrically surrounded by six chlorine neighbours, and stands in no privileged

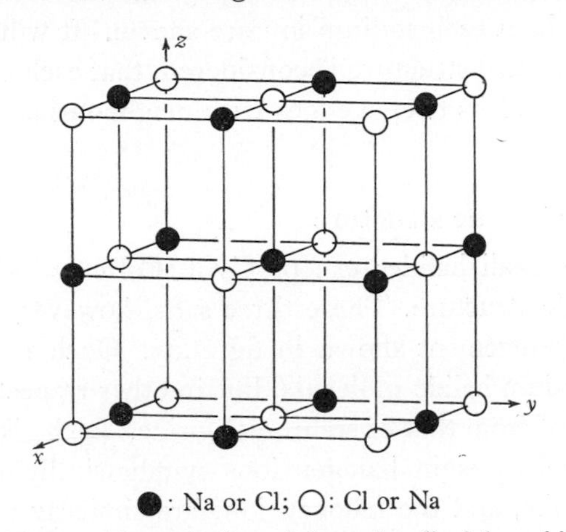

Fig. 3.01. Clinographic projection of the unit cell of the cubic structure of sodium chloride, NaCl.

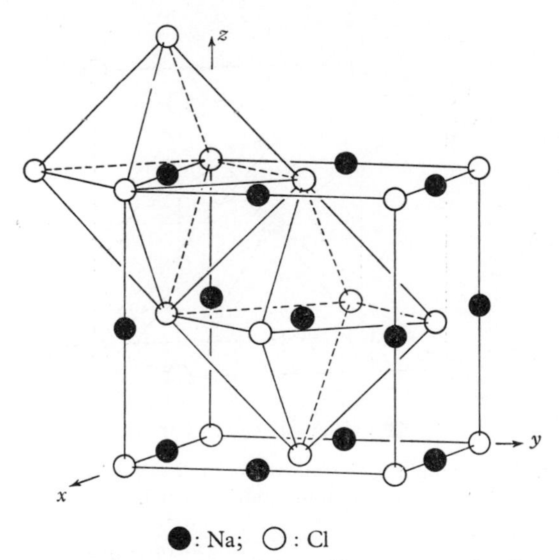

Fig. 3.02. Clinographic projection of the cubic structure of sodium chloride, NaCl, showing the co-ordinating octahedra of anions round the cations.

position relative to any one of them, while similarly each chlorine ion is symmetrically surrounded by six sodium neighbours. We say in such a case that the *co-ordination* of the sodium and chlorine ions is sixfold, or alternatively that it is octahedral in that the six neighbours of any ion are disposed about it at the corners of a regular octahedron. This aspect of the structure is made clearer in fig. 3.02, where the octahedra of chlorine ions about each sodium ion are shown. It will be seen, if the indefinitely extended structure is considered, that each of the octahedra shares every one of its twelve edges with neighbouring octahedra.

The caesium chloride structure

3.05. All the alkali halides except CsCl, CsBr and CsI possess the sodium chloride structure. These three salts, however, have the alternative cubic arrangement shown in fig. 3.03. Once again we see that there is no caesium halide molecule, but in other respects the structure is quite different from that of sodium chloride. Each alkali metal ion is now surrounded by eight halogen ions symmetrically arranged at the corners of a cube, and the halogen ions are similarly co-ordinated by eight caesium ions; the co-ordination is therefore eightfold or cubic.

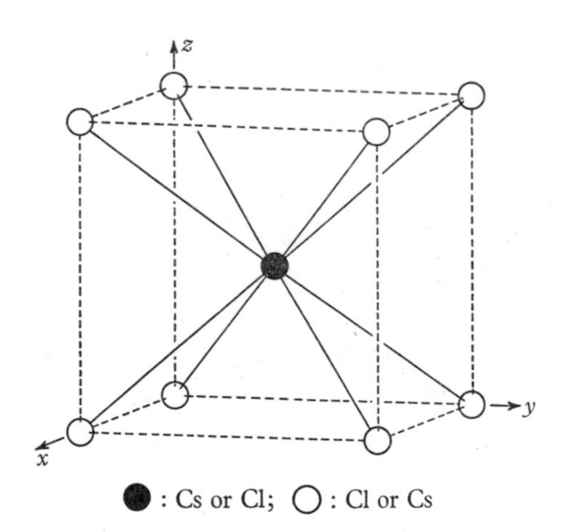

Fig. 3.03. Clinographic projection of the unit cell of the cubic structure of caesium chloride, CsCl.

The zincblende structure

3.06. Yet a third simple cubic structure, not found among the alkali halides, occurs nevertheless in a limited number of ionic compounds of composition AX. This is the zincblende arrangement, shown in fig. 3.04, in which the co-ordination is fourfold or tetrahedral, each ion being symmetrically surrounded by four ions of the opposite sign disposed at the corners of a regular tetrahedron.*

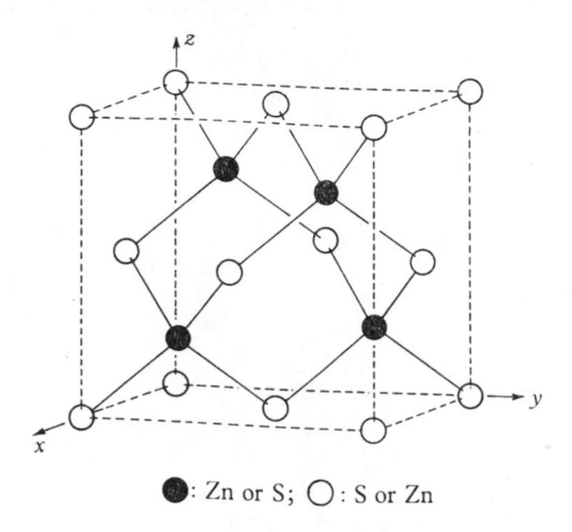

●: Zn or S; ○: S or Zn

Fig. 3.04. Clinographic projection of the unit cell of the cubic structure of zincblende, ZnS.

IONIC RADII

Interionic distances in the alkali halides

3.07. Although all the alkali halides possess either the sodium chloride or the caesium chloride structure the cell dimensions for the different salts are found to be different. From the observed cell dimensions it is clearly only a matter of simple geometry to derive the distance of closest approach between adjacent alkali and halogen ions. When this is done it is found that there is a progressive change in A–X distance in passing from the fluoride to the iodide of any one metal, or again in passing from the lithium to the caesium salt of any one halogen. This point is illustrated in table 3.01, in which the A–X distance for each of the alkali halides is recorded. A study of this table reveals that the change in interatomic distance accompanying a progressive substitution in the series is a regular one, and that, for example, the difference between the A–X distances in two halides with the same cation is very nearly the same whichever of the alkali metals that cation may be, while conversely, for the halides of a given pair of alkali metals, this difference is nearly

* Zincblende, ZnS, is not itself ionic but its name is given to the structure as it is the commonest compound in which this geometrical arrangement occurs.

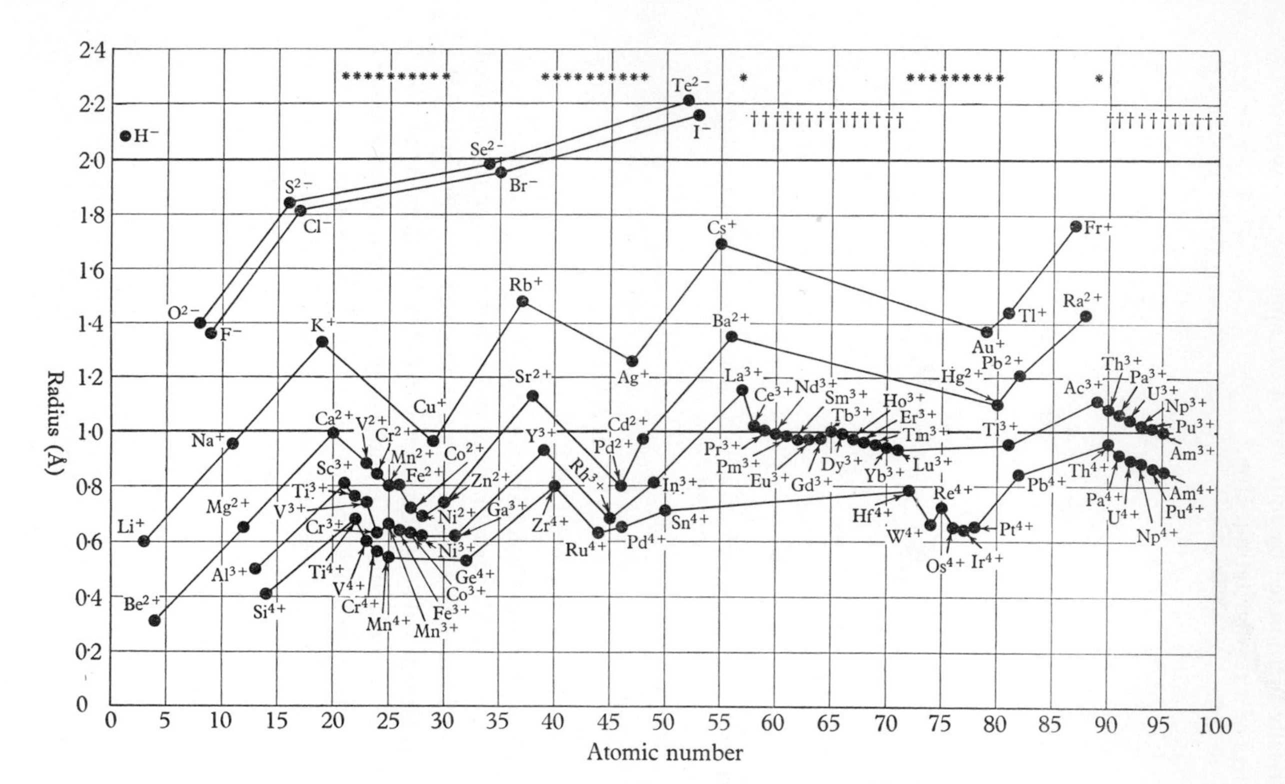

Fig. 3.05. The ionic radii of the elements. The values are those appropriate to 6-co-ordination.

independent of the particular halogen in combination. These regularities at once lend weight to the conception of the crystal structure as composed of a set of spherical atoms or ions, each of characteristic size, packed together in contact, and in fact in these data we have the first experimental confirmation of the views earlier expressed by Barlow and Pope (§1·03).

Table 3.01. *Interatomic distances in the alkali halides*

Values are in Ångström units

	Li	Δ	Na	Δ	K	Δ	Rb	Δ	Cs
F	2·01	0·30	2·31	0·35	2·66	0·16	2·82	0·18	3·00
Δ	0·56	—	0·50	—	0·48	—	0·45	—	0·56
Cl	2·57	0·24	2·81	0·33	3·14	0·13	3·27	0·29	3·56
Δ	0·18	—	0·17	—	0·15	—	0·16	—	0·15
Br	2·75	0·23	2·98	0·31	3·29	0·14	3·43	0·28	3·71
Δ	0·25	—	0·25	—	0·24	—	0·23	—	0·24
I	3·00	0·23	3·23	0·30	3·53	0·13	3·66	0·29	3·95

Ionic crystal radii

3.08. It is clear that from the observed interionic distances we can deduce only the sum of two ionic radii, but that if any one radius is known then other radii may be found. Various independent methods are available for estimating the radii of certain ions, and the values so determined, taken in conjunction with data from the crystal structures not only of the alkali halides but also of many other compounds, lead to the semi-empirical ionic crystal radii shown in table 3.02 and in fig. 3.05. The interpretation of the radii given in this table is subject to a number of qualifications which will be discussed below. For the present, however, it is sufficient to treat the radii as constant and characteristic of the ions concerned. For the alkali halides with the sodium chloride structure it will be seen that the interatomic distances quoted in table 3.01 are given with fair accuracy as the sum of the corresponding radii from table 3.02.

Certain general points which emerge from the data in this table call for explicit mention:

(*a*) The radii of comparable ions increase with atomic number, e.g.

Ion	Li^+	Na^+	K^+	Rb^+	Cs^+
Z	3	11	19	37	55
Radius (Å)	0·60	0·95	1·33	1·48	1·69

Table 3.02. *Ionic radii*

Values are in Ångström units and correspond to 6-co-ordination

The representative and transition elements

	1	2	3	4	5	6	7	8	0
1							H^{-} 1·54		He —
2	Li^{+} 0·60	Be^{2+} 0·31	B —	C —	N —	O^{2-} 1·40	F^{-} 1·36		Ne —
3	Na^{+} 0·95	Mg^{2+} 0·65	Al^{3+} 0·50	Si^{4+} 0·41	P —	S^{2-} 1·84	Cl^{-} 1·81		A —
4	K^{+} 1·33	Ca^{2+} 0·99	Sc^{3+} 0·81	Ti^{3+} 0·76 Ti^{4+} 0·68	V^{2+} 0·88 V^{3+} 0·74 V^{4+} 0·60	Cr^{2+} 0·84 Cr^{3+} 0·63 Cr^{4+} 0·56	Mn^{2+} 0·80 Mn^{3+} 0·66 Mn^{4+} 0·54	Fe^{2+} 0·80 Co^{2+} 0·72 Ni^{2+} 0·69 Fe^{3+} 0·64 Co^{3+} 0·63 Ni^{3+} 0·62	
	Cu^{+} 0·96	Zn^{2+} 0·74	Ga^{3+} 0·62	Ge^{4+} 0·53	As —	Se^{2-} 1·98	Br^{-} 1·95		Kr —
5	Rb^{+} 1·48	Sr^{2+} 1·13	Y^{3+} 0·93	Zr^{4+} 0·80	Nb —	Mo —	Tc —	Ru^{4+} 0·63 Rh^{3+} 0·68 Pd^{2+} 0·80 Pd^{4+} 0·65	
	Ag^{+} 1·26	Cd^{2+} 0·97	In^{3+} 0·81	Sn^{4+} 0·71	Sb —	Te^{2-} 2·21	I^{-} 2·16		Xe —
6	Cs^{+} 1·69	Ba^{2+} 1·35	La^{3+} 1·15	Hf^{4+} 0·78	Ta —	W^{4+} 0·66	Re^{4+} 0·72	Os^{4+} 0·65 Ir^{4+} 0·64 Pt^{4+} 0·65	
	Au^{+} 1·37	Hg^{2+} 1·10	Tl^{+} 1·44 Tl^{3+} 0·95	Pb^{2+} 1·21 Pb^{4+} 0·84	Bi —	Po —	At —		Rn —
7	Fr^{+} 1·76	Ra^{2+} 1·43	Ac^{3+} 1·11						

The lanthanide elements

Ce^{3+} 1·02	Pr^{3+} 1·00	Nd^{3+} 0·99	Pm^{3+} 0·98	Sm^{3+} 0·97	Eu^{3+} 0·97	Gd^{3+} 0·97
Tb^{3+} 1·00	Dy^{3+} 0·99	Ho^{3+} 0·97	Er^{3+} 0·96	Tm^{3+} 0·95	Yb^{3+} 0·94	Lu^{3+} 0·93

The actinide elements

Th^{3+} 1·08	Pa^{3+} 1·06	U^{3+} 1·04	Np^{3+} 1·02	Pu^{3+} 1·01	Am^{3+} 1·00
Th^{4+} 0·95	Pa^{4+} 0·91	U^{4+} 0·89	Np^{4+} 0·88	Pu^{4+} 0·86	Am^{4+} 0·85

Other ions

NH_4^{+} 1·48 OH^{-} 1·53

The increase, however, is by no means proportional to *Z*. This is readily understood, for although Cs^{+}, say, has many more extranuclear electrons than Na^{+} these electrons are under the attractive influence of a much greater nuclear charge.

(*b*) In a series of positive ions with the same number of extranuclear electrons the radius decreases rapidly with increasing charge:

Ion	Na^{+}	Mg^{2+}	Al^{3+}	Au^{+}	Hg^{2+}	Tl^{3+}	Pb^{4+}
Radius (Å)	0·95	0·65	0·50	1·37	1·10	0·95	0·84

Here again the effect can be attributed to the influence of the increasing nuclear charge as the atomic number increases. A second factor, however, also operates, for as the charge increases so also does the force of attraction between the cation and its co-ordinating anions, thus still further reducing its effective radius.

(*c*) In pairs of negative ions with the same number of extranuclear electrons, on the other hand, the radius increases with increasing negative charge:

Ion	F^{-}	O^{2-}	Cl^{-}	S^{2-}	Br^{-}	Se^{2-}
Radius (Å)	1·36	1·40	1·81	1·84	1·95	1·98

Here it is to be expected that O^{2-}, say, would be larger than F^{-}, owing to the influence of the smaller nuclear charge. That the difference is so small is to be attributed to the compensating effect of the increased force of attraction between the more highly charged anion and its co-ordinating cations.

(*d*) In the lanthanide or rare-earth series of elements the ionic radii actually *decrease* with increasing atomic numbers, the radii of the Ce^{3+} and Lu^{3+} ions being 1·02 and 0·93 Å, respectively. This effect can again be attributed to the influence of the increasing nuclear charge, since the differentiating electrons are deeply buried in the extranuclear structure and all the elements possess substantially the same configuration in the outermost shell. As a result of this so-called *lanthanide contraction*, ions immediately following the rare-earth elements have radii little or no larger than members of the same group in the period preceeding these elements. Since similarities in crystal structure, as we shall see, are closely associated with similarities on ionic size, the lanthanide contraction is largely responsible for the remarkable chemical resemblances between such pairs of elements as zirconium and hafnium, or niobium and tantalum.

A similar decrease in ionic radii with increasing atomic number is found in the limited number of ions of the actinide series of which the radii are known.

(*e*) Few common cations exceed 1 Å in radius, and the great majority are substantially smaller than this. Many common anions, however, including in particular O^{2-} and Cl^-, are much larger. In simple ionic crystals containing positive and negative ions in association the greater part of the volume is, therefore, occupied by negative ions, and it is primarily the arrangement of these which determine the structure.

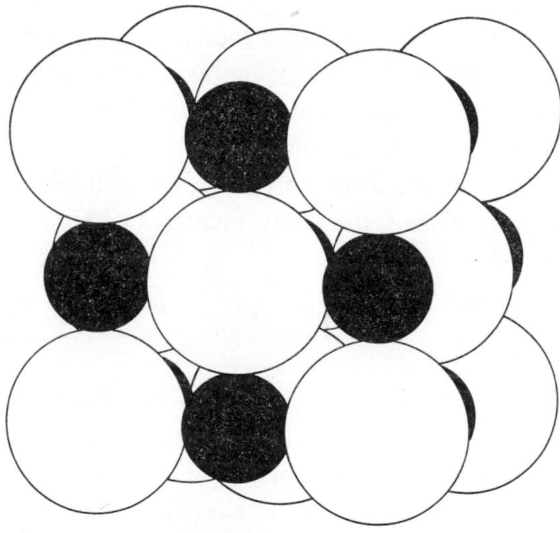

Fig. 3.06. Clinographic projection of the unit cell of the cubic structure of sodium chloride showing the ions in their correct relative sizes for NaCl. The solid circles represent sodium ions. The diagram is reproduced on the same scale as fig. 3·01.

3.09. The fact that ionic structures are composed of ions of characteristic and often very different radii emphasizes a shortcoming of the conventional crystal-structure diagrams such as fig. 3.01: these diagrams show clearly the positions of the ionic centres but give no indication of the relative sizes of the ions. Fig. 3.06 is an alternative presentation of the sodium chloride structure in which the ions are shown with approximately the correct relative radii. It will be seen that although the diagram gives a more faithful representation of the ionic packing the representation of the co-ordination of the ions is far less clear than in fig. 3.01. Diagrams showing only atomic centres are therefore generally preferred, but in studying them the relative sizes of the component atoms must always be kept in mind.

THE GEOMETRICAL BASIS OF MORPHOTROPY*

The effect of radius ratio

3.10. It is natural to enquire why different AX compounds should possess different structures, and, in particular, why CsCl, CsBr and CsI should have a structure different from that of the other alkali halides. We can answer this question if we consider fig. 3.07*a*, which represents a section through the caesium chloride unit cell on a vertical diagonal plane. The ions in this diagram are shown in their correct relative sizes for Cs^+ and Cl^-, and anions and cations are seen to be in contact at the points P. Now let us suppose that the cations are replaced by others of

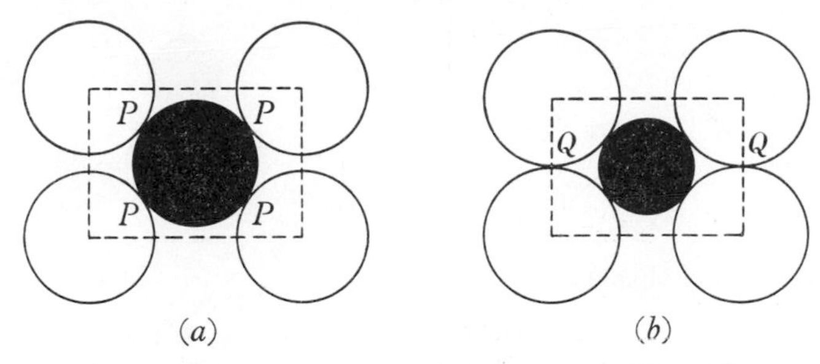

Fig. 3.07. Section through a unit cell of the caesium chloride structure on a vertical diagonal plane. The solid circles represent the cations. In (*a*) the ions are shown in their correct relative sizes for CsCl; (*b*) corresponds to the critical radius ratio for anion–anion contact.

progressively smaller radius, the anions remaining unchanged. As this is done the A–X distance between cation and anion decreases, the potential energy of the system falls and the structure becomes more stable. There is, however, a limit to this process, for there comes a situation, as shown in fig. 3.07*b*, when the anions are themselves in contact at Q. Any further reduction in the size of the cation will make no difference to the A–X distance or to the energy of the system; all that will happen is that the cations will be free to 'rattle' in the interstices between the negative ions. It is clear from the geometry of fig. 3.07*b* that this transition will occur when the ratio of the radii of cations and anions is given by

$$r^+/r^- = \sqrt{3}-1 = 0{\cdot}732.$$

* Morphotropy may be defined as a progressive change in crystal structure brought about by systematic chemical substitution.

Thus the variation of the energy of the structure as a function of radius ratio will be of the form represented by curve (*a*) in fig. 3.08, with a discontinuity at this critical value of r^+/r^-.

Now let us consider the sodium chloride structure in a similar way. Fig. 3.09 represents a section through this structure on a plane parallel to one face of the cubic unit cell with the ions in their correct relative sizes for Na^+ and Cl^-. The anions and cations are in contact at *P*. As before, let us now suppose that the sodium ions are replaced by others of

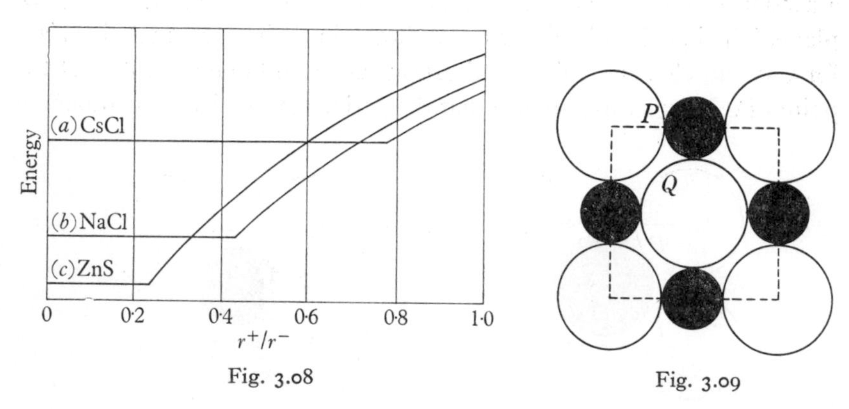

Fig. 3.08

Fig. 3.09

Fig. 3.08. The electrostatic component of the lattice energy of the caesium chloride, sodium chloride and zincblende ionic structures as a function of the radius ratio r^+/r^- (r^- assumed constant). The energy values are negative on an arbitrary scale, the zero of energy being above the top of the figure.

Fig. 3.09. Section through a unit cell of the sodium chloride structure on a plane parallel to a cube face. The solid circles represent the cations, and the ions are shown in their correct relative sizes for NaCl.

progressively smaller radius. The *A–X* distance will decrease, and with it the energy of the structure, but again there is a limit to this reduction when the anions come into mutual contact at *Q*. This will occur at a radius ratio given by

$$r^+/r^- = \sqrt{2}-1 = 0{\cdot}414.$$

For reasons which will appear shortly, the energy of the sodium chloride structure is slightly greater than that of the caesium chloride arrangements for large values of the ratio r^+/r^-. The variation of the energy of the sodium chloride structure with radius ratio is therefore as represented by curve (*b*) in fig. 3.08, with a discontinuity at $r^+/r^- = 0{\cdot}414$. It will be seen from this figure that the caesium chloride structure is the more stable of the two (but only slightly so) for radius ratios exceeding about 0·7, but that below this value the sodium chloride structure is by far the more stable.

We can continue this argument by considering the zincblende structure. In this case the critical radius ratio, corresponding to anion–anion contact, can be readily shown to be given by

$$r^+/r^- = \tfrac{1}{2}\sqrt{6}-1 = 0{\cdot}225.$$

For large values of this radius ratio the energy of the structure is considerably greater than that of the sodium chloride arrangement. The variation of energy with radius ratio is therefore as represented by curve (*c*) in fig. 3.08, and it will be seen that the zincblende arrangement is the most stable of the three structures discussed for values of r^+/r^- less than about 0·3.

The radius ratio r^+/r^- for each of the alkali halides is shown in table 3.03. Consideration of these values reveals that CsCl, CsBr and CsI would, indeed, be expected to have the caesium chloride structure, and that the majority of the remaining halides would be expected to show the sodium chloride arrangement. There are, to be sure, a number of halides with $r^+/r^- > 0{\cdot}7$ which, nevertheless, have the sodium chloride rather than the caesium chloride structure. Fig. 3.08, however, emphasizes that energetically there is little difference between these two structures when the radius ratio is large, and there are in any case other factors contributing to the lattice energy which we have so far ignored in our discussion.

Table 3.03. *The radius ratio r^+/r^- for the alkali halides*

	Li	Na	K	Rb	Cs
F	0·44	0·70	0·98	0·92*	0·80*
Cl	0·33	0·52	0·74	0·82	0·93
Br	0·31	0·49	0·68	0·76	0·87
I	0·28	0·44	0·62	0·69	0·78

* In these salts the cations are the larger ions; the radius ratios quoted are therefore r^-/r^+.

3.11. We may thus summarize our conclusions by saying that in simple ionic structures the atomic arrangement is determined primarily by geometrical considerations. Subject to the ions being present in the proper relative numbers to achieve electrical neutrality, the structure will be the one in which each of the cations (assuming these to be the smaller) is in contact with the maximum number of anions geometrically possible. Chemical considerations are unimportant, so that chemically similar compounds may have quite different crystal structures, while chemically dissimilar compounds may have the same structural arrangement.

Variation of ionic radii

Variation with co-ordination number

3.12. We have explained above that the concept of an ion as of characteristic and strictly constant size is subject to certain qualifications. In the first place it is found that the radius of a given ion is a function of the co-ordination in which it occurs; thus the radius of the Cl^- ion in sodium chloride is not precisely the same as in caesium chloride. This effect may be attributed to the polarization of the ions, polarization being crudely pictured as the deformation of the electronic structure of an ion by the electric field of its neighbours, as a result of which the interionic distance is reduced. It may be readily seen in a general way that this deformation may be expected to be most marked in structures of low or irregular co-ordination, and indeed it is found that in the three common AX structures described above a small but significant decrease in the effective radius of a given ion takes place with decreasing co-ordination number. If the radii corresponding to the sodium chloride structure are taken as standard, those in the 8-co-ordinated caesium chloride structure are systematically about 3 per cent larger and those in the 4-co-ordinated zincblende structure are about 5 per cent smaller. Thus the variation of radius with co-ordination number may be represented as follows:

Co-ordination number	8	6	4
Radius	1·03	1·00	0·95

It is clearly desirable that ionic radii should be expressed in a form appropriate to some particular co-ordination, and that of the sodium chloride structure is conventionally chosen as standard. The crystal radii quoted in table 3.02 and fig. 3.05 are therefore those for 6-co-ordination, and must be appropriately corrected for use in structures in which the co-ordination is different.

Variation with radius ratio

3.13. A second factor which influences ionic radii is the value of the radius ratio r^+/r^-. If we consider the two caesium chloride structures shown in fig. 3.07 it is clear that as the radius ratio approaches the critical value for anion–anion contact the repulsion between these ions will progressively increase and will have the effect of distending the structure, so increasing the apparent radii of the component ions. This effect is commonly ignored but it is by no means always negligible. Thus the radii in table 3.02 and fig. 3.05 apply strictly only to 6-co-ordinated structures in which r^+/r^- is approximately 0·7. For values of r^+/r^- appreciably less than this the *A–X* distances in structures of the sodium chloride type are significantly greater than the sum of the radii given, as is shown by the following figures for the alkali bromides:

Halide	RbBr	KBr	NaBr	LiBr
r^+/r^-	0·76	0·68	0·49	0·31
Sum of radii (Å)	3·43	3·28	2·90	2·55
Observed *A–X* distance (Å)	3·43	3·29	2·98	2·75

It will be seen from these figures that the observed *A–X* distance approximates closely to the sum of the radii when the radius ratio is in the neighbourhood of 0·7, but that as the ratio falls considerable discrepancies arise.

LATTICE THEORY OF IONIC CRYSTALS

Lattice energy

3.14. In our discussion of the factors governing the occurrence of the caesium chloride, sodium chloride and zincblende structures we have introduced the conception of lattice energy. For an ionic structure this is the work which must be done to disperse the crystal into an assembly of widely separated ions, or conversely the energy liberated when the crystal is formed from such an assembly of ions. The elementary physical picture of the ionic bond as arising from the electrostatic attraction between oppositely charged ions makes it possible to calculate this quantity in the case of a number of simple structures, and as early as 1918 Born and Landé developed a quantitative theory applicable to the structures of the alkali halides.

The electrostatic potential energy of a pair of ions considered as point charges $z_1 e$ and $z_2 e$ at a distance r apart is given by

$$u = z_1 z_2 e^2/r.$$

In a crystal, however, the potential energy for pairs of ions will not be given by this expression, for it will be necessary to take account of the mutual potential energies of all the charges in the structure. When this is done the electrostatic energy per pairs of ions is found to be given by the relation

$$u = Az_1z_2e^2/r, \tag{3.01}$$

where A is a numerical quantity, of the order of unity, termed the *Madelung constant*. The value of this constant depends upon the particular structure considered, and its exact calculation presents some difficulty. It has, however, been evaluated for a number of simple structures, with the following results:

Structure	Madelung constant
Caesium chloride	1·763
Sodium chloride	1·748
Zincblende	1·638

It is the differences between these values of the Madelung constant which account for the fact that the lattice energies of the three structures differ slightly at large values of the radius ratio r^+/r^-, as shown in fig. 3.08.

The greater part of the early work on lattice theory was undertaken with reference to the alkali halides, and it will simplify our discussion if we develop our further arguments in terms of these salts. For univalent ions, carrying charges $+e$ and $-e$, equation (3.01) becomes

$$u = -Ae^2/r. \tag{3.02}$$

Effect of repulsive forces

3.15. It is clear that equation (3.02) alone cannot represent completely the lattice energy of a crystal, for it is necessary to postulate not only an electrostatic attraction but also a force of repulsion between the ions in order that they may be in equilibrium at a distance apart corresponding to their characteristic radii. This repulsive force, unlike the force of attraction, cannot be described in terms of any simple physical picture, and we shall not discuss its origin here. All that it is necessary to say at this stage is that the fact that the component ions in ionic crystals behave to a close approximation as rigid spheres of characteristic constant radius implies that the force, whatever its nature, must be one which varies very rapidly with interatomic distance. The same conclusion can be drawn from the low compressibility of ionic crystals.

It has been found that the contribution of the interionic repulsive forces to the lattice energy can be well represented by a term of the form $B \exp[-r/\rho]$, where B and ρ are constants for any one crystal. The expression for the lattice energy therefore becomes

$$u = -Ae^2/r + B \exp[-r/\rho], \tag{3.03}$$

in which the two terms represent the effects of the attractive and of the repulsive forces, respectively. We can eliminate B from this equation by expressing the condition that the equilibrium distance r_0 between adjacent ions is that at which the potential energy is a minimum. This condition is

$$\left(\frac{\partial u}{\partial r}\right)_{r=r_0} = \frac{Ae^2}{r^2} - \frac{B}{\rho} \exp[-r/\rho] = 0,$$

whence $$B \exp[-r_0/\rho] = Ae^2 \rho/r_0^2,$$

so that $$(u)_{r=r_0} = -\frac{Ae^2}{r_0}\left(1 - \frac{\rho}{r_0}\right). \tag{3.04}$$

The value of ρ/r_0 can be deduced experimentally from compressibility measurements, for we can see in a general way from equation (3.03) that a small value of this quantity represents a rapid variation of repulsive force with distance and a correspondingly small compressibility. Values of ρ/r_0 for the alkali halides so obtained vary between about 0·09 and 0·12, confirming the rapid variation with distance which we have presupposed and indicating, from equation (3.04), that the contribution of the repulsive forces to the lattice energies of these salts is of the order of 10 per cent of that due to the forces of electrostatic attraction.

Lattice energies of the alkali halides computed from equation (3.04) are given in column (3) of table 3.04.

Effect of van der Waals forces

3.16. There is yet a third component of the lattice energy which must be considered, namely, that due to the van der Waals forces of attraction between the ions. These forces, like those of interionic repulsion, cannot be described in simple physical terms, but they, too, have been the subject of much recent work to which we shall have occasion to refer in chapter 6. The effect of these forces cannot be treated in isolation, for it can be shown that its inclusion involves a corresponding modification in the estimate of the effect of the repulsive forces. In consequence it is found that the necessary correction to the lattice energy so far computed is very small, and for the alkali halides never exceeds 3 kcal/mole.

Table 3.04. *Lattice energies of the alkali halides*

Values are in kcal/mole

(1) Halide	(2) *A–X* distance (Å)	(3) Lattice energy from equation (3.04)	(4) Lattice energy computed by more refined methods	(5) Observed lattice energy	(6) Lattice energy from Born–Haber cycle
LiF	2·01	−254	−244	—	−244
LiCl	2·57	−196	−200	—	−202
LiBr	2·75	−184	−189	—	−192
LiI	3·00	−169	−176	—	−180
NaF	2·31	−220	−215	—	−217
NaCl	2·81	−182	−184	−181	−185
NaBr	2·98	−173	−176	−176	−176
NaI	3·23	−159	−164	−166	−166
KF	2·66	−193	−193	—	−193
KCl	3·14	−166	−168	—	−168
KBr	3·29	−158	−161	−160	−161
KI	3·53	−148	−152	−154	−153
RbF	2·82	−183	−183	—	−185
RbCl	3·27	−159	−162	—	−164
RbBr	3·43	−152	−156	−151	−158
RbI	3·66	−143	−148	−146	−150
CsF	3·00	−175	−176	—	−173
CsCl	3·56	−150	−153	—	−158
CsBr	3·71	−143	−150	—	−152
CsI	3·95	−135	−143	−142	−145

Column (4) of table 3.04 shows the total lattice energies of the alkali halides as computed by more recent and refined methods. These values include the effect of the van der Waals forces and also include a contribution representing the zero-point energy. It will be seen that they do not differ greatly from the values in column (3) derived on the elementary theory.

The Born–Haber cycle

3.17. The lattice energy which we have discussed above is the amount of work which must be expended to disperse a crystal into an assemblage of widely separated ions. As such it cannot be immediately compared with any readily measurable quantity, and, in particular, is not to be identified either with the heat of sublimation, which is the energy necessary to disperse the crystal into a molecular gas, or with the chemical heat of formation, which is the energy released when the crystal is formed from metal atoms and diatomic halogen molecules. In some cases lattice energies have been measured directly, and values so obtained are shown in column (5) of table 3.04. In other cases indirect estimates can be made in terms of more readily measurable quantities by means of a cyclical process, proposed by Born and Haber, which we may illustrate by reference to the specific example of sodium chloride (fig. 3.10).

Consider one gram molecule of sodium chloride in the crystalline state. By the expenditure of an amount of work equal to $-u_{NaCl}$ this may be converted into a gas of ions Na^+ and Cl^-. The conversion of these ions into neutral atoms involves the expenditure of an amount of

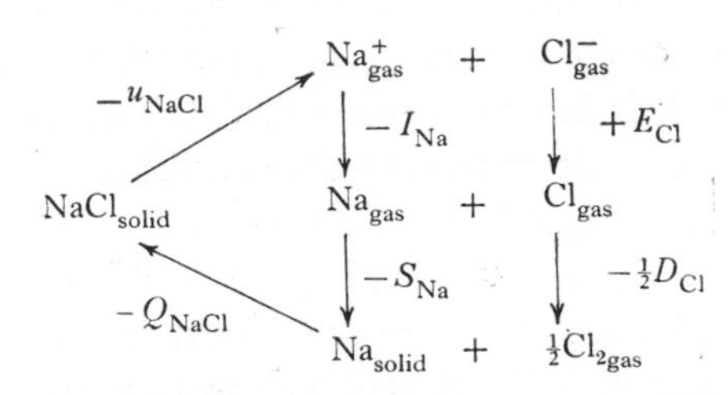

Fig. 3.10. The Born–Haber cycle applied to sodium chloride.

energy E_{Cl}, the electron affinity of the chlorine atom, and the recovery of an amount of energy I_{Na}, the ionization energy of sodium. If the sodium now be allowed to condense to the solid, and the chlorine atoms to associate to Cl_2, a further amount of energy $S_{Na}+\frac{1}{2}D_{Cl}$ is recovered. Here S_{Na} is the heat of sublimation of sodium and D_{Cl} the heat of dissociation of chlorine per gram molecule. Finally, if the sodium and chlorine thus obtained are allowed to combine chemically the heat of formation Q_{NaCl} is released.

In performing this cycle of operations the total energy interchange must clearly be zero, so that we have

$$-u_{NaCl}+E_{Cl}-I_{Na}-S_{Na}-\tfrac{1}{2}D_{Cl}-Q_{NaCl} = 0,$$

or

$$u_{NaCl} = E_{Cl}-I_{Na}-S_{Na}-\tfrac{1}{2}D_{Cl}-Q_{NaCl}. \quad (3.05)$$

In this equation all the quantities on the right-hand side are known, so that u_{NaCl} can be deduced. Introducing numerical values, we obtain

$$u_{NaCl} = 85{\cdot}8-117{\cdot}9-26{\cdot}0-28{\cdot}8-98{\cdot}3 = -185{\cdot}2 \text{ kcal/mole.} \quad (3.06)$$

Values of lattice energies derived in this way are shown in column (6) of table 3.04. It will be seen that the agreement between the calculated values in column (4) and the experimental values in columns (5) and (6) is on the whole very satisfactory.

A study of table 3.04 shows, as of course follows immediately from equation (3.02), that lattice energy increases as interionic distance decreases.* Other properties also show a systematic dependence on this distance. Thus, as the distance between the ions increases and the lattice energy is reduced the melting point and hardness of the crystals fall progressively, while conversely the coefficients of thermal expansion and of compressibility increase.

3.18. A number of oxides AO and sulphides AS of divalent metals are also essentially ionic in character and crystallize with the sodium chloride structure, so that their lattice energies can be calculated by a process exactly analogous to that described above. Values for some of these compounds are shown in table 3.05. Again we see an increase of energy with decreasing interionic distance, but we also note that the absolute values of the energies are much greater than those of the alkali halides, owing to the increase in ionic charge. The melting points and hardness of the oxides and sulphides are correspondingly much higher than those of the alkali halides.

Table 3.05. *Lattice energies of some oxides and sulphides*

Values are in kcal/mole

Oxide	A–X distance (Å)	Lattice energy	Sulphide	A–X distance (Å)	Lattice energy
MgO	2·10	−940	MgS	2·60	−778
CaO	2·40	−842	CaS	2·84	−722
SrO	2·57	−791	SrS	2·94	−687
BaO	2·76	−747	BaS	3·18	−656

Application of lattice theory to other properties

3.19. The theory of ionic crystals discussed above in its application to lattice energy has also been applied to other physical properties of these crystals. We have already seen how the compressibility of a crystal may be deduced, and have shown that experimental values for this quantity

* Lattice energy is, of course, negative since the zero of energy corresponds to the state in which the crystal is dispersed into widely separated ions. In speaking of an increase in energy we adopt the conventional usage and imply an increase in the arithmetical value of this negative quantity.

are in fact required in estimating the magnitude of the repulsive forces between ions. Born early calculated other elastic constants and also the characteristic frequency of lattice vibration, corresponding to the infra-red *Reststrahlen* frequency, and later deduced coefficients of thermal expansion and breaking strengths of simple ionic crystals. In all except the last case, values in satisfactory agreement with experiment were obtained. Breaking strength is now known to be a 'structure sensitive' property profoundly influenced by the history and precise condition of the specimen.

The structural significance of lattice theory

3.20. The structural importance of the quantitative lattice theory of ionic crystals discussed above lies in the light which it throws on the stability of crystal structures and the conditions which determine the appearance of different structures in substances chemically closely related, on the types of binding which occur in different structures, and on questions of solubility. We may consider these several points separately.

The stability of ionic structures

3.21. We have already seen that among the alkali halides the primary factor which determines the appearance of one structure or another is the radius ratio r^+/r^-. This is only another way of saying that the most stable structure is the one with the largest (negative) lattice energy, and the curves of fig. 3.08 are in fact curves showing the variation with r^+/r^- of the electrostatic component of the lattice energy for the three structure types under discussion; at any particular value of r^+/r^- the structure which occurs will be that corresponding to the lowest curve. Conversely, if the radius ratio happens to approximate to a value at which two curves cross (about 0·7 for the caesium chloride and sodium chloride structures, or about 0·3 for the sodium chloride and zincblende structures) we might expect the compound to be dimorphous and to exist with both structural arrangements. This does not, in fact, happen with the alkali halides at normal temperatures and pressures, but we shall later encounter many examples of polymorphism which can be attributed to just this cause.

3.22. Considerations of radius ratio enable us to predict which of several possible structural arrangements a given ionic compound will assume; they do not, however, enable us to predict whether or not a given compound can exist as an ionic crystal. To consider this problem we must return to equation (3.05), which we can write in the alternative form

$$Q_{NaCl} = -(I_{Na}+S_{Na}+\tfrac{1}{2}D_{Cl}-E_{Cl})-u_{NaCl}. \tag{3.07}$$

From this we see that the heat of formation of sodium chloride in the solid state may be regarded as made up of two components. The terms in parentheses on the right-hand side of equation (3.07) represent the work which must be done to produce a dispersed assembly of Na^+ and Cl^- ions from metallic sodium and chlorine gas; the final term represents the work recovered when these ions condense together to form the crystal. If $|u_{NaCl}|$ exceeds $|(I_{Na}+S_{Na}+\tfrac{1}{2}D_{Cl}-E_{Cl})|$ the whole process is exothermic, and we may expect sodium and chlorine to react together to form a stable crystal of sodium chloride. This, in fact, is of course the case, as we can see by introducing numerical values into equation (3.07):

$$\begin{aligned} Q_{NaCl} &= -(117{\cdot}9+26{\cdot}0+28{\cdot}8-85{\cdot}8)+185{\cdot}2 \\ &= -86{\cdot}9+185{\cdot}2 \\ &= +98{\cdot}3 \text{ kcal/mole.} \end{aligned} \tag{3.08}$$

These figures show that the process of forming the ions is endothermic, but that when the ions condense to form a crystal an amount of energy is released exceeding that previously expended. The overall process is therefore exothermic and the crystal is stable. The same conclusion is reached if we consider any other of the alkali halides.

The figures in equation (3.08) enable us to see at once what are the most important factors in determining the stability of these structures. Of the terms in parentheses the largest is seen to be the ionization energy of the cation, and the stability will be greater the smaller this quantity. Conversely, it is necessary that the lattice energy should be as large as possible, as will be the case (for a given state of ionization) if the ions approach as closely as possible. In other words, as is naturally to be expected, ionic structures will be formed most readily if the ions themselves can be readily produced and if their radii are not unduly large. These conditions are satisfied by all the alkali halides, and the close agreement between the calculated lattice energies of these salts and those derived experimentally is strong evidence that in these compounds the assumption of ionic bonding is in fact correct. Conversely, a wide discrepancy between calculated and observed energies would argue that the type of bonding which actually occurs is different from the ionic bonding assumed in the calculation.

The stability of hypothetical compounds

3.23. We may apply considerations of lattice energy to explain not only the existence of actual structures but also the non-existence of hypothetical compounds. The atom of neon can be ionized to form the ion Ne^+, and we might imagine an ionic crystal of composition NeCl with the sodium chloride structure. The ionization energy of neon is much larger than that of sodium, namely 496 kcal/mole. Introducing this value in the equation corresponding to (3.07), we have

$$\begin{aligned} Q_{\text{NeCl}} &= -(496+0+28{\cdot}8-85{\cdot}8)-u_{\text{NeCl}} \\ &= -439-u_{\text{NeCl}}. \end{aligned}$$

Thus NeCl will be stable only if the lattice energy exceeds 439 kcal/mole. This, however, is an impossibly large value, for it would imply an Ne–Cl separation of about 1·2 Å, a distance considerably less than the radius of the Cl^- ion alone. In other words, it is the high ionization energy of neon which prevents it from forming ionic compounds.

Solution

3.24. The lattice energy of an ionic crystal is an important factor in determining its solubility. In an ionizing solvent the solute occurs as dispersed ions, and solution will be possible only if an amount of work equal to the lattice energy is made available in the process. In the case of solution in water there is ample evidence that the ions exist co-ordinated by water molecules, to which they are relatively strongly attached owing to the polar character of these molecules. It is the energy of hydration of the ions which provides the work necessary to disperse them into solution. Thus ionic solution is possible only if the solvent is polar and the lattice energy is reasonably small. Sodium chloride is therefore insoluble in benzene because this is not a polar solvent, and magnesium oxide is insoluble in water because its lattice energy is too large.

An alternative approach to the problem of the solution of ionic salts is to regard it as arising from the high dielectric constant of water, which reduces the interionic attraction and the lattice energy. At first sight this treatment may seem different from that just described, but in fact it is equivalent to it; it is just because the molecules are strongly polar that water has a high dielectric constant.

3.25. The brief account which we have given of the structural importance of lattice theory illustrates in a general way some of the directions in which it may be applied, but it also emphasizes some of the inevitable shortcomings of the quantitative treatment. We have discussed the theory in its application to the relatively very simple structures of the alkali halides, and even for these we have seen that refined calculations are necessary if values of the lattice energy are to be obtained in close accord with experiment. The vast majority of structures to which the methods of X-ray analysis have been applied are, however, of far greater complexity, and for these the problem of precise treatment is as yet unsolved. Moreover, we have considered only ionic structures, and the problem of those in which other types of bonding occur is even less tractable. For the present, therefore, we must unfortunately be content with a purely qualitative account of the properties of the great majority of known structures.

4 THE COVALENT BOND AND SOME COVALENT STRUCTURES

INTRODUCTION

Early theory of the covalent bond

4.01. The electron transfer from an electropositive to an electronegative element which occurs in ionic compounds, and by means of which each atom achieves a stable electron configuration, clearly cannot be responsible for the formation of such molecules as those of hydrogen, H_2, and fluorine, F_2. The two atoms in these molecules can, however, both attain a stable inert gas configuration by a *sharing* of two electrons in a way which may be symbolized thus:

$$\mathrm{H}\underset{\times}{\overset{\bullet}{}}\mathrm{H} \qquad\qquad \begin{array}{ccccc} & \cdot\,\cdot & & \times\,\times & \\ \cdot & & \cdot & & \times \\ & \mathrm{F} & & \mathrm{F} & \\ \cdot & & \times & & \times \\ & \cdot\,\cdot & & \times\,\times & \end{array}$$

where the dots and crosses represent the valency electrons of the outermost shell. If the two shared electrons, one from each atom, are regarded as belonging to both atoms, each atom has in effect acquired an extra electron in its extranuclear structure. Such a mechanism for the formation of the covalent or homopolar bond was proposed in 1916 by Lewis, whose work may be regarded as the foundation of modern valency theory.

The conception of a covalent bond as due to a sharing of electrons between the atoms which it binds together immediately gives rise to an important distinction between it and the ionic bond. In an ionic crystal the number of neighbours to which an ion may be bound is determined primarily by geometrical considerations, and may vary for a given ion from one structure to another. In the case of the covalent bond, however, it is clear that the number of such bonds by which a given atom can be linked to others is limited and is determined by quite different considerations. Thus, in the molecules of hydrogen and fluorine each atom is able to achieve a stable configuration by sharing one of its electrons with the other atom, but once this state has been realized no further covalent bonds can be formed. In forming a single covalent bond an atom effectively increases by one the number of electrons in its extranuclear structure, and the covalency of an atom is therefore determined by the extent to which its electronic configuration falls short of that corresponding to a stable outermost shell. For this reason oxygen, with six electrons in the L shell, is divalent, nitrogen, with five such electrons, is trivalent, and so on.

Later theory

4.02. The more recent treatment of the covalent bond, based on the application of the principles of wave mechanics, has developed in two distinct forms, usually termed the *valence-bond* and *molecular-orbital* theories, respectively. Although ultimately there is no inconsistency between these two theories, they do in fact approach the problem of chemical binding from different points of view, and we shall generally find that for our purposes the valence-bond treatment is the more suitable. This theory starts from concepts already familiar to the chemist and its conclusions can usually be expressed verbally in qualitative terms; the molecular-orbital theory, on the other hand, is more mathematical in its approach and lends itself less readily to such an interpretation. We shall, therefore, first discuss the valency-bond theory, and refer only briefly to the molecular-orbital treatment later in the chapter.

THE VALENCE-BOND THEORY

4.03. The valence-bond theory of covalent binding involves the reinterpretation of the Lewis picture of the chemical bond in terms of the more detailed description of the electronic structure of the elements discussed in chapter 2, and immediately new considerations arise. The bond is still conceived as due to a sharing of two electrons, or more precisely to an overlapping of their orbitals, but now sharing is possible only between electrons of opposite spin. It follows at once that electrons already paired in orbitals can play no part in covalent binding, so that for any given element it becomes important to consider which particular orbitals will be available for bond formation.

Valency in the first two periods

4.04. In the hydrogen atom the single electron in the s orbital is of course unpaired, and it is a pairing of the two electrons in two atoms which is responsible for the H–H bond in the hydrogen molecule; such a bond may therefore be termed an s–s bond. In fluorine, with the configuration $1s^2$, $2s^2 2p_x^2 2p_y^2 2p_z^1$, all the electrons are paired except that in the $2p_z$ orbital, and this electron alone is therefore available for bond formation. The F–F bond in the fluorine molecule, arising from the pairing of the $2p_z$ electrons in the two atoms, is accordingly a p–p bond. In hydrogen fluoride, HF, the H–F bond represents a pairing of the $1s$ electron in the hydrogen atom with the $2p_z$ electron in the fluorine atom, and is therefore an s–p bond. This distinction of bond type according to the orbitals involved does not arise in the Lewis picture but is of importance because the strengths of the different bonds are found to be different and to decrease in the order

$$p\text{–}p > s\text{–}p > s\text{–}s.$$

4.05. Oxygen has the configurations $1s^2$, $2s^2 2p_x^2 2p_y^1 2p_z^1$ and there are now two unpaired electrons, namely those in the $2p_y$ and $2p_z$ orbitals. Oxygen is therefore divalent, as in the Lewis picture, but here yet another new feature arises for we have seen that the p_x and p_z orbitals represent *directed* distributions of electron density mutually disposed at right angles (fig. 2.01 *b*). Accordingly, the two covalent bonds from an oxygen atom are directed in this characteristic way and we may expect the molecule of water, for example, to have the configuration

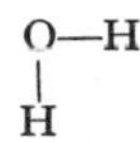

This distinctive spatial distribution of the bonds from an atom is a most important characteristic of the covalent link and one by means of which it may often be recognized in crystal structures.

4.06. In nitrogen, with the configuration $1s^2$, $2s^2 2p_x^1 2p_y^1 2p_z^1$, there are three unpaired electrons; accordingly, this element is trivalent with its three valencies mutually disposed at right angles, and we may expect a molecule such as that of ammonia, NH_3, to be pyramidal in form with the nitrogen atom at the apex and the hydrogen atoms at the corners of an equilateral triangular base.

Hybridization

4.07. Carbon, with the configuration $1s^2$, $2s^2 2p_x^1 2p_y^1$, presents an apparent difficulty: although there are four electrons in the L shell, two of these are already paired in the $2s$ orbital and only two are available for bond formation, suggesting that carbon should be divalent. To account for the well-known quadrivalency of carbon it is necessary to postulate that the paired $2s^2$ electrons are uncoupled and that one of them is promoted to the vacant $2p_z$ orbital to give four unpaired electrons in the L shell with the resultant configuration $1s^2$, $2s^1 2p_x^1 2p_y^1 2p_z^1$. Energy will be required for this process, but the energy will be available from the heat of reaction released when the bonds are formed. Even so, it would still appear that four electrons available are not equivalent, three being in p orbitals and one in an s orbital. By a process of *hybridization*, however, these four orbitals are combined to form four precisely equivalent orbitals, and the *hybrid bonds* arising from these new orbitals are directed towards the corners of a regular tetrahedron. Moreover, by this process of hybridization the bonds acquire additional strength so that they are stronger than the p–p bonds already discussed. Bonds arising from hybridization of the type just described are termed sp^3 bonds since one s and three p electrons are involved.*

4.08. Hybridization must similarly be invoked to account for the known valencies of boron and beryllium. The boron atom in its ground state has the configuration $1s^2$, $2s^2 2p_x^1$, with only one unpaired electron available for bond formation. If, however, the $2s^2$ electrons are uncoupled and one is promoted to the $2p_y$ orbital we obtain the configuration $1s^2$, $2s^1 2p_x^1 2p_y^1$, and if now the three unpaired electrons are hybridized we obtain three precisely equivalent sp^2 orbitals, the bonds corresponding to which lie in a plane directed to the corners of an equilateral triangle. In this way the trivalent state of boron is accounted for. The distinction between the planar arrangement of the sp^2 hybrid bonds in boron and the pyramidal disposition of the three pure p bonds in nitrogen should be noted.

In beryllium ($1s^2$, $2s^2$) there are in the ground state no unpaired electrons. Uncoupling of the $2s^2$ electrons, however, gives the configuration $1s^2$, $2s^1 2p_x^1$, and if this is followed by hybridization two equivalent sp orbitals are produced, the bonds from which are oppositely directed

* It may well seem that the introduction of the concept of hybridization to account for the valency of the carbon atom is both arbitrary and artificial, and this would indeed be the case if it had no other basis. As with so many other aspects of the application of quantum mechanics to valency theory, however, the justification is mathematical, and here, where we are concerned only with those aspects of valency theory which have a direct bearing on crystal structure, we shall unfortunately often be compelled to quote conclusions without seeking a rigorous justification; for the mathematical arguments the reader is referred to the many works on the theory of valency already available.

along a straight line. Again we note the distinction between the linear arrangement of the *sp* hybrid bonds in beryllium and the right-angled distribution of the two pure *p* bonds in oxygen.

Valency in the later periods

4.09. In the elements of the second period just discussed the maximum possible number of unpaired electrons in the *L*-shell, after uncoupling if necessary, is clearly four, namely, one *s* electron and three *p* electrons; the covalency of elements in this period therefore cannot exceed 4. When, however, we come to consider elements of higher atomic number new possibilities arise because now *d* (and ultimately *f*) electrons may be involved in covalent bonds. These electrons do not generally occur in the outermost shell of an atom and therefore do not normally take part in 'pure' covalent binding. We have seen, however, that in many elements, and especially in those of the transition series, the incompletely filled *d* orbitals of the penultimate shell differ little in energy from the *s* and *p* orbitals of the outermost shell (fig. 2.02). If one or more of the *d* electrons is promoted to the outermost shell hybridization of the $(n-1)d$, *ns* and *np* electrons may occur. Such a hybrid is termed a $d^x s^y p^z$ hybrid, where the superscripts indicate the number of electrons of each type involved and the sequence of orbitals in the symbol implies that the energy of the *d* orbital is lower than that of the *s*. Alternatively, *s* and *p* electrons may be promoted to vacant *d* orbitals in the same shell.

Table 4.01. *Some common hybrid orbitals*

Hybrid	Number of bonds	Distribution of bonds	Ref.*	Examples
sp	2	Linear	(*a*)	Cu^{I}, Ag^{I}, Au^{I}
sp^2	3	Planar to corners of equilateral triangle	(*b*)	—
dsp^2	4	Planar to corners of square	(*c*)	Cu^{II}, Ag^{II}, Au^{III}; Ni^{II}, Pd^{II}, Pt^{II}
sp^3	4	To corners of regular tetrahedron	(*d*)	Cu^{I}, Ag^{I}
d^2sp^3	6	To corners of regular octahedron	(*e*)	Fe^{II}, Fe^{IV}, Co^{II}, Co^{III}, Ni^{II}, Ni^{III}, Pd^{IV}, Pt^{IV}
sp^3d^2	6	To corners of regular octahedron	(*e*)	Ti^{IV}, Zr^{IV}
d^4sp	6	To corners of trigonal prism	(*f*)	Mo^{IV}

* References are to the parts of fig. 4.01.

In this case hybridization would involve the *ns*, *np* and *nd* electrons and would give rise to $s^x p^y d^z$ hybrids.

For elements of high atomic number the number of possible hybrids is large, and unfortunately quantum mechanical calculations are rarely sufficiently precise to predict with certainty which particular state of hybridization a given element will adopt. Generally, therefore, it is necessary for us to determine experimentally the number and disposition

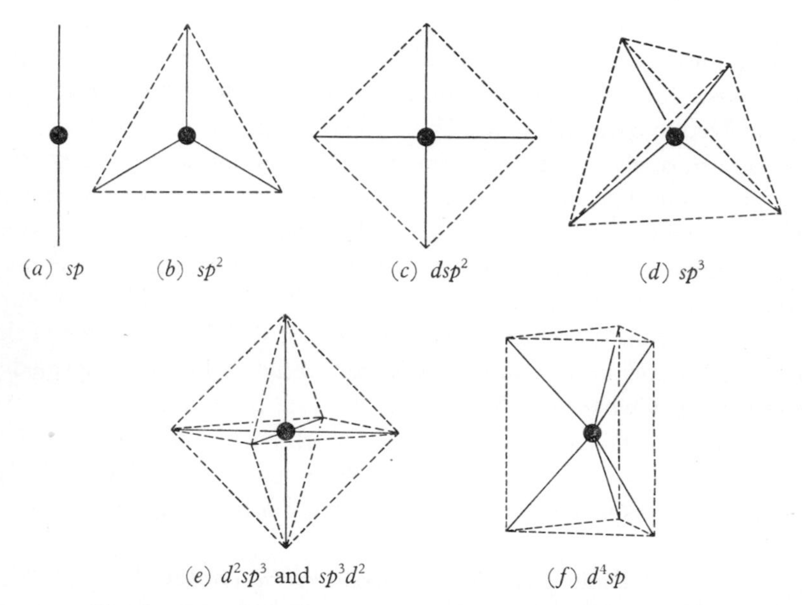

Fig. 4.01. The bond configurations corresponding to some simple hybrid orbitals. (*a*) *sp*—linear; (*b*) sp^2—to corners of an equilateral triangle; (*c*) dsp^2—to corners of a square; (*d*) sp^3—to corners of a regular tetrahedron; (*e*) d^2sp^3 and sp^3d^2—to corners of a regular octahedron; (*f*) d^4sp—to corners of a trigonal prism.

of the bonds about the atom under consideration and from these observations to deduce the hybrid involved. Table 4.01 records most of the hybrids of common occurrence in crystal structures, and fig. 4.01 shows the corresponding bond distributions. Configurations of particularly common occurrence in crystals (apart from the *sp*, sp^2 and sp^3 hybrids already discussed) are (1) the dsp^2 hybrid, giving rise to four coplanar bonds directed to the corners of a square, and (2) the d^2sp^3 and sp^3d^2 hybrids, giving rise to bonds directed to the corners of a regular octahedron.

4.10. Also shown in table 4.01 are a number of examples of metallic elements which have been found experimentally to adopt the states of hybridization indicated. It will be seen that for these elements the valency state has been quoted; the reason for this will be made clear when we come later to consider the crystal structures of some of the compounds of these elements in detail. At this point, however, we may draw attention to the fact that some elements (e.g. Ni^{II}) can occur in several alternative states of hybridization, some (e.g. Pt^{II}) are found only in one, and for some the state of hybridization depends on the valency (e.g. Cu^{II} is always dsp^2, Cu^{I} may be sp or sp^3 hybridized).

SOME SIMPLE COVALENT STRUCTURES

4.11. The characteristic properties of the covalent bond discussed above impose severe restrictions on the possible types of crystal structure in which such bonds may occur. Since the number of neighbours to which a given atom may be bound by covalent bonds is limited to the covalency of that atom, and since this covalency is usually small, the vast majority of structures containing the covalent bond are molecular structures in which covalent forces occur within discrete molecules but in which the molecules are bound to one another by forces of a different kind. There are, however, a number of structures in which the binding throughout is due to covalent forces, and of these the simplest is that of carbon in the form of diamond.

The diamond structure

4.12. The cubic unit cell of the diamond structure is illustrated in fig. 4.02. Here it will be seen that every carbon atom is surrounded by only four others and that these neighbours are symmetrically arranged at the corners of a regular tetrahedron. The C–C distance is 1·54 Å. Geometrically the structure may be regarded as the zincblende structure (§3.06) but with carbon atoms replacing those of both zinc and sulphur. Such a comparison, however, obscures a fundamental difference. The zincblende structure occurs (among ionic AX structures) because the relative sizes of the ions are such that it represents the tightest possible packing of these ions in equal numbers. In diamond the position is quite different. Here the atoms are all of the same size and if they were packed as closely as is geometrically possible each would be surrounded by twelve neighbours, an arrangement found in the crystal structures of many metals. The fact that in diamond the co-ordination is not twelvefold, but only fourfold, is of course an expression of the quadrivalency of the carbon atom in the sp^3 hybrid state. Similarly, the arrangement of neighbours about any one atom reflects the characteristic spatial distribution of the sp^3 orbitals.

It immediately follows from this discussion that the interatomic forces throughout the diamond structure are covalent or 'chemical' in nature, exactly analogous to those responsible for the formation of, say, the molecule of fluorine, F_2. Thus, if we regard F_2 as a molecule of fluorine

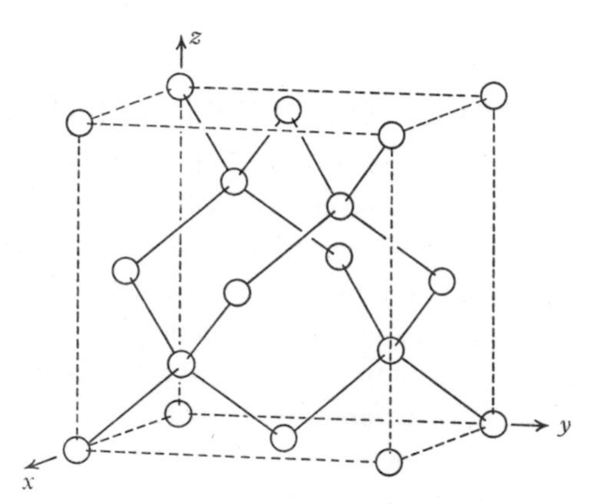

Fig. 4.02. Clinographic projection of the unit cell of the cubic structure of diamond, C.

we must also regard the whole crystal of diamond as an indefinitely large molecule of carbon, its physical coherence being therefore attributed to 'chemical' forces. This merely emphasizes that there is no fundamental distinction between chemical and physical forces, and that 'chemical' forces are in fact 'physical' in origin.

The zincblende and wurtzite structures

4.13. We have so far discussed the zincblende structure as an ionic arrangement, and it is certainly true that it is found in a limited number of ionic *AX* structures in which the radius ratio r^+/r^- is sufficiently small. A very much larger number of covalent *AX* compounds, however, also possess this structure or the closely related wurtzite arrangement illustrated in fig. 4.03. This latter structure is hexagonal but, as in zincblende, each atom is surrounded by four neighbours at the corners of a regular tetrahedron and it is only when next-nearest neighbours are also considered that the distinction between the two arrangements is seen.

Among the covalent *AX* compounds which crystallize with either the zincblende or wurtzite structure are ZnS itself, ZnO, AlN, AlP, HgS, CuCl and many more. At first sight it would seem that such an arrangement is inconsistent with the known covalencies of the atoms involved. In ZnS, for example, the zinc and sulphur atoms, with two and six electrons, respectively, in the outermost shell, would be expected to be limited to a covalency of 2 and therefore to be capable of covalent

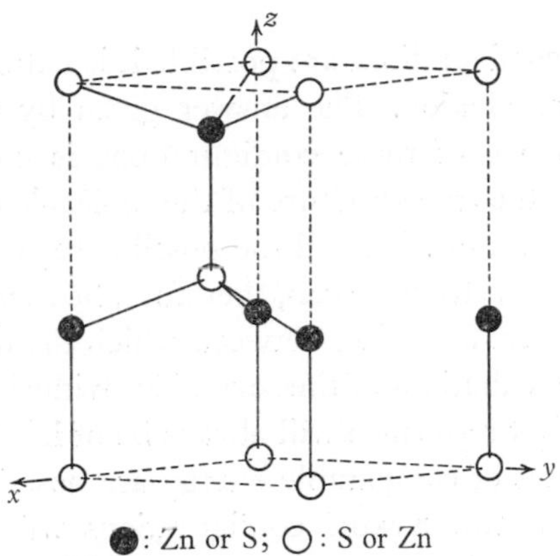

●: Zn or S; ○: S or Zn

Fig. 4.03. Clinographic projection of the unit cell of the hexagonal structure of wurtzite, ZnS.

bonding with only two other atoms. We may, however, regard the structure as one in which the covalent bonding is not between atoms but between ions carrying the formal charges Zn^{2-} and S^{2+}. If two electrons are removed from such sulphur atom and transferred to a zinc atom, both atoms effectively acquire an electron configuration in the outermost shell analogous to that of carbon, and so achieve a covalency of 4. Analogous arguments apply to the other *AX* compounds with the zincblende and wurtzite structures. We shall later find many further examples of this concept of formal charges on covalently bonded atoms.

RESONANCE

4.14. It is well known that there are many molecules whose structure can be formulated in more than one way. We are not here referring to different isomers, or to tautomerism between them, but to configurations in which the atomic nuclei retain their positions although the patterns of bond distribution are different. A familiar example is benzene, whose molecule may be formulated in terms of the two Kekulé configurations (I) and (II), and the three Dewar configurations (III), (IV) and (V).

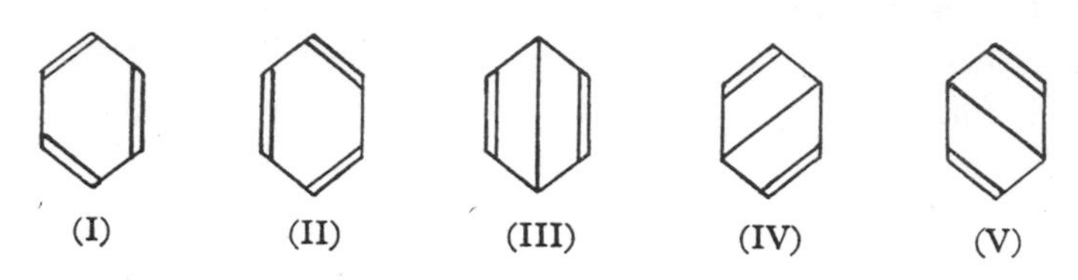

When alternative configurations are possible it is natural to ask which is the 'correct' representation. The answer given by the valence-bond theory is that no one of these configurations is correct or has any existence, and that the true structure of the molecule corresponds to an intermediate condition to which all the possible separate configurations contribute. The actual structure can, therefore, be represented only by a combination of several formulae, between which the molecule is said to 'resonate'. The contributions of the several individual structures are not necessarily equal, but resonance will always be of importance whenever alternative structures of comparable energy are posssible. Thus, in the case of benzene the two Kekulé configurations will make by far the largest contribution to the resonance structure. If we ignore the contribution of the Dewar configurations altogether our picture of the benzene molecule becomes a regular hexagon of carbon atoms with the C–C bonds all equivalent and all intermediate in character between single and double bonds. This is in complete accord with the chemical and crystallographic evidence that the benzene ring is a regular plane hexagon with a C–C distance of 1·39 Å, a value intermediate between the lengths 1·54 and 1·34 Å characteristic of C–C single and double bonds, respectively.

4.15. A second very important point which emerges from the valence-bond theory is that resonance confers additional stability on a covalent bond so that a resonance structure is energetically more favoured than any of the single configurations which contribute to it. Indeed, this is the fact which accounts for the very existence of the covalent bond, as we can illustrate by considering the simplest possible example, namely, the hydrogen molecule. If we distinguish the two hydrogen atoms by the subscripts $_A$ and $_B$, and represent the two electrons by the superscripts

[1] and [2] we can consider a hydrogen molecule as arising from either of the states

$$H_A^1 + H_B^2 \quad \text{or} \quad H_A^2 + H_B^1.$$

These two states are clearly of equal energy, resonance between them is possible, and it is this resonance which stabilizes the covalent H–H bond. In the particular case of benzene, the structure is stabilized by resonance to the extent of 37 kcal/mole.

4.16. In many cases we must give consideration also to configurations showing formal charges. The molecule of nitrous oxide, N_2O, is known to be linear with the atoms arranged in the sequence N–N–O. This arrangement cannot be explained in terms of covalent bonds between uncharged atoms, but if we entertain the possibility that the atoms may carry formal charges the configurations (I) and (II) must be considered.*

$$\underset{\text{(I)}}{\overset{\ominus}{N}{=}\overset{\oplus}{N}{=}O} \qquad\qquad \underset{\text{(II)}}{N{\equiv}\overset{\oplus}{N}{-}\overset{\ominus}{O}}$$

* In such cases the reader may find it helpful to consider separately the number of electrons in the *L* shell of each of the charged atoms concerned, thus:

	$\overset{\oplus}{N}$	$\overset{\ominus}{N}$	$\overset{\ominus}{O}$
Charged atom	N	N	O
Number of electrons	4	6	7
Valency of charged atom	4	2	1

These charged atoms will clearly have the valencies shown, conforming correctly to the bond patterns (I) and (II) above.

An alternative approach is to consider, say, the N–O bond in structure (II) as due to the sharing of two electrons, both of which are contributed by the neutral nitrogen atom; in this way the oxygen atom achieves a stable shell of eight electrons but also effectively acquires one unit of negative charge from the nitrogen atom. Such a covalent bond is sometimes termed a *dative bond* and represented by an arrow indicating the direction of transfer of the negative charge. In terms of this symbolism the structures (I) and (II) would be written

$$\underset{\text{(I)}}{N \overset{\leftarrow}{=} N{=}O} \qquad\qquad \underset{\text{(II)}}{N{\equiv}N \rightarrow O}$$

In this notation it is no longer necessary to indicate the charge distribution since this is implied by the arrows.

It is important to realize, however, that, once formed, a dative bond is in all respects identical with an 'ordinary' covalent bond. For this reason some authors prefer not to use the arrow symbolism but to represent all bonds in the same way, so that the above structures would be written simply as

$$\underset{\text{(I)}}{N{=}N{=}O} \quad \text{and} \quad \underset{\text{(II)}}{N{\equiv}N{-}O}$$

Here again the charge distribution, although not explicitly shown, is implicit in the seemingly abnormal valencies of the atoms concerned.

Both of these configurations are clearly strongly polar, and if resonance implied that the resultant configuration was a *mixture* of the component structures nitrous oxide would show a large dipole moment. In fact, the moment is very small, as would be expected if the two structures, the polarities of which are opposed, contributed nearly equally to the resonance configuration.

4.17. Even when a molecule can be assigned a plausible structure on the basis of neutral atoms it may, nevertheless, still be necessary to countenance possible alternative structures in which formal charges are involved. Thus, to revert to the example of the hydrogen molecule, it is possible to envisage a structure in which the binding is ionic between a proton H^+ and an H^- ion. Here again two possibilities will rise, namely

$$\overset{\oplus}{H_A} + \overset{\ominus}{H_B}\!: \quad \text{and} \quad :\!\overset{\ominus}{H_A} + \overset{\oplus}{H_B},$$

resonance between which will yield a non-polar bond. These polar configurations are less stable than the covalent structures discussed above, but calculation shows that their contribution is by no means negligible and that the bonding in the hydrogen molecule must therefore be treated as a resonance hybrid between the two covalent and the two ionic configurations.

Covalent/ionic resonance

4.18. The great importance of resonance theory from our point of view arises from the fact that it emphasizes that there is no absolute distinction between ionic and covalent bonding and that a continuous transition from one to the other is possible. In the hydrogen molecule, as we have just seen, there is an appreciable ionic component in the H–H bond, but this does not, of course, imply any dipole moment because the two possible ionic configurations contribute equally. Suppose, however, that we now consider the hydrogen fluoride molecule, HF. Here, again, a covalent configuration (I) is possible, but we must also take into account possible ionic configurations (II) and (III).

$$\text{H}\,\colon\!\ddot{\underset{..}{\text{F}}}\colon \qquad \overset{\oplus}{\text{H}} + \colon\!\overset{\ominus}{\ddot{\underset{..}{\text{F}}}}\colon \qquad \overset{\ominus}{\text{H}}\colon + \overset{\oplus}{\ddot{\underset{..}{\text{F}}}}\colon$$

(I) (II) (III)

We shall not now expect these two ionic configurations to contribute equally, for (II) is far more stable than (III) since fluorine is so much more strongly electronegative than hydrogen. Our picture of the HF molecule is therefore a resonance hybrid to which configurations (I) and (II) make the chief contributions. Such a hybrid will be polar, the magnitude of the dipole moment depending on the relative contributions of these two configurations. Exactly similar arguments will apply to the other hydrogen halides, but we shall expect the ionic component of the binding to diminish as fluorine is replaced by the successively less and less electronegative elements chlorine, bromine and iodine.

4.19. In favourable cases we can form a quantitative estimate of the contribution of the ionic component in a bond by comparing the observed dipole moment with that which would be expected on the basis of a purely ionic link. In HF, for example, the ratio of these quantities is about 0·45, from which we may say that the bond has 45 per cent ionic character. In HCl, HBr and HI the corresponding figures are about 17, 14 and 5 per cent, respectively.

The degree of ionic character; electronegativity

4.20. Since the percentage ionic character of a bond is determined by the difference between the electronegativities of the two atoms concerned, it is tempting to endeavour to express the relationship quantitatively. Several alternative formulae have been proposed, but here we shall describe the treatment given by Pauling. This author ascribes to each atom a number, x, as a measure of its electronegativity. Then if two atoms A and B are bonded together, the resonance energy arising from the ionic component of the binding is proportional to $(x_A - x_B)^2$ and the percentage ionic character of the bond is given by the empirical relation

$$p = 16|x_A - x_B| + 3{\cdot}5|x_A - x_B|^2. \tag{4.01}$$

Values of x as given by Pauling and others are shown in table 4.02, and some of these values are also shown in fig. 4.04; the expression (4.01) is represented graphically in fig. 4.05.

We may illustrate the application of this treatment by reverting to the example of hydrogen fluoride. The electronegativities of the hydrogen and fluorine atoms are 2·1 and 4·0, respectively, so that we have

$$p = 16(4{\cdot}0 - 2{\cdot}1) + 3{\cdot}5(4{\cdot}0 - 2{\cdot}1)^2 = 43\,\%,$$

Table 4.02. *The electronegativities of the elements*

H																
2.1																
Li	Be											B	C	N	O	F
1·0	1·5											2·0	2·5	3·0	3·5	4·0
Na	Mg											Al	Si	P	S	Cl
0·9	1·2											1·5	1·8	2·1	2·5	3·0
K	Ca	Sc	Ti	V	Cr	Mn	Fe	Co	Ni	Cu	Zn	Ga	Ge	As	Se	Br
0·8	1·0	1·3	1·5	1·6	1·6	1·5	1·8	1·8	1·8	1·9	1·6	1·6	1·8	2·0	2·4	2·8
Rb	Sr	Y	Zr	Nb	Mo	Tc	Ru	Rh	Pd	Ag	Cd	In	Sn	Sb	Te	I
0·8	1·0	1·2	1·4	1·6	1·8	1·9	2·2	2·2	2·2	1·9	1·7	1·7	1·8	1·9	2·1	2·5
Cs	Ba	La	Hf	Ta	W	Re	Os	Ir	Pt	Au	Hg	Tl	Pb	Bi	Po	At
0·7	0·9	1·1	1·3	1·5	1·7	1·9	2·2	2·2	2·2	2·4	1·9	1·8	1·8	1·9	2·0	2·2
Fr	Ra	Ac														
0·7	0·9	1·1														

The lanthanide and actinide elements

Ce–Lu	Th	Pa	U	Np–No
1.1–1·2	1·3	1·5	1·7	1·3

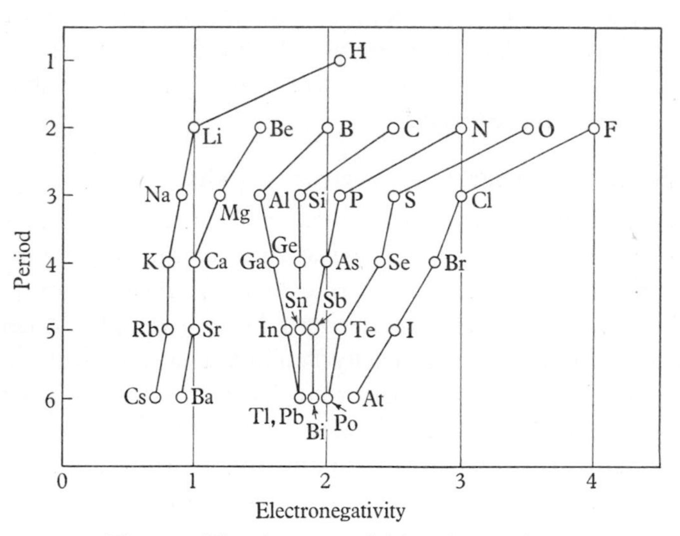

Fig. 4.04. The electronegativities of some elements.

a value in very satisfactory agreement with that determined experimentally. We must not, however, expect that the agreement will always be equally satisfactory, and it must be understood that the application of equation (4.01) to give a quantitative measure of bond character is subject to many qualifications which cannot be discussed here. It will suffice for our purposes to confine ourselves to its qualitative interpretation.

4.21. Many conclusions can be drawn from the data presented in table 4.02 and fig. 4.04, and we shall have frequent occasion to refer to these electronegativity values in our later discussions of bond type in crystal structures. It is convenient, however, at this point to summarize certain of these conclusions:

(*a*) An electronegativity difference of about 2·1 corresponds to a bond which is 50 per cent ionic. Bonds with a larger electronegativity

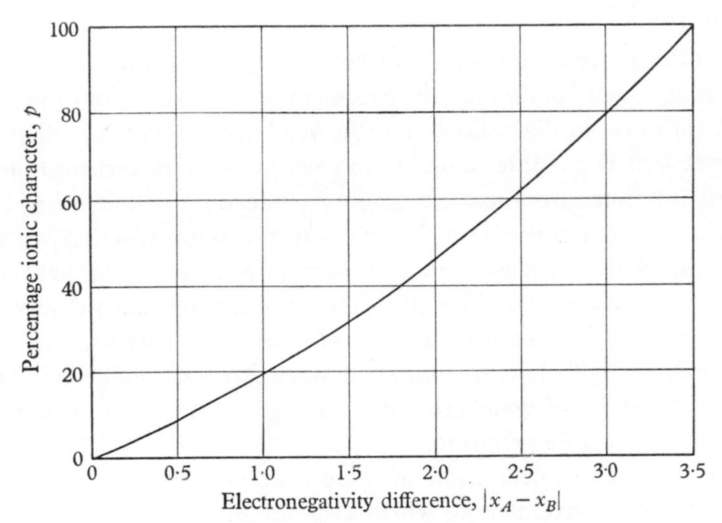

Fig. 4.05. Ionic character of a bond A–B as a function of the electronegativity difference $|x_A - x_B|$.

difference are therefore primarily ionic, those with a smaller difference are primarily covalent.

(*b*) The bonds from the alkali metals, magnesium and the alkaline earths on the one hand to fluorine, oxygen and other strongly electronegative non-metallic elements on the other are essentially ionic.

(*c*) Beryllium and aluminium have the same electronegativity. Bonds from these elements to fluorine are essentially ionic, to other elements essentially covalent.

(*d*) Boron forms with hydrogen a bond which is almost purely covalent. The bond between boron and other more electronegative elements is partially ionic, but in all cases the covalent component predominates.

(*e*) Bonds between silicon and all other elements (except perhaps fluorine) are primarily covalent. The Si–O bond has about 40 per cent ionic character, a value which we shall find to be of importance when we come to discuss silicate structures.

(*f*) Fluorine and oxygen are the most electronegative elements. Bonds between these elements and all the metals have considerable ionic character, and in most cases are predominantly ionic.

COVALENT BOND LENGTHS

Covalent radii

4.22. In the above discussion of the covalent bond we have frequently had occasion to refer to characteristic interatomic distances; for example, the C–C distance in diamond is 1·54 Å. We have already seen that in ionic crystals it is possible to assign to each *ion* a characteristic radius such that interionic distances are given very nearly by the sum of the appropriate radii, and it is therefore natural to enquire whether we are similarly justified in assigning to *atoms* characteristic radii such that their sum will give covalent bond lengths. This problem has been discussed by many authors but, subject to a number of qualifications, some of which are discussed below, we may say at once that such radii can in fact be assigned and are of great value in enabling bond lengths to be predicted with reasonable precision.

Values for the covalent radii of a number of atoms are given in table 4.03, the arrangement of which calls for some explanation. The first section (*a*) gives the *normal radii* of atoms. These are the radii applicable when an atom forms the number of covalent bonds appropriate to its place in the Periodic Table—one bond for fluorine, two for oxygen, and so on. Different values are given corresponding to the different lengths of single, double and triple bonds. The remaining sections of the table give the radii of atoms involved in hybrid bond formation. Thus the *tetrahedral radii* (*b*) are applicable when sp^3 hybridization takes place, as in the very large number of AX compounds having the zincblende and wurtzite structures. For elements in group 4 these radii are the same as the normal single-bond radii since such elements, as we have already seen, can be quadrivalent only in the sp^3 hybrid state; for other elements the radii are slightly smaller than the normal radii. The *octahedral radii* (*c*) correspond to the d^2sp^3 and sp^3d^2 hybrids, which give rise to six bonds directed to the corners of a regular octahedron, and the *square radii* (*d*) are those which apply when a dsp^2 hybrid is formed. These latter radii are numerically indistinguishable from the octahedral radii but have been tabulated separately because in some cases the valency states corresponding to the types of co-ordination are not the same; for example, Pt^{IV} forms a d^2sp^3 hybrid whereas Pt^{II} is dsp^2 hybridized. Finally, the *linear radii* (*e*) are those appropriate to the limited number of elements which form two *sp* hybrid bonds oppositely directed along a straight line.

4.23. We have already remarked that the concept of characteristic covalent radii is subject to a number of qualifications. The first qualification which we would make is that the covalent radius of an atom (unlike its ionic radius) must not be interpreted as implying that the atom is a sphere of that size. The covalent radii are applicable only to

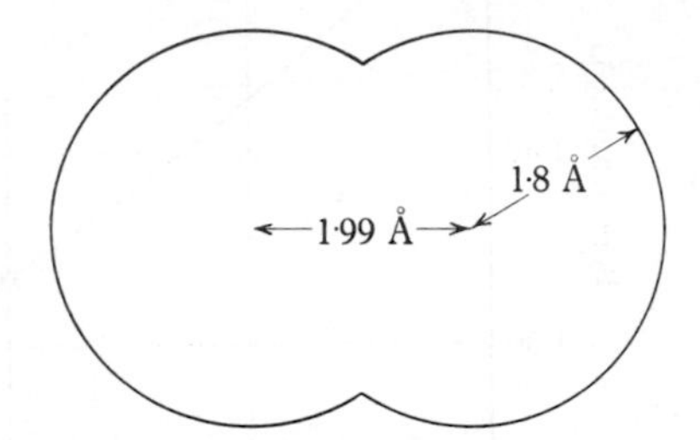

Fig. 4.06. The dimensions of the chlorine molecule.

the calculation of interatomic distances between atoms joined by covalent bonds, and tell us nothing about the distances between atoms of the same kind when not so united. Thus the single-bond radius of chlorine is shown in table 4.03 as 0·99 Å, and this is in perfect agreement with the observed Cl–Cl distance of 1·988 Å in the chlorine molecule. In the crystal structure of solid chlorine, however, the interatomic distance between contiguous chlorine atoms of different molecules, which are bound together only by van der Waals forces, is about 3·6 Å. Thus the van der Waals radius of the chlorine atom is about 1·8 Å (nearly double the covalent radius) and we must picture the Cl_2 molecule as having the form shown in fig. 4.06.

A second point which must be made is that bond length depends upon bond number: triple bonds are shorter than double bonds, which in their turn are shorter than single bonds. For this reason separate radii are required for these three types of bond, and it will be seen from the values given in table 4.03 that triple-bond radii are about 0·17 Å less and double-bond radii about 0·10 Å less than the corresponding single-bond radii. We must, however, also remember that a bond will not necessarily have an integral bond number and that this number may be fractional if the bond involves resonance. Thus in benzene the C–C bond has a bond number 1·5, and the bond length of 1·39 Å is intermediate between the lengths 1·54 and 1·34 Å characteristic of single and double C–C bonds. It might be expected that such intermediate bond lengths could be predicted by interpolation between the lengths corresponding to integral bond numbers, but this procedure must be applied with caution. In fig. 4.07 the points represent the bond lengths corresponding to pure single, double and triple C–C bonds, and a smooth

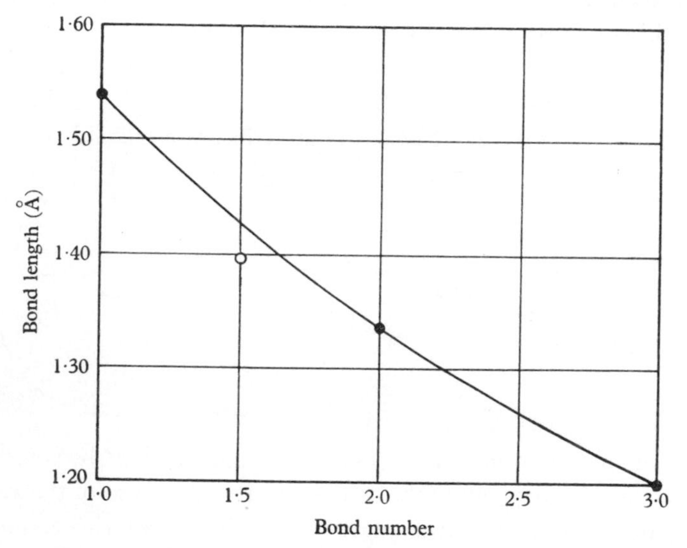

Fig. 4.07. The lengths of the single, double and triple C–C bonds.

curve has been drawn through these points. The circle representing the C–C distance in benzene is seen to be appreciably below this curve, the reason being that resonance has stabilized the structure and conferred additional strength on the bond, thus reducing its length. Pauling stresses this point by distinguishing between the bond *number* of 1·5 and the bond *order* of about 1·66, a figure which emphasizes that although each C–C bond in benzene is a resonance hybrid to which single and double bonds contribute equally the resultant bond, in length and strength, is nearer to a double than to a single bond. When, as here, two bonds of different order contribute equally to a resonance hybrid the bond length is found to be less than that deduced by direct interpolation.

A third qualification to be taken into account in employing the radii recorded in table 4.03 is that we must consider resonance not only between covalent bonds of different bond number but also between covalent and ionic bonds, for, as we have already seen, many covalent bonds do in fact have considerable ionic character. The effect of this covalent/ionic resonance may be considerable, and in some instances reduces the length of a bond by as much as 0·1 Å or more. Attempts have been made to express the necessary correction as a function of the difference in electronegativity of the atoms involved. Unfortunately, however, it cannot be claimed that these attempts have met with great success, for in some cases the 'corrections' so derived increase the discrepancy between predicted and observed bond lengths. For our purpose it seems best to ignore the effect of covalent/ionic resonance in deriving bond lengths from the radii given in table 4.03 but to remember that such lengths may be appreciably too large in cases where the ionic component of the bond is considerable.

A fourth and final point which we wish to make concerning the calculation of bond length from atomic radii is that even if a bond is purely covalent its character and strength are not uniquely determined unless we know the orbitals involved in its formation. We must therefore use different effective radii in calculating the lengths of different types of hybrid bond. It is for this reason that the radii for some common states of hybridization are shown separately in table 4.03.

4.24. We have discussed the question of atomic radius in covalent compounds at some length, but even so many other factors remain to be considered if bond lengths are to be predicted with precision. It is found, for example, especially in crystal structures, that seemingly equivalent bonds differ significantly in length and that interbond angles depart from the values to be expected on the basis of the highly symmetrical configurations associated with the various patterns of hybridization. Some of these points will arise when we come to consider individual crystal structures, but it would be inappropriate to discuss them in further detail at this point. It is sufficient for our purposes to say that the radii given in table 4.03 will enable us to predict bond lengths with reasonable accuracy (say to 0·1 Å or better) and that the concept of hybridization shows us why a given atom can sometimes display several different covalencies each with its own characteristic distribution of bonds in space.

The reader wishing to pursue these questions in greater detail should consult some of the works quoted in appendix 1.

Table 4.03. *Covalent radii*

Values are in Ångström units

(*a*) Normal radii

	H	C	N	O	F
Single bond	0·30	0·77	0·74	0·74	0·72
Double bond	—	0·67	0·62	0·62	—
Triple bond	—	0·60	0·55	—	—
		Si	P	S	Cl
Single bond		1·17	1·10	1·04	0·99
Double bond		1·07	1·00	0·94	—
Triple bond		1·00	0·93	—	—
		Ge	As	Se	Br
Single bond		1·22	1·21	1·17	1·14
Double bond		1·12	1·11	1·07	—
		Sn	Sb	Te	I
Single bond		1·40	1·41	1·37	1·33
Double bond		1·30	1·31	1·27	—

(*b*) Tetrahedral radii

	Be	B	C	N	O	F
	1·06	0·88	0·77	0·70	0·66	0·64
	Mg	Al	Si	P	S	Cl
	1·40	1·26	1·17	1·10	1·04	0·99
Cu	Zn	Ga	Ge	As	Se	Br
1·35	1·31	1·26	1·22	1·18	1·14	1·11
Ag	Cd	In	Sn	Sb	Te	I
1·52	1·48	1·44	1·40	1·36	1·32	1·28
Au	Hg	Tl	Pb	Bi		
1·50	1·48	1·47	1·46	1·46		

(*c*) Octahedral radii

d^2sp^3 hybrids

$\mathrm{Fe^{II}}$	$\mathrm{Co^{II}}$	$\mathrm{Ni^{II}}$	$\mathrm{Ru^{II}}$			$\mathrm{Os^{II}}$			
1·23	1·32	1·39	1·33			1·33			
	$\mathrm{Co^{III}}$	$\mathrm{Ni^{III}}$		$\mathrm{Rh^{III}}$			$\mathrm{Ir^{III}}$		
	1·22	1·30		1·32			1·32		
$\mathrm{Fe^{IV}}$		$\mathrm{Ni^{IV}}$			$\mathrm{Pd^{IV}}$			$\mathrm{Pt^{IV}}$	$\mathrm{Au^{IV}}$
1·20		1·21			1·31			1·31	1·40

sp^3d^2 hybrids

$\mathrm{Ti^{IV}}$	$\mathrm{Zr^{IV}}$	$\mathrm{Sn^{IV}}$	$\mathrm{Te^{IV}}$	$\mathrm{Pb^{IV}}$
1·36	1·48	1·45	1·52	1·50

(*d*) Square radii

$\mathrm{Ni^{II}}$	$\mathrm{Pd^{II}}$	$\mathrm{Pt^{II}}$	$\mathrm{Au^{III}}$
1·39	1·31	1·31	1·40

(*e*) Linear radii

$\mathrm{Cu^{I}}$	$\mathrm{Ag^{I}}$	$\mathrm{Hg^{II}}$
1·18	1·39	1·29

THE MOLECULAR-ORBITAL THEORY

4.25. Our treatment of the valence-bond theory of the covalent binding between two atoms A and B has involved two essentially distinct steps: first, we have built up the electronic structure of the atoms A and B by feeding electrons into the atomic orbitals of these atoms, and then we have considered the interaction of these orbitals to form the bond. The alternative molecular-orbital theory combines these two stages in one. In treating the molecule AB we now start with the nuclei of the two atoms and feed all the electrons of the molecule into *molecular* orbitals associated with the molecule as a whole. Just as the orbitals of an atom are distinguished as s, p, d, etc., so those of a molecule are distinguished as σ, π, δ, etc. When this procedure is treated analytically it is found that the orbitals of all but the valency electrons are in fact atomic orbitals, so that the inner electrons of each atom remain associated with that atom as in the valence-bond theory. Moreover, the molecular orbitals which give rise to bonding represent a concentration of electron density between the two atoms just as in the simpler theory.

In principle the molecular-orbital method could be applied to any molecule whether simple or complex and whether finite (as in benzene) or infinite (as in diamond). In practice the mathematical difficulties are insuperable and we must adopt a compromise solution. For all but the simplest diatomic molecules, therefore, we first form the atomic orbitals of the individual atoms in the usual way and then, by a linear combination of the relevant bonding orbitals, we deduce the molecular orbitals of the molecule or crystal. We may illustrate this point by considering the molecules of ethylene, $H_2C{=}CH_2$, acetylene, $HC{\equiv}CH$, and benzene, C_6H_6.

Ethylene

4.26. We have already seen that in the ground state the carbon atom has the configuration $1s^2$, $2s^2 2p_x^1 2p_y^1$ but that in order to explain the quadrivalency of carbon we must assume that one of the $2s^2$ electrons is promoted to the vacant $2p_z$ orbital, giving the configuration $1s^2$, $2s^1 2p_x^1 2p_y^1 2p_z^1$. Hybridization of the one $2s$ orbital and the three $2p$ orbitals then gives an sp^3 hybrid with the four bonds tetrahedrally disposed. It is not, however, satisfactory to assume this state of hybridization in the ethylene molecule, for it would imply that the two C–C bonds arose from pairs of orbitals steeply inclined to one another and would not account for the fact that the interbond angles are known to be all close to 120°.

There is, however, another type of hybridization possible in the carbon atom in which the 2*s* orbital is hybridized with only two of the 2*p* orbitals (say the p_x and p_y). The three sp^2 hybrid bonds will then lie in a plane symmetrically inclined at 120°, leaving the unaltered p_z orbital perpendicular to this plane, as shown in fig. 4.08*a*. The three hybrid bonds of each carbon atom will then be responsible for the two bonds between it and the hydrogen atoms with which it is associated and also for one bond between it and the other carbon atom. This picture explains the observed interbond angles of 120° but does not

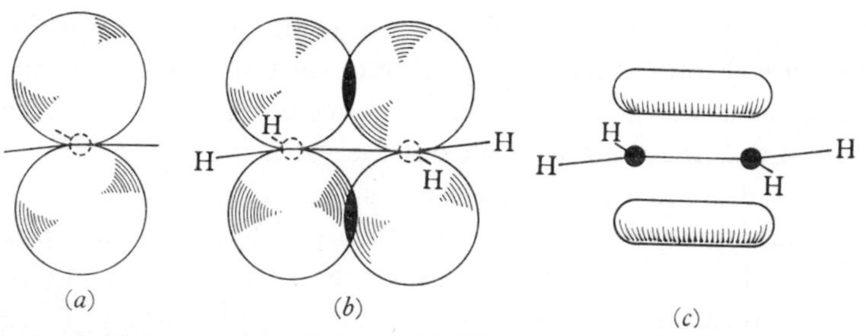

Fig. 4.08. Orbital overlap in ethylene. (*a*) The sp^2 bonds of a single carbon atom, with the unaltered p_z orbital perpendicular to the plane of these bonds. (*b*) Overlap of the p_z orbitals of the two carbon atoms. (*c*) The resultant π bond.

explain the coplanar character of the molecule or the absence of free rotation about the C–C bond. In terms of the molecular-orbital theory, however, there will also be lateral interaction between the two p_z orbitals provided that their axes are parallel (fig. 4.08*b*). This interaction will give rise to a molecular orbital represented by two cigar-shaped regions of electron density parallel to the line joining the carbon atoms but symmetrically displaced on either side of it (fig. 4.08*c*). The bond arising in this way is termed a π bond, whereas the other bonds of each carbon atom are designated σ bonds, a characteristic of the σ bond being that it corresponds to a circularly symmetrical distribution of electron density about the bond in question. The σ bonds are stronger than π bonds so that in the molecular-orbital picture of the ethylene molecule the simple double bond between the carbon atoms is replaced by two bonds of different strengths. We can if we wish represent this distinction by writing the molecule as

```
H       H
 \     /
  C:::C
 /     \
H       H
```

We see that this picture explains the planar character of the molecule, for if the carbon atoms were twisted about their common axis overlap of the p_z orbitals would be reduced and the π bond would be broken.

Acetylene

4.27. Similar arguments apply to the acetylene molecule. Here we must assume that in each carbon atom hybridization takes place between the $2s$ orbital and only one of the $2p$ orbitals (say p_x), giving rise to two collinear sp or σ bonds. The unaltered p_y and p_z orbitals stand perpendicular to these bonds and to each other. In the molecule the σ bonds will be responsible for the bonding H–C–C–H, but again there

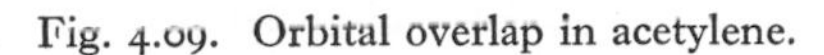

Fig. 4.09. Orbital overlap in acetylene.

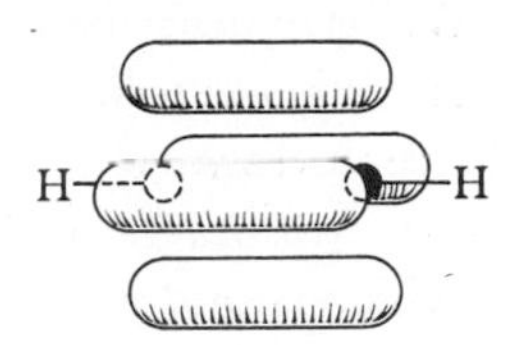

will be additional bonds between the carbon atoms due to interaction between the p_y and p_z orbitals, provided, as before, that these are parallel in the two atoms. There will therefore now be two C–C π bonds represented by four cigar-shaped regions of electron density parallel to the C–C axis, as shown in fig. 4.09. Of the three C–C bonds one is therefore a strong σ bond and two are weaker π bonds, a state of affairs which we can represent by writing the molecule as

H—C⋮⋮⋮C—H

Benzene

4.28. As a final example of the molecular-orbital treatment we may consider the molecule of benzene. Here we assume initially that the carbon atoms are sp^2 hybridized, as in ethylene. The resulting σ bonds will then immediately give rise to a planar, regular hexagonal ring as is, of course, observed. We have, however, still to consider the unaltered p_z orbitals of each carbon atom standing perpendicular to the plane of the ring, and here a new point arises. The p_z orbital of any one atom will overlap the corresponding orbitals of both its neighbours so that the electron density associated with the π bond arising from this overlap will no longer be localized as two cigar-shaped figures between a single pair of carbon atoms; instead it will take the form of two delocalized ring-shaped figures above and below the plane of the benzene ring and related symmetrically to all six carbon atoms, as shown in fig. 4.10.

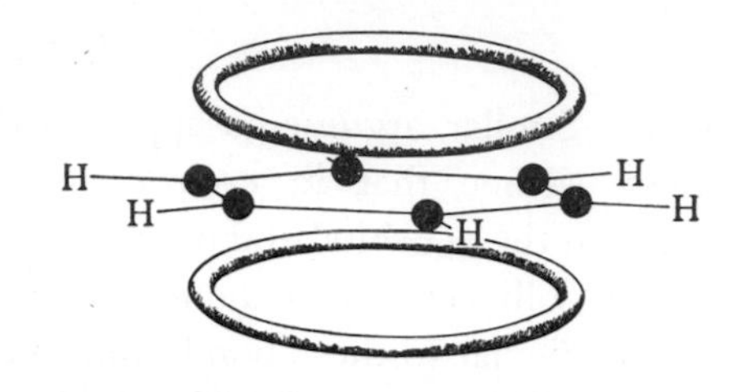

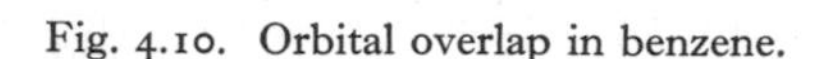
Fig. 4.10. Orbital overlap in benzene.

Comparison with the valence-bond theory

4.29. We have described the molecular-orbital theory only very briefly and in a greatly over-simplified form, the reason being that in discussing crystal structures we shall generally find that the valence-bond theory is better suited to our needs. This is not to belittle the molecular-orbital theory, which is undoubtedly conceptually the simpler and aesthetically the more satisfying; it is merely an expression of the fact that the concepts of the molecular-orbital theory can generally be expressed only in mathematical terms, whereas those of the valence-bond theory lend themselves to verbal expression in terms already familiar to the chemist. In ultimate analysis the two theories converge very closely, and any theoretical prediction cannot be regarded as satisfactory unless it is consistent with both. So long, however, as valency theory can be applied with precision only to the very simplest of diatomic molecules the value of any theory of interatomic binding must lie primarily in the qualitative insight which it gives rather than in its quantitative application. It is in this respect that the valence-bond theory has an advantage, and it is for this reason that we shall generally interpret the interatomic forces in crystals in valence-bond terms. If, however, in particular cases the molecular-orbital theory appears to have advantages we shall not hesitate to employ it.

5 THE METALLIC BOND AND THE STRUCTURES OF SOME METALLIC ELEMENTS

INTRODUCTION

5.01. Metals are sharply distinguished from other types of solid matter by a number of physical properties, of which the high electrical and thermal conductivity and the optical opacity are among the most obvious. About three-quarters of all the known elements are metallic, but in spite of this fact, and in spite of the great technological importance of these elements, the study of metals and of intermetallic systems has received relatively little attention from chemists. In part, no doubt, this is because metal systems have proved to be an intractable field for study; alloys do not generally display the constancy of composition characteristic of chemical compounds, and even when a phase does display a distinctive composition it is rarely one conforming to the accepted principles of chemical valency. In the last few decades, however, there have been notable advances in our understanding of the metallic state, and it is significant that these advances have stemmed largely from a study of the crystal structures of the metals, so that from our present point of view they are of particular importance. Finality has certainly not been reached, but already we have a picture of interatomic binding in metals which not only enables us to understand many of the physical properties of these elements and of alloy systems but also correlates our picture of the metallic bond with those of the ionic and covalent bonds discussed in previous chapters.

The Drude–Lorentz theory of the metallic bond

5.02. The first successful theory of the metallic state may be said to have arisen from the work of Drude and Lorentz in the early years of the present century. On this theory a metal is to be regarded as an assemblage of positive ions immersed in a gas of free electrons. A potential gradient exists at the surface of the metal to imprison the electrons, but within the metal the potential is uniform. Attraction between the positive ions and the electron gas gives the structure its coherence, and the free mobility of this electron gas under the influence of electrical or thermal stress is responsible for the high conductivity. In terms of this picture many of the physical properties of metals can be given a very satisfactory qualitative and even quantitative description, but there are nevertheless many properties of which no such explanation seems possible.

The Drude–Lorenz theory of metals was, of course, developed before the crystal structures of these elements were known, and it paid relatively little attention to the problem of the nature of the interatomic forces in the metallic state. Nevertheless, it is possible even in terms of this simple theory to draw certain important conclusions on this point. The picture of the coherence of a metal as arising from the attraction between positive ions and the electron gas confers neither spatial nor numerical limitations on the metallic bond: the bonds from any one atom must be regarded as spherically distributed and as being capable of acting on as many neighbours as can be packed round that atom. In a covalent structure the bonds from any one atom are severely limited both in number and in spatial distribution, while in an ionic crystal the demands of electrical neutrality impose limitations on the types of structure which can occur. In metal systems, however, none of these factors arises, and we would expect the structure to be determined by geometrical considerations alone. In the majority of metals this is in fact the case, and many metallic elements display one or other of two *close-packed structures* which represent the geometrically most compact arrangement of spheres in space. This picture immediately accounts for the high densities of metals and also for their remarkable mechanical behaviour (notably their ductility), characteristics which distinguish metals from non-metallic solids no less decisively than their electrical and optical properties.

SOME SIMPLE METAL STRUCTURES

Close-packed structures

5.03. If an assemblage of equal spheres is packed together as tightly as possible in two dimensions the arrangement shown in fig. 5.01 *a* results. The spheres can be seen to be centred at the points A of an equilateral triangular network, and each is in contact with six neighbours at the corners of a regular hexagon. This same arrangement is shown again in fig. 5.01 *b*, where, however, the spheres are drawn smaller for clarity. The upper surface of such a single layer of spheres is seen to display a series of triangular depressions or 'dimples', and these may be conveniently divided into two sets, centred about B and C, respectively, although inversion of the diagram shows that there is no fundamental distinction between the two. It will be noted that the pattern of B's alone, or of C's alone, is exactly the same (except for a translation) as that of the A's alone.

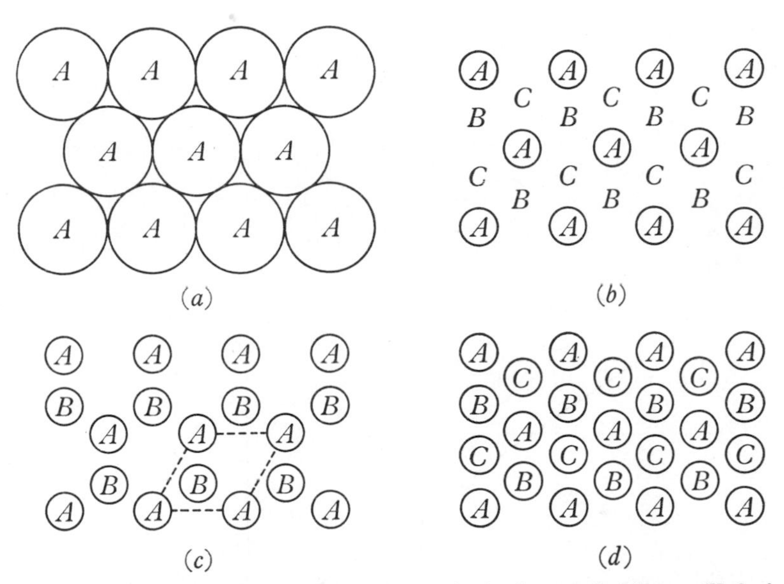

Fig. 5.01. The close packing of spheres. (*a*) A single close-packed layer. (*b*) A single close-packed layer showing the two alternative sets of 'dimples' into which the spheres of a second close-packed layer can fit. (*c*) Hexagonal close packing. The layers repeat in the sequence $A\ B\ A\ B\ A\ldots$. The plan of the unit cell is outlined by broken lines. (*d*) Cubic close packing. The layers repeat in the sequence $A\ B\ C\ A\ B\ C\ A\ldots$.

If now a second close-packed layer is superimposed as tightly as possible on the first, the spheres in the second layer will rest in one or other of these two sets of dimples, say the B's. (They could, of course, equally well occupy the dimples C, but it is clear that in an indefinitely extended structure no difference between these arrangements exists.) Each sphere in this second layer is now in contact with nine neighbours, six in its own layer and three in the layer beneath, and, as before, the upper surface of the layer displays two sets of dimples, now centred about A and C. Whereas, however, for a single layer these two sets would be indistinguishable, for the assemblage of two layers they are different: the dimples A lie vertically above spheres in the first layer while the dimples C do not. It follows, therefore, that a third layer may be added in two fundamentally distinct ways according as to whether the spheres in this third layer are placed at A or at C.

If the former arrangement is adopted the third layer is identical with the first, and if further layers are added they will form the sequence $ABABAB$... indefinitely continued. This arrangement (fig. 5.01 *c*)

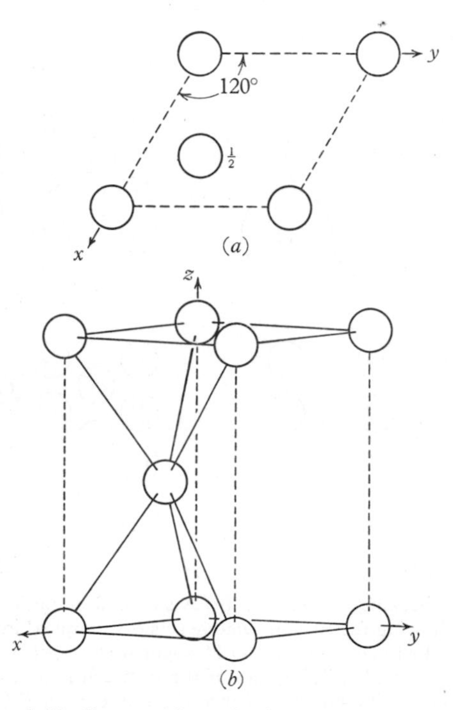

Fig. 5.02. (*a*) Plan and (*b*) clinographic projection of the unit cell of hexagonal close packing. In (*b*) the atom at a height $\frac{1}{2}c$ is co-ordinated by the six atoms at the same height in the adajcent cells as well as by the six atoms shown.

possesses hexagonal symmetry and is therefore termed *hexagonal close packing*. The unit cell is outlined in fig. 5.01 *c* and is shown separately in plan and clinographic projection in fig. 5.02. The c/a ratio is 1·633. It will be noted that in the three-dimensional structure each sphere has twelve neighbours, six in its own layer, three in the layer below and three in the layer above. No more highly co-ordinated packing of equal spheres is geometrically possible.

If the alternative arrangement of the third layer is adopted this layer is different from the first, and the three layers form the sequence ABC.

A fourth layer added so that it, similarly, is different from the second and third must fall vertically above the first, and if further layers are added they will form the sequence $ABCABC...$ indefinitely continued (fig. 5.01 *d*). It is not by any means immediately apparent that this arrangement has cubic symmetry and is, in fact, an arrangement of spheres at the corners and face centres of a cubic unit cell (fig. 5.03 *a*), so that it constitutes *cubic close packing*. This point, however, is made clear by fig. 5.03 *b*, which represents 27 unit cells of the structure. Here the face-centred unit cells are apparent by a comparison with fig. 5.03 *a*, but the close-packed nature of the layers normal to one of the cube

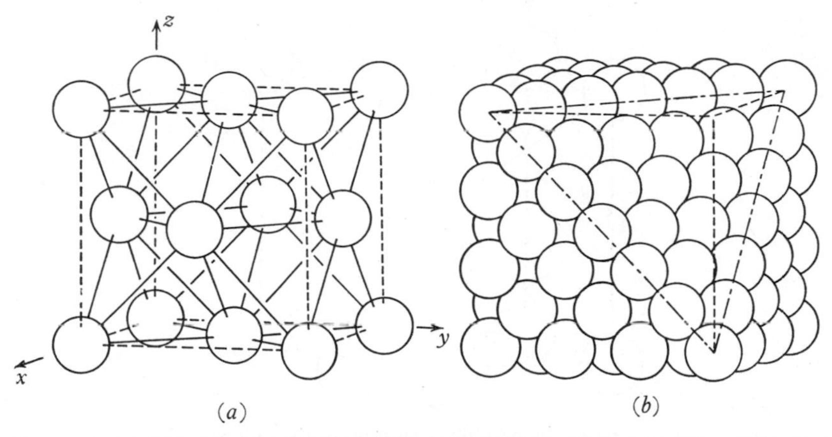

Fig. 5.03. (*a*) Clinographic projection of the unit cell of cubic close packing. The atoms occupy the corners and face centres of the cubic cell. (*b*) Clinographic projection of 27 unit cells of cubic close packing, showing interatomic contacts. Some atoms have been omitted to reveal the close-packed layers perpendicular to one of the cube diagonals.

diagonals has been emphasized by the omission of a number of spheres. As in hexagonal close packing, each sphere is co-ordinated by twelve neighbours.

It is clear that the two close-packed structures can differ little in lattice energy. It is also clear that they represent an equally economical filling of space, and it may be readily shown that if the radius of each sphere is a the volume of the structure per sphere is in both cases $5{\cdot}66a^3$; the proportion of space occupied is therefore $\frac{4}{3}\pi a^3/5{\cdot}66a^3 = 74{\cdot}1$ per cent. The two structures do, however, differ in one important geometrical respect, namely the disposition of the close-packed layers. In hexagonal close packing there is only one direction normal to which the spheres are arranged in individually close-packed sheets, whereas in the cubic structure such sheets occur in four directions normal to the four cube diagonals. This distinction gives rise to important differences between the mechanical properties associated with the two structures.

The body-centred cubic structure

5.04. Yet a third structure common among the metallic elements is the very simple cubic body-centred arrangement (fig. 5.04), in which each atom is surrounded by only eight neighbours arranged at the corners of a cube. In addition to these eight neighbours, however, each atom has

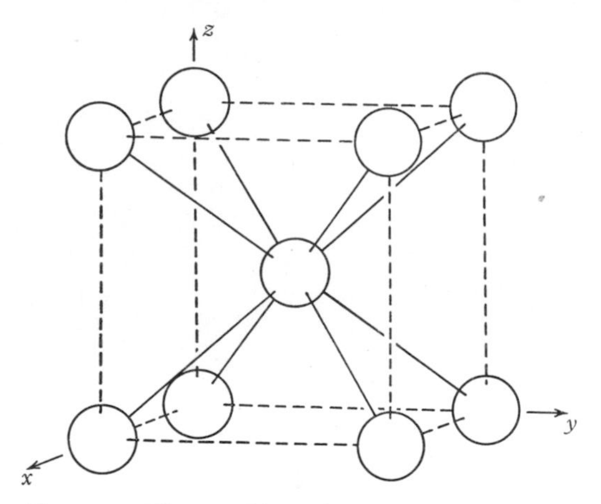

Fig. 5.04. Clinographic projection of the unit cell of the cubic body-centred structure.

also six others only slightly more remote (namely, for the atom at the centre of any one cell, the atoms at the centres of the adjacent cells). The packing density is therefore not greatly different from that in the close-packed structures: the volume of the structure per sphere is $6{\cdot}16a^3$ and the proportion of space occupied is 68·1 per cent.

The structures of the metallic elements

5.05. The cubic close-packed, the cubic body-centred and the hexagonal close-packed structures are conventionally represented by the symbols A_1, A_2 and A_3, respectively, and the structures of the metallic elements are indicated in table 5.01 by these symbols. The symbol X implies the existence of one or more different and more complex structures.

It will be seen from table 5.01 that few of the B sub-group metals have simple structures, but that among the more truly metallic elements the three structures described above are very common. Many of the metals are polymorphous, showing two or more different structures, and these may co-exist or may be stable at different temperatures. In such cases no attempt has been made to show in the table the stability ranges of the various phases, and it must not be assumed that the simplest phase is that which is stable at room temperature; for example, the A_2 structure of uranium is found only above 760 °C., and below that temperature two other more complex structures obtain.

Table 5.01. *The structures of the metallic elements*

The representative and transition elements

	1*A*	2*A*	3*A*	4*A*	5*A*	6*A*	7*A*	8			1*B*	2*B*	3*B*	4*B*	5*B*	6*B*
2	Li A_2 A_1 A_3	Be A_2 A_3 X														
3	Na A_2 A_3	Mg A_3											Al A_1			
4	K A_2	Ca A_1 A_3 X	Sc A_1 A_3	Ti A_3 A_2	V A_2	Cr A_2 A_3 X	Mn X A_1 A_2	Fe A_2 A_1	Co A_3 A_1	Ni A_1 A_3	Cu A_1	Zn X	Ga X	Ge X	As X	Se X
5	Rb A_2	Sr A_1 A_3 A_2	Y A_2	Zr A_3 A_2	Nb A_2	Mo A_2 A_3	Tc A_3	Ru A_3	Rh A_1	Pd A_1	Ag A_1	Cd X	In X	Sn X	Sb X	Te X
6	Cs A_2	Ba A_2	La A_1 A_3	Hf A_3 A_2	Ta A_2	W A_2	Re A_3	Os A_3	Ir A_1	Pt A_1	Au A_1	Hg X	Tl A_3 A_2	Pb A_1	Bi X	Po X
7	Fr —	Ra —	Ac A_1													

The lanthanide elements

Ce	Pr	Nd	Pm	Sm	Eu	Gd	Tb	Dy	Ho	Er	Tm	Yb	Lu
A_1 A_3	A_1 X	X	—	X	A_2	A_3	A_3	A_3	A_3	A_3	A_3	A_1	A_3

The actinide elements

Th	Pa	U	Np	Pu	Am	Cm	Bk	Cf	E	Fm	Mv	No
A_1 A_2	X	X A_2	X	X A_1 A_2	—	—	—	—	—	—	—	—

A_1, cubic close packing.
A_2, cubic body-centred structure.
A_3, hexagonal close packing.
X, other more complex structures.

THE ATOMIC RADII OF THE METALS

5.06. Measurement of the cell dimensions of the crystal structures of the metals enables characteristic atomic radii to be derived for these elements. These radii are shown in table 5.02 and in fig. 5.05, but the values there given call for some explanation and qualification before their full significance can be discussed. In the case of metals with close-packed structures the atomic radius is clearly half the distance of closest approach. In structures of lower co-ordination, such as the cubic body-centred arrangement of the alkali metals, a similar definition of atomic radius is applicable, but the values so deduced are not immediately comparable with those derived from close-packed structures. It is found, both from a study of polymorphous metals having several structures of different co-ordination, and from alloy systems in which metal atoms often occur in a state of co-ordination different from that which obtains in the pure element, that a small but systematic decrease of atomic radius takes place with decreasing co-ordination. The extent of this change of radius was studied by Goldschmidt, who summarized observations on a series of elements and alloy systems by expressing the radius in 8-, 6- and 4-co-ordination corresponding to unit radius in a close-packed structure as follows:

Co-ordination	12	8	6	4
Radius	1·00	0·97	0·96	0·88

It will be seen from these figures that the change in radius may be considerable so that a comparison of radii of different elements is of significance only when the state of co-ordination is known.

The same difficulty arises in a still more acute form in discussing the radii of the *B* sub-group metals, for in these elements the co-ordination is so irregular that the concept of a precise radius ceases to have any very definite meaning. Even in these cases, however, we can define the radius as half the distance of closest approach, but the values thus obtained will be of significance only in the structure of the element and will have no immediate influence on its behaviour in alloy systems. Values of the radius derived from a study of alloys are therefore often of more general utility than those deduced from the elements themselves, and values can, moreover, be so obtained corresponding to degrees of co-ordination not found in the pure metal. For this reason two radii are in general given for each element in table 5.02 and fig. 5.05. One of

Table 5.02. *The atomic radii of the metals*

Values are in Ångström units. The upper value is half the distance of closest approach; the lower value is that corresponding to 12-co-ordination

The representative and transition elements

1*A*	2*A*	3*A*	4*A*	5*A*	6*A*	7*A*	8			1*B*	2*B*	3*B*	4*B*	5*B*	6*B*
Li	Be														
1·52	1·12														
1·56	1·12														
Na	Mg											Al			
1·85	1·60											1·42			
1·91	1·60											1·42			
K	Ca	Sc	Ti	V	Cr	Mn	Fe	Co	Ni	Cu	Zn	Ga	Ge	As	Se
2·31	1·96	1·60	1·46	1·31	1·25	1·12	1·23	1·25	1·24	1·28	1·33	1·21	1·22	1·25	1·16
2·38	1·96	1·60	1·46	1·35	1·28	1·36	1·27	1·25	1·24	1·28	1·37	1·35	1·39	—	—
Rb	Sr	Y	Zr	Nb	Mo	Tc	Ru	Rh	Pd	Ag	Cd	In	Sn	Sb	Te
2·46	2·15	1·81	1·60	1·43	1·36	1·35	1·33	1·34	1·37	1·44	1·48	1·62	1·40	1·45	1·43
2·53	2·15	1·81	1·60	1·47	1·40	1·35	1·33	1·34	1·37	1·44	1·52	1·67	1·58	1·61	—
Cs	Ba	La	Hf	Ta	W	Re	Os	Ir	Pt	Au	Hg	Tl	Pb	Bi	Po
2·62	2·17	1·87	1·58	1·43	1·37	1·37	1·35	1·35	1·38	1·44	1·50	1·71	1·74	1·55	1·68
2·70	2·24	1·87	1·58	1·47	1·41	1·37	1·35	1·35	1·38	1·44	1·55	1·71	1·74	1·82	—

The lanthanide elements

Ce	Pr	Nd	Pm	Sm	Eu	Gd	Tb	Dy	Ho	Er	Tm	Yb	Lu
1·82	1·82	1·81	—	—	1·98	1·78	1·77	1·75	1·76	1·73	1·74	1·93	1·73
1·82	1·82	1·81	—	—	2·04	1·78	1·77	1·75	1·76	1·73	1·74	1·93	1·73

The actinide elements

Th	Pa	U	Np	Pu	Am	Cm	Bk	Cf	E	Fm	Mv	No
1·80	1·60	1·38	1·30	1·64	—	—	—	—	—	—	—	—
1·80	1·63	1·54	1·50	1·64	—	—	—	—	—	—	—	—

these corresponds to the distance of closest approach in the structure of the element and the other, determined indirectly if necessary, is that appropriate to 12-co-ordination. For elements with close-packed structures the two values are, of course, identical, but in other cases they differ. For the *B* sub-group metals, in particular, the difference is often considerable.

5.07. Certain general features of table 5.02 and fig. 5.05 call for emphasis:

(*a*) The atomic radii are very much larger than the corresponding

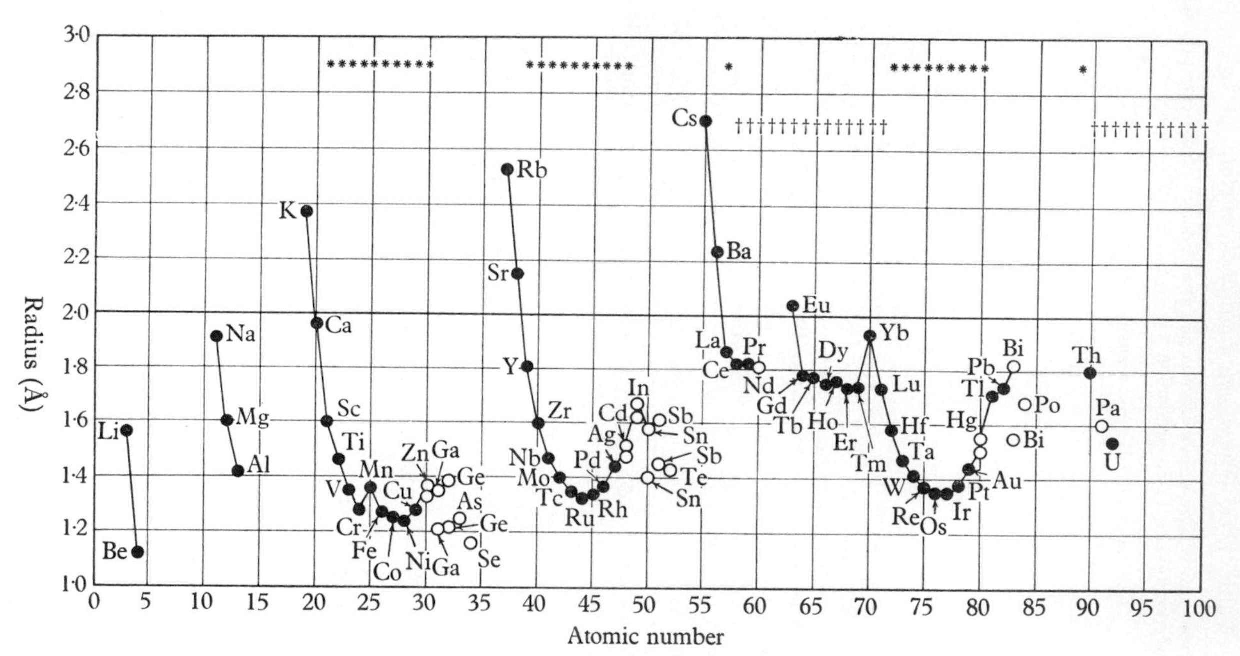

Fig. 5.05. The atomic radii of the metallic elements. Dots refer to metals with one of the three structures A_1, A_2 or A_3, and represent the radii appropriate to 12-co-ordination. Circles refer to metals having only more complex structures. For these metals, if a single radius is shown it is that corresponding to the distance of closest approach in the structure; if two radii are shown the smaller corresponds to the distance of closest approach and the larger is that appropriate to 12-co-ordination.

ionic radii (table 3.02), owing to the greater number of electrons in the extranuclear structure, e.g.

Atom	Li	Cs	Be	Ba
Atomic radius (Å)	1·56	2·70	1·12	2·24
Ionic radius (Å)	0·60	1·69	0·31	1·35

On the other hand the atomic radii are not very different from the covalent radii given in table 4.03, and in a number of cases are very nearly the same. We shall return to this point later in the chapter (§ 5.24).

(*b*) The extreme values of the radii of the metals differ by a factor little greater than 2, and the increasingly complex extranuclear structure of the elements of high atomic number does not give rise to any very marked increase in radius, owing to the compensating influence of the greater nuclear charge. If the alkali and alkaline earth metals are excluded, the radii of all the other metallic elements lie within the range 1·2–1·9 Å.

(*c*) In each period the alkali metal has by far the largest radius, and with increasing valency in the *A* groups the addition of electrons to the extranuclear structure is accompanied by a marked decrease in radius owing to the increased nuclear charge. This decrease in radius is arrested in the middle of each family of transition metals, so that all these elements within each period have very roughly the same radius. In each of the families of *B* sub-group elements a pronounced increase in radius takes place.

(*d*) The systematic increase in the radius of the transition elements which takes place in passing from the fourth to the fifth period is not observed in passing from the fifth to the sixth, so that the radii of corresponding elements in these last two periods are very nearly the same, e.g.

Period 4	Atom	Cr	Fe	Ni	Cu
	Radius (Å)	1·28	1·27	1·24	1·28
Period 5	Atom	Mo	Ru	Pd	Ag
	Radius (Å)	1·40	1·33	1·37	1·44
Period 6	Atom	W	Os	Pt	Au
	Radius (Å)	1·41	1·35	1·38	1·44

This is another manifestation of the lanthanide contraction already discussed in connexion with ionic radii (§ 3.08).

THE SOMMERFELD FREE-ELECTRON THEORY

5.08. The simple Drude–Lorentz theory described earlier in this chapter pictures the valency electrons in a metal as free to move in a potential well of the form shown in fig. 5.06. Within the metal, from B to C, the potential is uniform, but at the surface a potential difference, V (of the order of 10 V), prevents electron escape. If the electrons are assumed to obey the laws of classical mechanics their energies will correspond to the Boltzmann distribution appropriate to the temperature of the specimen. At room temperatures a quite negligible fraction of the electrons will have energies sufficient to surmount the potential

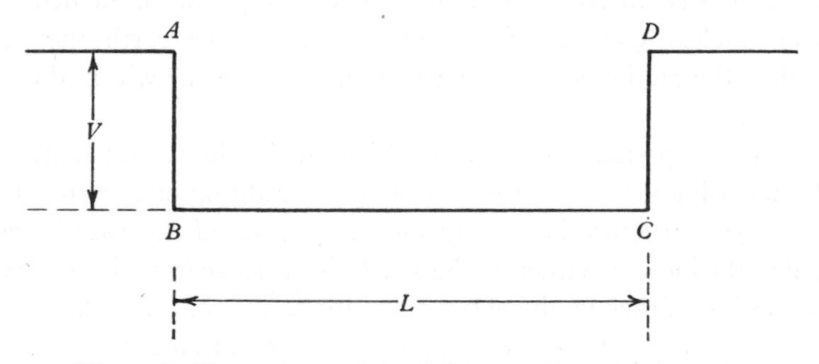

Fig. 5.06. Form of potential well in a one-dimensional metal on the basis of the free-electron theory.

barrier V, but as the temperature is raised the fraction will increase and a number of electrons will escape in the form of a thermionic current. There is, however, a grave objection to this argument, for if the electrons have a Boltzmann distribution of energies they should contribute to the specific heat, a conclusion contrary to experimental observation.

5.09. This difficulty was resolved by Sommerfeld, who treated the electrons in a metal as obeying the laws of the quantum theory instead of those of classical mechanics. In terms of wave mechanics an electron in motion has a wave-like character and its behaviour must be associated with that of a wave which is propagated in the same direction as the electron and whose wavelength is given by

$$\lambda = h/mv, \tag{5.01}$$

where h is Planck's constant, m is the mass of the electron and v is its velocity. Alternatively we can specify the wave associated with the electron in terms of a *wave number*, k, defined by the relation*

$$k = 2\pi/\lambda, \tag{5.02}$$

in which case we have, from (5.01),

$$k = \frac{2\pi}{h} mv. \tag{5.03}$$

The kinetic energy of an electron in motion is given by

$$E = \tfrac{1}{2}mv^2 = \frac{h^2}{8\pi^2 m} k^2, \tag{5.04}$$

so that the relationship between E and k is parabolic, as shown in fig. 5.07*a*. It must be realized that, strictly speaking, k is a vector in that it can be taken to represent the velocity of the electron in both magnitude and direction, and it is for this reason that both positive and negative values of k are shown.

Permissible energy levels

5.10. So far the wave mechanical theory has added nothing to our picture of the electrons in a metal, but we must now consider the important consequence of that theory that not all energy values given by equation (5.04) are permissible. We can best illustrate this point by supposing that fig. 5.06 represents the potential in a one-dimensional metal specimen of length L. In this case the waves representing the electrons must have nodes at the potential barriers B and C, for unless this is so the electrons, no matter how low their energy, will have a finite probability of penetrating the potential barrier and of escaping from the metal. The only permissible wavelengths are therefore those given by the relation

$$\lambda = 2L/n, \tag{5.05}$$

where $n = 1, 2, 3, \ldots$, etc., and the corresponding permissible k values are given by

$$k = \pi n/L. \tag{5.06}$$

Equation (5.04) still represents the relationship between E and k, and this relationship is still parabolic, as shown in fig. 5.07*a*, but this figure no longer represents a continuous variation of E; now E can assume only

* Some authors define the wave number as $k = 1/\lambda$.

the values corresponding to those values of k which satisfy equation (5.06). Since, however, the intervals between permitted k values are inversely proportional to L, we can say that in any specimen of appreciable size these intervals will be extremely small, so that the relationship between E and k can be regarded as quasi-continuous.

If we insert in equation (5.04) the k values given by (5.06) we have for the permitted energies

$$E = \frac{h^2}{8mL^2} n^2. \tag{5.07}$$

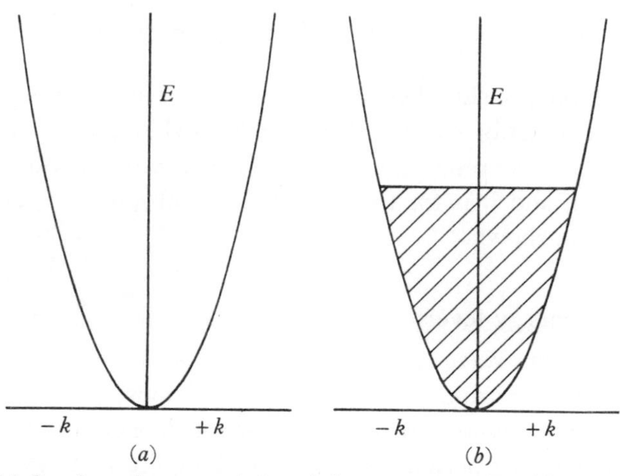

Fig. 5.07. (*a*) Quasi-continuous variation of electron energy, E, with wave number, k, on the basis of the free-electron theory; (*b*) as (*a*) but with the occupied energy levels indicated by shading.

Analogous arguments applied to the case of a three-dimensional specimen in the form of a cube of side L give for the corresponding permitted energies the expression

$$E = \frac{h^2}{8mL^2} (n_x^2 + n_y^2 + n_z^2), \tag{5.08}$$

where n_x, n_y and n_z independently assume the values 1, 2, 3 ..., etc.

5.11. Having thus found the permitted energy levels we can proceed to consider which of these will actually be occupied by the electrons in the metal. Here the argument is very similar to that which we applied in building up the ground state of an isolated atom. If we consider our metal to be at 0 °K the first electron introduced will enter the lowest possible state, of energy $h^2/8mL^2$. A second electron, provided it has opposite spin, can have the same energy, but once these two electrons are present the lowest energy level is fully occupied and the exclusion principle demands that the next pair of electrons (again of opposite spin) shall enter the next level of energy $4h^2/8mL^2$. Successive pairs of electrons will enter successively higher energy levels, so that if N is the total number of free electrons the $\frac{1}{2}N$ states of lowest energy will be occupied and all those of higher energy will be vacant. It is easy to show that the energy corresponding to the highest occupied state is given by

$$E_{\text{max.}} = \frac{h^2}{8m}\left(\frac{3N}{\pi V}\right)^{\frac{2}{3}}, \tag{5.09}$$

where V is the volume of the metal. Although we have developed our argument in terms of a specimen in the form of a cube it can be shown that $E_{\text{max.}}$ is in fact independent of the shape of the specimen; it will be seen that it is also independent of volume, since N is itself directly proportional to that quantity.

We can represent the state of affairs in a metal at 0 °K in the manner shown in fig. 5.07*b*, where shading indicates occupied energy levels and those unshaded are vacant. It is, however, generally more convenient to represent this condition in a slightly different way, which we now proceed to describe.

The $N(E)/E$ relationship

5.12. At 0 °K the number of electron states with an energy less than any given value E can be shown to be

$$\frac{4\pi V}{3h^3}(2mE)^{\frac{3}{2}}. \tag{5.10}$$

If we represent by $N(E)\,dE$ the number of states with energies in the range E to $(E+dE)$ we have at once, by differentiation,

$$N(E)\,dE = \frac{2\pi V}{h^3}(2m)^{\frac{3}{2}}E^{\frac{1}{2}}\,dE. \tag{5.11}$$

The relation between $N(E)$ and E is therefore parabolic, and for a metal at 0 °K takes the form shown in fig. 5.08*a*. Below $E_{\text{max.}}$ the value of $N(E)$ is finite, above $E_{\text{max.}}$ it is zero.

If we evaluate $E_{\text{max.}}$ from equation (5.09) we reach the seemingly surprising conclusion that this energy corresponds to a temperature of the order of 30,000 °K. Expressed in other words, this means that even at the absolute zero of temperature some of the electrons in a metal are moving with velocities characteristic of a gas at this very high temperature. In this conclusion we have an immediate explanation of why the free electrons make little or no contribution to the specific heat. As a metal is heated, additional energy can be acquired by the electron only by promotion to vacant states of energy higher than $E_{max.}$ to give a distribution such as that shown in fig. 5.08*b*. This, however, will be

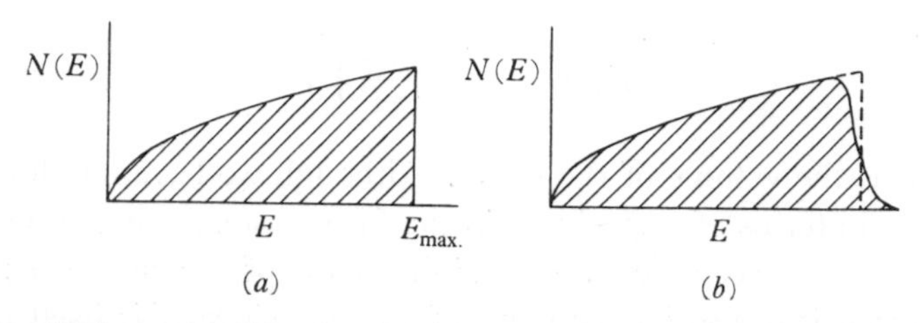

Fig. 5.08. $N(E)$ as a function of E on the basis of the free-electron theory (*a*) at the absolute zero and (*b*) at a higher temperature. Occupied energy levels are indicated by shading.

possible only at very high temperatures, so that at room temperature the energy distribution will differ little from that at the absolute zero. Except at high temperatures, very few of the electrons can absorb thermal energy and accordingly they contribute little to the specific heat.

THE BLOCH THEORY

Effect of crystal field

5.13. The theory which we have outlined above is usually termed the free-electron theory because the electrons in the metal are treated as moving in a field-free space. As we have seen, it contributes notably to our understanding of the metallic state, but it still cannot be regarded as a satisfactory description of the conditions prevailing in a metal crystal. The reason for this is that the postulate of a uniform potential throughout a metal crystal is untenable; instead we must imagine that the periodic structure of the crystal imposes a corresponding periodicity on the potential, so that in our simple one-dimensional picture the potential field of fig. 5.06 must be replaced by that shown in fig. 5.09. The effect of this is profound, for we no longer find that the energy E varies quasi-continuously with k as shown in fig. 5.07; instead, certain values of E, corresponding to discrete values of k, are absolutely forbidden. How this comes about can best be explained by continuing to consider the one-dimensional case.

Brillouin zones

5.14. Suppose that the points in fig. 5.10*a* represent atoms in a one-dimensional metal crystal and that they are regularly spaced with a separation a. The electrons moving in the metal will have wavelengths given by equation (5.01), and the corresponding electron waves may experience Bragg reflexion from crystal planes, just as X-rays may be reflected. In this particular case the direction of propagation of the waves is parallel to XX and the crystal planes in question can be

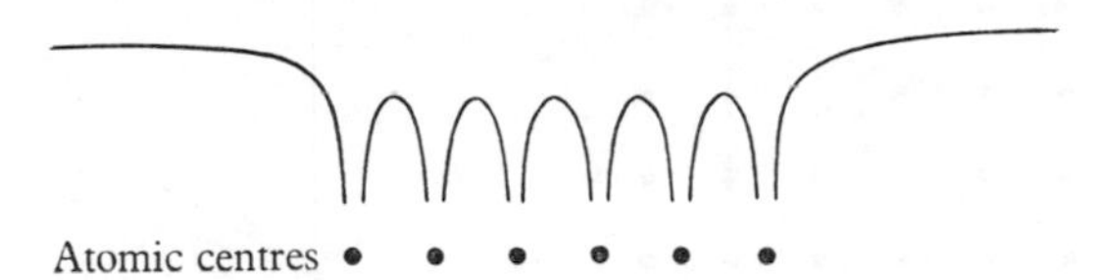

Fig. 5.09. Form of potential well in a one-dimensional metal on the basis of the Bloch theory.

considered to be those perpendicular to this direction and having a spacing a. The waves are therefore incident normally on these planes, and the wavelengths which satisfy the Bragg condition for reflexion are given by

$$n\lambda = 2a, \tag{5.12}$$

where $n = 1, 2, 3, \ldots$, etc. Expressing this in terms of k we have

$$k = \frac{\pi}{a} n. \tag{5.13}$$

Electrons whose k values satisfy equation (5.13) cannot travel through the crystal since they suffer reflexion in the atomic planes. Physically this means that such electrons have no existence in the crystal and that the corresponding energy values are missing from the energy distribution.

A convenient way to represent this state of affairs is to plot a diagram in k space on which are shown the forbidden values of k. Such a diagram, corresponding to our one-dimensional model, is shown in fig. 5.10*b*. O is the origin, and the points $k_1, k_2, \ldots$ represent the values of k given by equation (5.13). It must be remembered that k is a vector, and such a diagram must therefore be interpreted in the sense that the line Ok_1, say, represents by its length the magnitude of k_1 and by its direction the direction of this quantity.

More refined arguments show that the energies of electrons whose

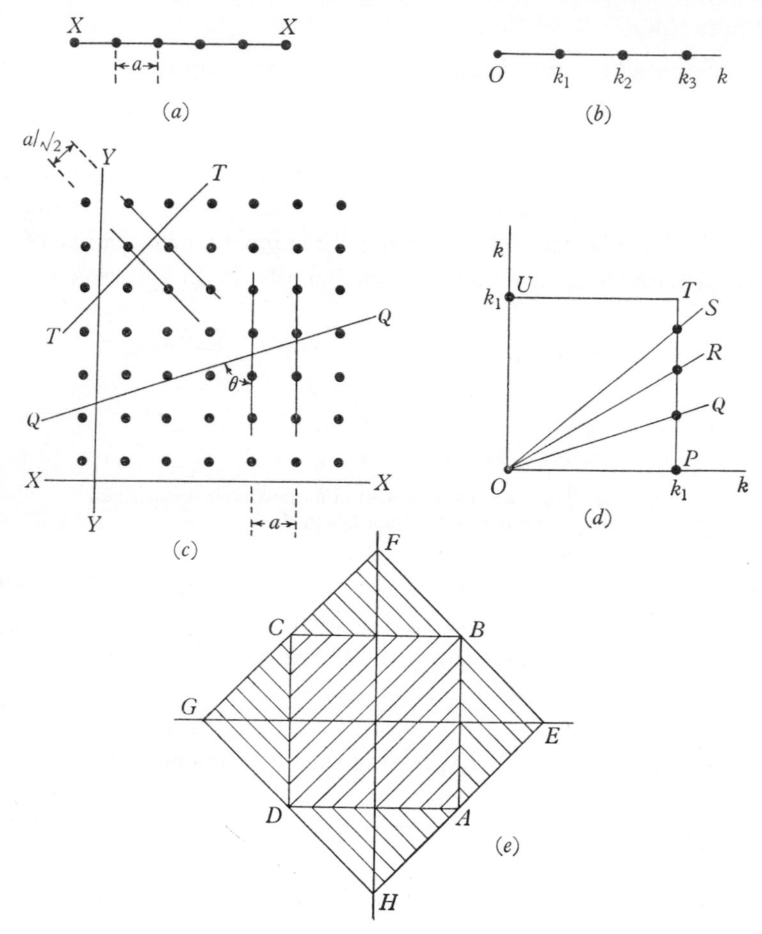

Fig. 5.10. (*a*) Atomic centres in a one-dimensional metal, and (*b*) the corresponding forbidden wave numbers. (*c*) Atomic centres in a two-dimensional metal, and (*d*) one quadrant of the corresponding first Brillouin zone. (*e*) The first and second Brillouin zones for the two-dimensional metal (*c*). The first zone is the square *ABCD*; the second zone consists of the four triangular shaded regions.

k values lie near the forbidden values are also affected: energies corresponding to k values just below the critical values are reduced, those corresponding to k values just above these critical values are increased. Remote from the critical k values the energies are those given by the free-electron theory. The effect of this is to modify the form of the relation between E and k from that shown in fig. 5.07 to that given in fig. 5.11, and it will be seen that there are now bands of forbidden energy corresponding to the k values given by equation (5.13). The first two of these are shown in the figure.

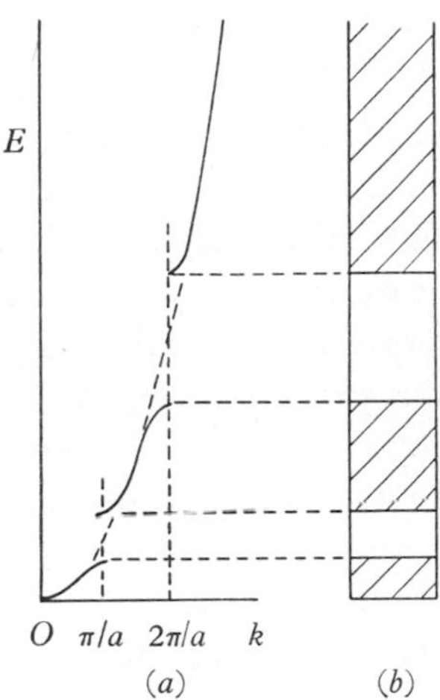

Fig. 5.11. (*a*) Variation of electron energy with wave number for a one-dimensional metal on the basis of the Bloch theory. The first two forbidden k values are shown and in (*b*) the corresponding bands of forbidden energy are indicated. The only permissible energy levels are those within the shaded regions.

5.15. We must now consider the extension of these arguments to the two-dimensional crystal shown in fig. 5.10*c*. For electrons travelling parallel to XX or YY conditions will be as before, and the corresponding forbidden k values can be plotted in k space (fig. 5.10*d*) along OP and OU (for simplicity only k_1 is now plotted and only one quadrant of the diagram is shown). Let us next consider electrons travelling in some arbitrary direction such as QQ. These electrons will now be incident on the family of planes normal to XX at an angle θ different from 90°, and equation (5.12) must be replaced by

$$n\lambda = 2a \sin \theta. \tag{5.14}$$

The wavelengths which suffer Bragg reflexion are reduced and the corresponding k values are increased so that in the direction OQ (fig. 5.10*d*) k_1 is in the position shown, at a greater distance from O than before. If we similarly consider other directions OR, OS, ... in k space we can plot the corresponding k_1 positions, and elementary geometry shows that these will fall on the straight line PT, as indicated; in the same way k_1 values corresponding to reflexions from families of planes normal to YY will lie on the straight line UT. We can therefore draw a new diagram of k space (fig. 5.10*e*) in which the square $ABCD$ is the locus of the k_1

positions and is such that as the k vector of an electron crosses the boundaries of this square the energy of that electron suffers a discontinuous increase.

The above argument was based on a consideration of families of planes in the crystal structure normal to XX and YY and having a spacing a. There are, however, many other families of planes, one of which is that perpendicular to TT of spacing $a/\sqrt{2}$. Similar arguments applied to this family give rise to a second square $EFGH$ in k space, which again has the property that the energy of an electron increases discontinuously when its k vector crosses the boundaries of the square. If we consider yet other families of planes we obtain a series of further larger and larger figures (not necessarily square) in k space all possessing this property. We may thus regard k space as divided into a number of zones such that as the representative point passes from one zone to the next a discontinuous increase in energy takes place. Such zones are termed *Brillouin zones*, and in the example we have given the square $ABCD$ is the first Brillouin zone, the four triangular regions ABE, etc., are the second, and so on.

In the three dimensions of a real crystal similar arguments again apply and we find that the Brillouin zones are now defined by a series of concentric polyhedra in k space.

5.16. In fig. 5.11 we represented for a one-dimensional crystal the energy changes which take place as the k value of an electron is increased. For two- and three-dimensional crystals this diagram requires modification because the energy changes which occur on crossing the boundary of a Brillouin zone are different in magnitude and occur at different k values in different directions. This is illustrated for the first Brillouin zone in fig. 5.12*a*, where k_a, k_b and k_c are the critical k_1 values for three different directions in space. For any one direction the energy curve (e.g. OA_aB_aC) follows the course of that shown in fig. 5.11, but in different directions the positions and magnitudes of the energy gaps A_aB_a, A_bB_b, etc., differ. For the purpose of our argument let us assume that k_a is the minimum and k_c the maximum critical k_1 value for the first Brillouin zone, corresponding in the two-dimensional case to the directions OP and OT in fig. 5.10*d*. Then in this case there is a finite energy gap A_cB_a between the first and second zones and these zones are separated by a band of forbidden energies, as shown in fig. 5.12*b*.

Suppose, however, that the E/k curves take the form shown in fig. 5.12*c*. It is still true that in any given direction the curve is as shown in fig. 5.11, but we now note that for certain directions in the first zone (e.g. k_c) the energy is greater than for other directions in the second zone (e.g. k_a). There is therefore now no gap of forbidden energy values and the two zones overlap (fig. 5.12*d*). This distinction is of profound importance, for it enables us to discuss what happens as the energy states in a metal are successively filled with electrons.

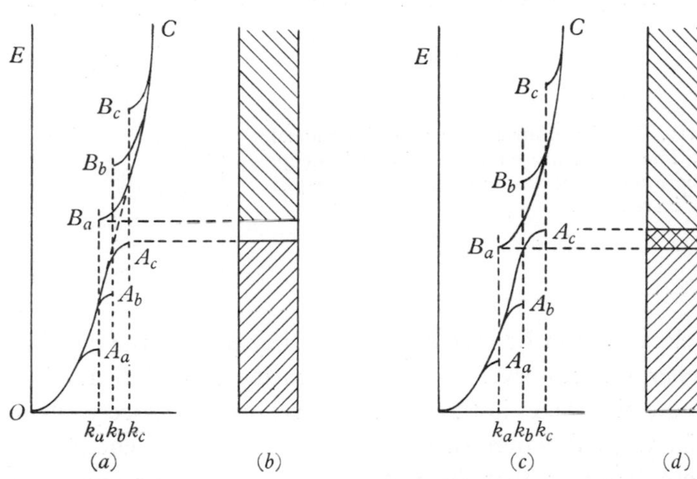

Fig. 5.12. (*a*) Variation of electron energy with wave number for a two-dimensional metal on the basis of the Bloch theory. k_a, k_b and k_c are values of the first forbidden wave number in different directions in the first Brillouin zone. The band of forbidden energy in (*b*) indicates that the energy levels of the first and second zones do not overlap. (*c*) As (*a*) but for the case in which there is overlap between the energy levels of the first and second zones, as shown in (*d*).

The Fermi surface

5.17. We have already seen that the permitted energy states are those which satisfy equation (5.08). We can thus represent each energy state by an element of volume in k space so that as more and more electrons are fed into the metal the occupied volume of k space increases. Initially energy is independent of the direction of k, so that the occupied volume, usually termed the *Fermi surface*, is spherical, and we can imagine a series of concentric spherical shells in k space corresponding to equal increments of energy. These shells will not be equally spaced since E is proportional to k^2. In two dimensions the corresponding figures are circles, as represented by the contours (1) and (2) in fig. 5.13.

As the Fermi surface begins to approach the boundaries of the first Brillouin zone, however, it will become distorted. In directions such as *OP* (fig. 5.10*d*) the energy is abnormally depressed, for a given value of *k*, and the Fermi surface is correspondingly distended (contour (3) in fig. 5.13). When the surface finally touches the boundary of the Brillouin zone further growth in that direction is arrested and the Fermi surface then extends only towards the corners of the zone (contours (4–6) in fig. 5.13). If the system is one corresponding to the condition shown in fig. 5.12*a* the zone will ultimately be fully occupied and

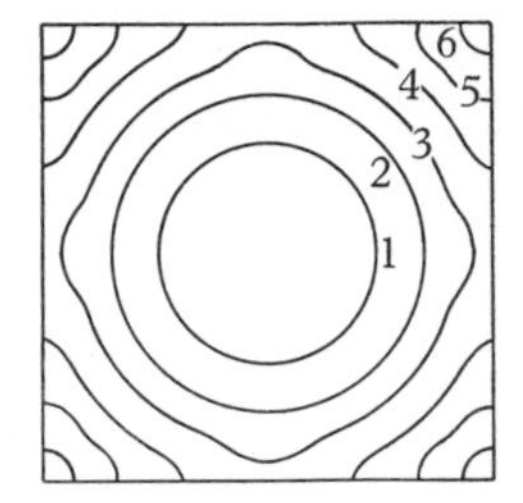

Fig. 5.13. The Fermi surface in a two-dimensional metal. The curves are contours of equal energy and the intervals between them represent equal energy increments.

further electrons can be accepted only if they possess the considerably higher energy necessary for promotion to the second Brillouin zone. These successive stages in the process are represented purely schematically in fig. 5.14*a–d*, where shading represents occupied energy levels.

Suppose, however, that the system is one of the type represented in fig. 5.12*c*. In this case the early stages will proceed exactly as before. Now, however, there is no gap of forbidden energy between the first and second Brillouin zones and electrons will therefore start to enter the second zone before the first is fully occupied. The first zone will ultimately fill up, but not before a considerable number of electrons have taken their place in the second zone. This case is represented in fig. 5.14*e–h*.

5.18. The distinction between these two cases, as we shall see shortly, is of such great importance that it is desirable to emphasize it in another way by considering the form of the corresponding $N(E)/E$ diagram. We have already shown the form of this diagram on the free-electron picture in fig. 5.08*a*, where we found a parabolic relation between $N(E)$ and E. In terms of our more refined treatment, the parabolic relation will still hold for low E values, but for higher E values departures from the relationship will occur as soon as the Fermi surface ceases to be spherical.

Let us consider first a system in which there is no overlap between the energy levels of the first two Brillouin zones. As the Fermi surface approaches the boundary of the first Brillouin zone the energy states become more crowded together (see fig. 5.13) and the number for a given energy increment will increase. The $N(E)$ curve thus rises, as shown at

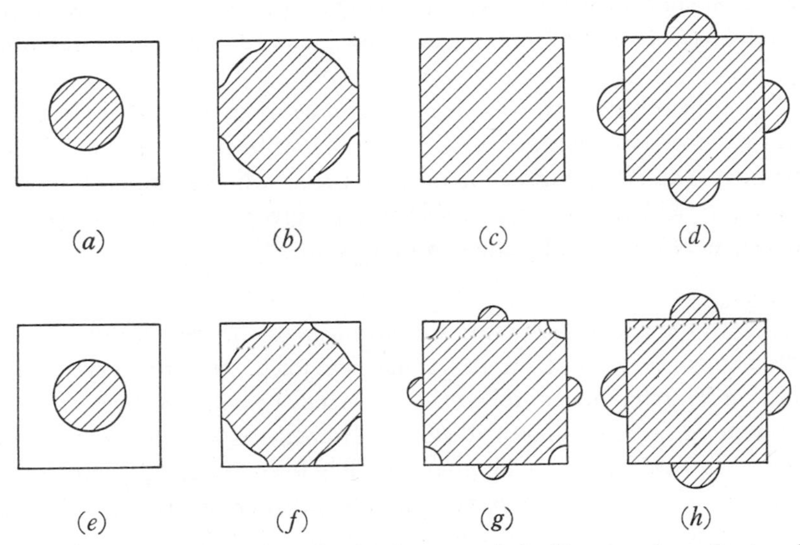

Fig. 5.14. Successive stages in the development of the Fermi surface of a two-dimensional metal: (*a–d*) when the energy levels of the first and second Brillouin zones do not overlap; (*e–h*) when there is overlap between these energy levels.

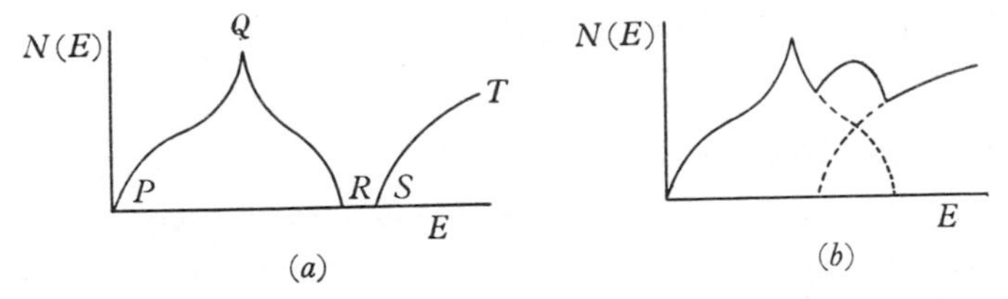

Fig. 5.15. $N(E)$ as a function of E for a metal (*a*) when the energy levels of the first and second Brillouin zones do not overlap, and (*b*) when there is overlap between these energy levels.

PQ in fig. 5.15*a*. This rise will continue until the Fermi surface touches the Brillouin zone in some direction, corresponding to the condition represented by the point A_a in fig. 5.12*a*. Thereafter, certain *k* directions are forbidden, and there will be more and more such directions as the filling of the zone proceeds. It is only electrons travelling in allowed *k* directions which can have higher energies, and the number of such electrons will rapidly dwindle. After Q the $N(E)$ curve will therefore fall steeply and will finally sink to zero at R, where the zone is full and where the energy corresponds to the point A_c in fig. 5.12*a*. A range RS of forbidden energies will now follow, and finally electrons will start to enter the second zone at S. So long as the number of electrons in this zone is small they will behave essentially as on the free-electron theory and the $N(E)$ curve will once again follow the parabolic form represented by ST.

If, on the other hand, we consider the case where the energy levels of the two zones overlap (fig. 5.12*c*) there will now be no range of forbidden energies. The $N(E)$ curves for the two zones will overlap, as shown in fig. 5.15*b*, and the resultant $N(E)$ will be given by adding these two curves, as represented by the continuous line.

5.19. In the above discussion we have treated the very simple case of only two zones, each of simple geometrical form. In actual crystals the position is often more complex: the zones may be bounded by more than one type of face at different distances from the origin in k space, and more than two zones may be involved. The form of the $N(E)$ curve will therefore in general be more complex than that shown in fig. 5.15, but we may expect that it will show a pattern of peaks and valleys the detailed feature of which will depend upon the crystal structure and upon the magnitude of the energy gaps at the surfaces of the zones. We shall find that the form of these distributions for different systems throws much light on their different properties.

Relation to molecular-orbital theory

5.20. It will be realized that the discussion of metal theory given above is in fact the application to metal systems of the molecular-orbital theory briefly described in chapter 4. It is an essential feature of that theory that the electrons concerned in interatomic binding are no longer to be associated with individual atoms but exist in molecular orbitals in which two or more atoms are involved; in the particular case of benzene, for example, we saw that some of the electrons had to be regarded as occupying delocalized orbitals embracing the whole molecule. This however, is precisely the state of affairs which exists in a metal, provided that we now regard the whole crystal as constituting a single giant molecule.

5.21. This point of view may be illustrated by a particular example. Suppose that we consider a crystal of sodium to be formed by the condensation of an array of widely dispersed sodium atoms. In the dispersed state the isolated atoms will have the configuration $1s^2$, $2s^2 2p^6$, $3s^1$, and the $3s$ electrons of all the atoms will have the same energy. As the atoms approach, however, interaction of the $3s$ orbitals will take place and it will no longer be possible, on account of the Pauli exclusion principle, for more than two electrons to possess the same energy. In consequence the energies of the $3s$ electrons are now spread over a finite range, and the closer the approach of the atoms the broader this range becomes, as indicated purely schematically in fig. 5.16.

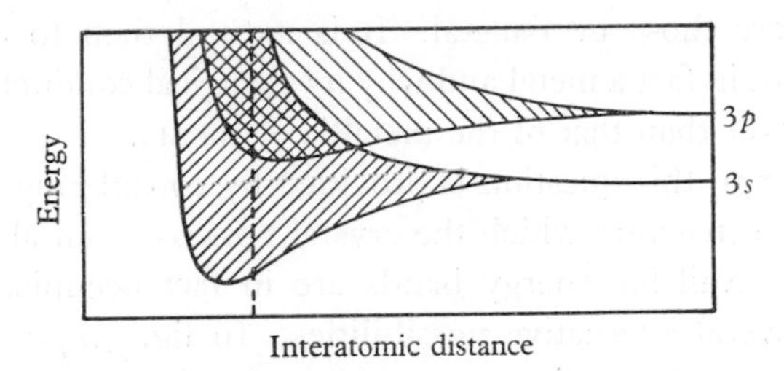

Fig. 5.16. Schematic representation of the spreading of the $3s$ and $3p$ energy levels of isolated sodium atoms as the atoms condense to form a crystal. The broken line represents the equilibrium interatomic distance in the solid.

We must, however, also consider the unoccupied $3p$ orbitals, the energy of which, in the isolated atom, is only slightly higher than that of the $3s$ orbital. These orbitals, too, will spread, and it may happen that at the equilibrium interatomic distance, represented by the broken line, overlap with the $3s$ levels takes place. There is then a broad quasi-continuous range of energy levels available to the valency electrons, and how far this range is filled will depend upon the number of these electrons. This is just another way of describing the range of permitted energies which we have already discussed in terms of the Bloch theory. It is also another way of describing the phenomenon of hybridization, for within the range of overlap it is impossible to identify electrons as $3s$ or $3p$ and we can describe them only as a hybrid of these orbitals. This emphasizes that the electronic configurations of the atoms given in table 2.03 are applicable only to *isolated atoms* in their ground states, and that when these atoms are closely associated with others quite different configurations of the outer electrons may be found. We shall find that this is particularly true of the transition metals, in which hybridization between several orbitals of comparable energy frequently occurs in the solid state.

Conductors, semi-conductors and insulators

5.22. The molecular-orbital treatment of metal theory is of particular value from our point of view because it serves to emphasize that there is no sharp distinction between the metallic and covalent bonds, just as we have already seen that there is no sharp distinction between covalent and ionic binding. In diamond, for example, we can regard the electrons responsible for the C–C bonds as existing in molecular orbitals which pervade the crystal as a whole, so that these electrons are as free to circulate as are those in a metal. It is natural then to enquire why diamond is not in fact a metal and why its electrical conductivity is some 10^{24} times lower than that of the metallic elements.

The answer to this question is provided by considering not only the Brillouin-zone structure which the crystal possesses but also the extent to which the available energy bands are in fact occupied. Fig. 5.17 represents several alternative possibilities. In fig. 5.17*a* the first two zones are separated by a finite energy gap and the number of electrons present (represented by shading) is insufficient to fill the first zone. If now an electric field is applied electrons will strive to acquire a drift velocity and so to increase their energy. Since vacant levels of higher energy are available to receive these electrons such a process is possible and conductivity will result. Conductivity will similarly take place under the conditions represented in figs. 5.17*b*–*d*, where the first two zones overlap and where again vacant energy levels are available for electron promotion. These conditions are those characteristic of most of the metallic elements, and fig. 5.17*b* represents the state of affairs in the univalent metals where the number of electrons available is only one-half of that required to fill the first zone. It should be noted that under the conditions represented in fig. 5.17*d* the electrons in the first zone can play no part in the conductivity process, so that although the total number of electrons is greater than in the preceding examples the number contributing to conductivity may well be smaller. In fig. 5.17*c* electrons have already entered the second zone, although the first is not yet full; in this case electrons in both zones are effective in the conductivity process.

In fig. 5.17*e* the zones are again shown as separated, but now the first zone is completely filled and the next is empty. Clearly in this case it will not be possible for the electrons to acquire additional energy in an electric field, since this would bring them into the forbidden energy range, and the substance is accordingly an insulator. Often, of course, the actual zone structure will be more complex than in this simple example, but nevertheless we can still say that in general an insulator is characterized by the fact that the electrons completely fill one or more

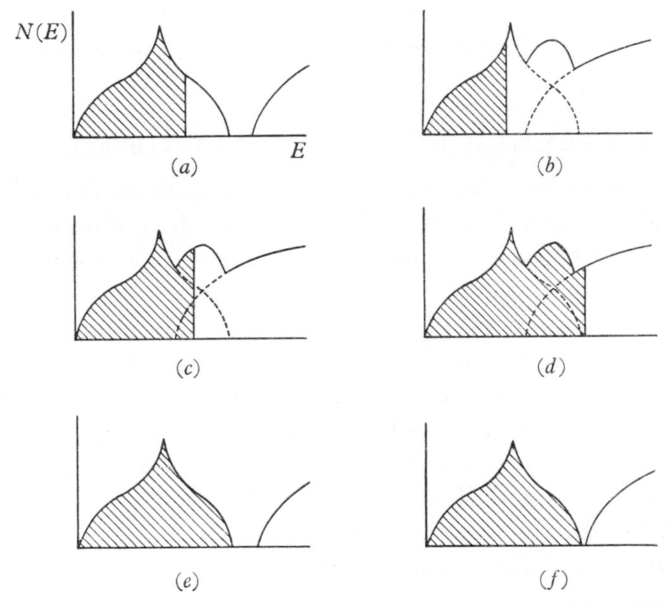

Fig. 5.17. Idealized representations of $N(E)$ as a function of E in different solids: (*a–d*) metals; (*e*) an insulator; (*f*) an intrinsic semi-conductor.

Brillouin zones, the outermost of which is separated from the next by a finite energy gap. The distinction between conductors and insulators is therefore not a distinction between free and bound electrons, but is a distinction between systems in which vacant states are and are not available for electron promotion.

If, under the conditions corresponding to fig. 5.17*e*, some of the electrons can be promoted to the second zone, conductivity will be possible since vacant energy levels will become available in both zones. This can be achieved if the crystal is illuminated with light whose quantum energy is sufficient to bridge the gap between the zones, and under these circumstances photoconductivity will be observed in what is ordinarily an insulating material. If the energy gap is very small, as shown in fig. 5.17*f*, the crystal will be an insulator at the absolute zero of temperature, but now even the energy of thermal agitation may be sufficient to promote some electrons into the second zone. This is the state of affairs in intrinsic semi-conductors, and it is easy to see why such substances will have a positive temperature coefficient of conductivity: as the temperature is raised more and more electrons are able to take part in the process of conduction. Even so, the absolute magnitude of the conductivity will still be much lower than that of a metal since the number of conductivity electrons remains relatively very small.

THE VALENCE-BOND THEORY OF METALS

5.23. So far we have discussed metal systems on the basis of the molecular-orbital theory. We now proceed to a brief discussion of the treatment of these systems in terms of the alternative valence-bond theory, as developed principally by Pauling.

It is well known that certain univalent metals (for example, lithium) form diatomic molecules in the vapour in which the interatomic binding is presumably covalent in character. On the valence-bond theory it is assumed that such bonds also operate in the solid state, but since the number of electrons available is inadequate to give rise to covalent bonds between each atom and all its neighbours (eight in lithium) resonance is assumed to take place throughout the solid in a way which may be symbolized, in two dimensions, thus:

```
Li—Li        Li   Li
        ↔    |    |
Li—Li        Li   Li
```

This mechanism involves the synchronized resonance of pairs of bonds, but much greater stability will result if we countenance the far more numerous structures arising from unsynchronized resonance between arrangements in which ions are also involved, thus:

```
Li—Li        Li—Li⁻
        ↔        |
Li—Li        Li⁺ Li
```

This process will be possible only if a vacant orbital is available in the lithium atom to accept the extra electron present in the Li^- ion. In fact this is so, for the three vacant 2*p* orbitals are energetically little above the 2*s* orbital occupied by the valency electron. We can thus regard the

Li^- ion as having the structure $1s^2$, $2s^1 2p^1$, with the $2s$ orbital and one of the $2p$ orbitals, probably hybridized, as available for the formation of two covalent bonds.

It is natural to enquire how metallic conductivity arises on this picture. The answer is, as on the molecular-orbital theory, that metallic conduction will be possible only if in addition vacant 'metallic' orbitals are also available to accept conductivity electrons. The essential requirement for the existence of a conducting metallic crystal structure is therefore that there shall be available a sufficiency of orbitals, of energy not very different from that of the outermost occupied orbitals, to produce a pattern of resonating covalent bonds, and in addition one or more wholly vacant orbitals, again of energy comparable with that of the occupied orbitals, to accept conductivity electrons. This condition is, indeed, satisfied in the example of lithium, given above, for two of the $2p$ orbitals are not involved in the covalent structure and so act as metallic orbitals.

5.24. The valence-bond theory of metals is of importance in that it enables us to form a qualitative picture of the properties of these elements. In the first place it explains the close relationship, already mentioned in §5.07, between metallic and covalent radii, since it treats metallic binding as being essentially covalent in origin. A second important feature is that it enables us to understand the distribution of the metallic elements in the Periodic Table, as we may now show.

The distribution of the metallic elements

The short periods

5.25. Let us begin by considering the elements of the second and third periods (Li–Ne and Na–A). In each of these the orbitals available for bond formation are limited to four in number, namely, one s and three p orbitals, so that if four or more electrons are present it will be possible for all the orbitals to be filled by stable groups of electrons, as in the crystals of diamond and silicon or in the molecules of nitrogen, oxygen, fluorine, etc. Metallic properties in these two short periods will therefore be confined to the elements of the first three groups.

The long periods

5.26. As soon, however, as we pass to the long periods we must consider not only the *s* and *p* orbitals of the outermost shell but also the *d* orbitals of the penultimate shell, since these now differ little in energy. In all, therefore, nine orbitals (five *d*, one *s* and three *p*) must be taken into account, and the properties of any particular element will be determined by the distribution of the available electrons among these orbitals. We may illustrate this point by some examples.

5.27. Let us consider the case of tin, a *B* sub-group element in group 4. One possible configuration for this atom is as shown for Sn_A below (where only the 4*d*, 5*s* and 5*p* electrons are represented), and by analogy with carbon the element would be expected to be non-metallic with a co-valency of 4. This in fact is true of one modification of tin (grey tin, stable below 18 °C), which is an insulator with the same structure as diamond. An alternative configuration, however, is that shown for Sn_B. Here only two orbitals are available for covalent bond formation, but there is in addition one wholly vacant 'metallic' orbital (represented by the symbol ○) to confer metallic properties. This is probably the configuration in white tin, the common metallic form of the element stable above 18 °C, although it is possible that both configurations make some contribution, thus accounting for the fact that white tin does not have the close-packed structure characteristic of a truly metallic element.

	4*d*	5*s*	5*p*	Co-valency	Number of metallic orbitals
Sn_A	↑↓ ↑↓ ↑↓ ↑↓ ↑↓	↑	↑ ↑ ↑	4	0
Sn_B	↑↓ ↑↓ ↑↓ ↑↓ ↑↓	↑↓	↑ ↑ ○	2	1

5.28. As a second example we may consider the element zinc, for which the three configurations shown below may be written. Zn_A has no vacant metallic orbitals and corresponds to a purely covalent non-metallic structure of covalency 6. Zn_B and Zn_C, however, both possess vacant orbitals. Separate zinc structures corresponding to these different configurations are not known, but the assumption that in the actual structure there is some contribution from each accounts elegantly for the fact that the crystal structure is not truly close packed. We shall revert to this point later when we discuss the metal structures in detail (§7.19).

	3*d*	4*s*	4*p*	Co-valency	Number of metallic orbitals
Zn_A	↑↓ ↑↓ ↑↓ ↑ ↑	↑	↑ ↑ ↑	6	0
Zn_B	↑↓ ↑↓ ↑↓ ↑↓ ↑	↑	↑ ↑ ○	4	1
Zn_C	↑↓ ↑↓ ↑↓ ↑↓ ↑↓	↑	↑ ○ ○	2	2

5.29. As a final example of the application of the valence-bond treatment to metal structures we may consider the transition metal nickel, and here we must introduce the idea that an unpaired electron in the *d* shell may occur either in a *bonding* role (as in Zn_A and Zn_B discussed above) or in a *non-bonding* role in which, although unpaired, it plays no part in interatomic bonding. The importance of the unpaired non-bonding electrons is that they will confer paramagnetic properties on the metal structure, whereas a structure in which all electrons are paired would be expected to be diamagnetic.

Two structures for nickel may be formulated, as shown in the diagram below where a broken arrow represents a single electron in a non-bonding orbital. Ni_A has no vacant metallic orbital and corresponds to a purely covalent non-metallic structure of covalency 6, which is paramagnetic on account of the two non-bonding orbitals. Ni_B on the other hand has a vacant metallic orbital but no non-bonding orbitals, and is therefore a diamagnetic metal. The actual structure must represent a combination of these configurations, and their relative contributions can be assessed from the magnitude of the observed paramagnetism.

	3*d*	4*s*	4*p*	Co-valency	Number of metallic orbitals	Number of non-bonding orbitals
Ni_A	↑↓ ⇡ ⇡ ↑ ↑	↑	↑ ↑ ↑	6	0	2
Ni_B	↑↓ ↑↓ ↑ ↑ ↑	↑	↑ ↑ ○	6	1	0

5.30. Similar arguments apply to the other transition metals. We shall not, however, pursue these discussions further at this point because we shall generally find that for our purposes a purely qualitative treatment of the binding in metallic systems will suffice. Moreover, it must be admitted that in its more detailed quantitative interpretation the valence-bond picture of the metallic state is by no means universally accepted. As with the covalent bond, the molecular-orbital and valence-bond treatments are to a large extent complementary. The former is undoubtedly the more aesthetically satisfying: it starts from a theoretical basis, it lends itself to quantitative calculation and it gives an elegant account of the conductivity and other properties of metals. The latter, on the other hand, is largely empirical in its approach: the electronic configuration in each element is deduced not on theoretical grounds but from a detailed interpretation of the observed interatomic distances, magnetic properties and other characteristics of each particular system. It has been particularly successful in interpreting atomic radii and correlating these radii with those observed in covalent systems, and it gives us a clear picture of the distribution of metallic and non-metallic properties among the elements of the Periodic Table. One day we may hope for a closer co-ordination of the two treatments, but that day is not yet.

6 THE VAN DER WAALS BOND

INTRODUCTION

6.01. None of the three types of interatomic bonding discussed in the preceding chapters can be responsible for the cohesion of the inert gases in the solid state or for the intermolecular forces in organic crystals. In all of these cases the forces involved are those due to the residual or van der Waals bond.

The fact that weak forces of attraction exist between atoms and molecules already chemically saturated is shown by the Joule–Thomson effect, and as early as 1873 van der Waals postulated such forces in deriving his equation of state for real gases, without, however, making any attempt to account for their origin or to explain their nature. The van der Waals bond must be regarded as operating between all atoms and ions in all solids, but it is so weak compared with the ionic, covalent and metallic bonds that its effect is largely masked in any structure in which it occurs in conjunction with any of these stronger bonds. In the qualitative discussion of such structures its effect may therefore often be ignored, although, as we have already seen, a quantitative treatment sometimes demands that it should be taken into account. The only solids in which the properties of the van der Waals bond can be studied in isolation are the inert gases in the solid state to which they condense at sufficiently low temperatures. The very low melting points of those gases, their large coefficients of thermal expansion and their low heats of sublimation reveal the weakness of the bond.

6.02. Structurally the van der Waals bond bears a close formal resemblance to the metallic bond as one which can link an atom to an indefinite number of neighbours, and one which is spatially undirected. In consequence of these characteristics the structural arrangement in crystals bound by these bonds is determined almost entirely by geometrical considerations, and we find, for example, that the crystals of the inert gases have the same close-packed structures as the true metals: helium at low temperatures and under high pressure forms hexagonal close-packed crystals and the remaining inert gases in the solid state are cubic close packed.

Similar considerations apply to the very much more important and far more numerous body of molecular structures in which discrete molecules, strongly bound within themselves, are bound to one another by van der Waals bonds. These structures comprise almost all the compounds of organic chemistry, as well as a considerable number of inorganic substances, and now we are concerned with the packing of molecules which are no longer spherical in shape and which often comprise different atoms of different sizes. Clearly the structure can no longer be close packed in the strict geometrical sense, but we shall, nevertheless, find when we discuss such structures in detail that the arrangement of the molecules is usually as compact as is consistent with their irregular shape. Moreover, the physical properties of such structures again reflect the weakness of the van der Waals bond. Although the strength of this bond may vary between fairly wide limits, so that many organic compounds are solid at room temperature, it is clearly true that broadly speaking organic and other molecular crystals are softer and have lower melting points and larger coefficients of thermal expansion than, say, metals and inorganic salts.

VAN DER WAALS RADII

6.03. Measurements of the cell dimensions of crystals of the inert gases enable characteristic van der Waals radii to be deduced for the atoms of these elements. In the same way van der Waals radii for other atoms may be derived from molecular structures by determining the closest distance of approach between adjacent molecules. Some radii found in this way are given in table 6.01. These radii cannot be regarded as having the same degree of constancy as the ionic, covalent and metallic radii already discussed, and variations of as much as 0·1 Å may be found in different molecular structures. It will be seen, however, that they are generally much larger than the corresponding covalent radii of the same elements (table 4.03), emphasizing once again that in molecular structures the effective radius of an atom within a molecule is quite different from its radius towards other molecules.

Table 6.01. *van der Waals radii*

Values are in Ångström units

		H 1·2	
N 1·5	O 1·4	F 1·35	Ne 1·6
P 1·9	S 1·85	Cl 1·8	A 1·9
As 2·0	Se 2·0	Br 1·95	Kr 2·0
Sb 2·2	Te 2·2	I 2·15	Xe 2·2

THEORY OF THE VAN DER WAALS BOND

van der Waals lattice energy

6.04. The origin of the van der Waals forces has been the subject of much work which we shall discuss only briefly here. In 1912 Keesom considered the case of uncharged molecules possessing a permanant dipole μ, and attributed the forces to the attraction between these dipoles. As a result of this so-called *orientation effect* he derived an expression for the interaction energy between atoms or molecules at a distance r apart of the form

$$u = -\frac{2}{3}\frac{1}{r^6}\frac{\mu^4}{kT}, \tag{6.01}$$

where k is the Boltzmann constant and T is the absolute temperature. Debye, however, showed that this expression was by itself inadequate on the grounds that the temperature variations which it predicted was not in accord with the experiment, and he added to the interaction energy a second term arising from the attraction between a permanent dipole and the moments induced by it in neighbouring molecules. This *induction effect* contributes to the energy a term

$$u = -2\frac{\alpha\mu^2}{r^6}, \tag{6.02}$$

where α is the polarizability of the molecule and measures the dipole induced in it by an applied electric field. Broadly speaking, we may say that the larger a particle and the more loosely bound its electronic structure the greater will be its polarizability.

While these two effects are undoubtedly relevant in solids in which molecules are polar, neither can play any part in the numerous structures, such as those of the inert gases, methane, hydrogen, benzene and many others, in which the atoms or molecules possess no permanent dipole moment. Moreover, even when a dipole moment does exist, the forces predicted are far smaller than those experimentally observed. Some other component of interatomic force must therefore exist, and it was London who first ascribed this to a *dispersion effect* associated with the dynamic polarization of an atom or molecule arising from its zero-point motion. Even if we consider a completely symmetrical system, with no permanent dipole moment, it will still be possible for such a system to possess an instantaneous moment, since at any given instant the electrons will not necessarily be distributed with the high symmetry corresponding to their distribution averaged over a long period of time. This instantaneous dipole moment of one atom or molecule will polarize its neighbours, just as in the induction effect discussed by Debye, and since the effect is proportional to μ^2 the average force will be finite even if the average moment is zero. Arguing in this way, London derived for the energy due to the dispersion effect the expression

$$u = -\frac{3}{4}\frac{h\nu_0\alpha^2}{r^6}, \tag{6.03}$$

where ν_0 is the frequency of the zero-point motion and h is Planck's constant.

6.05. The relative magnitudes of the three components of the van der Waals energy for a number of molecular structures are shown in table 6.02. It will be seen that in all cases the influence of the dispersion effect is considerable and that in most cases this is by far the largest contribution. Only in very strongly polar molecules, such as NH_3 and H_2O, is the orientation effect important, and even in HCl its contribution is not large. The induction effect is very small in almost every case. Within any one group of comparable structures, such as those of the hydrogen halides, the dispersion energy is seen to increase with particle size because the influence of increased interparticle distance in equation (6.03) is more than compensated by the increase in polarizability.

Comparison with experiment

6.06. The van der Waals lattice energy of a crystal is the energy required to disperse the structure into an assemblage of widely separated molecules and may therefore be directly compared with its heat of sublimation. Some values of this latter quantity are given in table 6.02 and it will be seen that on the whole agreement is satisfactory, especially when it is remembered that we have taken no account of the effect of the repulsive forces which must operate to confer on molecules their characteristic sizes.

6.07. A comparison of the lattice energies given in table 6.02 with those of ionic structures (tables 3.04 and 3.05) reveals that the van der Waals bond has a strength one or more orders of magnitude smaller than that of the ionic link. Its weakness compared with the covalent bond is similarly revealed by a comparison of the van der Waals energy of molecular structures with the heat of dissociation of the same molecules. A few values of this quantity are also given in table 6.02, and it will be seen, for example, that the van der Waals energy of a nitrogen crystal is 1·7 kcal/mole, whereas the heat of dissociation of the nitrogen molecule (corresponding to the rupture of the covalent bonds) is one hundred times greater. Ammonia and water are exceptional owing to the strongly polar character of these molecules; water alone, of the substances cited in the table, has a boiling point above room temperature.

Table 6.02. *van der Waals energy*

Values are in kcal/mole

	Orientation effect	Induction effect	Dispersion effect	Total energy	Heat of sublimation	Dissociation energy
Ne	—	—	—	−0·6	0·6	—
A	−0·00	−0·00	−2·03	−2·0	2·0	—
Kr	—	—	—	−2·8	2·8	—
H_2	—	—	—	—	0·2	103
N_2	—	—	—	−1·7	1·7	170
O_2	—	—	—	−1·7	2·0	117
CO	−0·00	−0·00	−2·09	−2·1	1·9	223
HCl	−0·79	−0·24	−4·02	−5·0	4·8	102
HBr	−0·16	−0·12	−5·24	−5·5	5·5	—
HI	−0·01	−0·03	−6·18	−6·2	6·2	—
NH_3	−3·18	−0·37	−3·52	−7·1	7·1	90
H_2O	−8·69	−0·46	−2·15	−11·3	11·3	118

THE CLASSIFICATION OF CRYSTAL STRUCTURES

6.08. We have now completed our reviews of the four principal types of interatomic force responsible for the cohesion of solids, and have illustrated by examples some characteristic properties of simple crystal structures in which these different types of force operate. These properties, and also some additional physical properties which we shall not discuss in detail, are summarized in table 6.03.

In the first edition of this book these four types of bond were used as the basis for a system of classifying crystal structures. It has now become abundantly clear, however, as we have repeatedly emphasized, that the distinction between these types of force is less real than was at one time imagined, and is by no means absolute; in many structures binding of an

Table 6.03. *Physical and structural properties associated with the four interatomic bonds*

Property	Ionic	Covalent	Metallic	van der Waals
Structural	Non-directed, giving structures of high co-ordination	Spatially directed and numerically limited, giving structures of low co-ordination and low density	Non-directed, giving structures of very high co-ordination and high density	Formally analogous to metallic bond
Mechanical	Strong, giving hard crystals	Strong, giving hard crystals	Variable strength. Gliding common	Weak, giving soft crystals
Thermal	Fairly high M.P. Low coefficient of expansion. Ions in melt	High M.P. Low coefficient of expansion. Molecules in melt	Variable M.P. Long liquid interval	Low M.P. Large coefficient of expansion
Electrical	Moderate insulators. Conduction by ion transport in melt. Sometimes soluble in liquids of high dielectric constant	Insulators in solid and melt	Conduction by electron transport	Insulators
Optical and magnetic	Absorption and other properties primarily those of the individual ions, and therefore similar in solution	High refractive index. Absorption profoundly different in solution or gas	Opaque. Properties similar in liquid	Properties those of individual molecules, and therefore similar in solution or gas

intermediate character is found and in many other structures different types of force operate between different constituent atoms. A classification based on bond type therefore suffers from the objection that it prejudges the question as to which particular type of force is operating in any particular structure and makes no provision for any re-interpretation of bond type which may be demanded by later developments in valency theory.

Alternative classifications based on purely geometrical considerations have been proposed. These, too, are not without objection, for while they are unambiguous they have the disadvantage of associating together structures which are geometrically analogous but between which no other close relationship exists; similarly they separate on purely geometrical grounds structures which in other respects are closely related.

Certain broad features of a classification can, nevertheless, be recognized. The metals and intermetallic systems can be distinguished from inorganic compounds containing non-metallic elements, and both can be distinguished from organic crystal structures in which well defined molecules have a discrete existence. Beyond this, however, it does not seem profitable to go. In the systematic discussion of crystal structures, to which the rest of this book is devoted, we shall therefore adopt no formal classification, but instead we shall consider structures roughly in order of increasing complexity.

PART II

SYSTEMATIC CRYSTAL CHEMISTRY

7 THE ELEMENTS

INTRODUCTION

7.01. It is the aim of our discussion of systematic crystal chemistry, to which the rest of this book is devoted, to investigate the relationships which exist between the structure and the chemical constitution of crystalline substances in the hope that we may be able to interpret as many as possible of the chemical and physical properties of these substances in terms of their crystal structures. Ideally we would wish not only to do this but also, conversely, to be in a position to propose, and even to create, structures having any desired combination of properties whatsoever. We are as yet a long way from this idealized goal, but X-ray methods have now been applied to a sufficiently wide range of bodies, representative of inorganic, organic, metallurgical, mineralogical and biological chemistry, for a number of general principles to have emerged, and we have already learned to recognize the structural significance of the properties of many of these substances.

It would be quite out of place in a work such as this to describe the detailed features of all, or even of many, of the several thousand crystal structures now known, especially as such descriptions are already available in the invaluable works of reference, cited in appendix 1, which crystallographers are fortunate in having at their disposal. Our concern with individual structures lies primarily in the principles which they illuminate, and in so far as detailed descriptions are necessary they will be confined to structures of common occurrence and general interest. Frequently the detailed features of a structure are unimportant or insignificant so that an idealized or simplified description will serve our purpose, and sometimes a purely qualitative account of only certain aspects of a structure will suffice to illustrate the general principles involved.

We now proceed to our systematic survey of crystal chemistry. For this purpose it is convenient to open with a discussion of the structures of the elements, for these structures in themselves furnish examples of many of the structure types and of many of the general principles outlined in the preceding chapters. We find among them both molecular and non-molecular structures, and structures in which covalent, metallic and van der Waals bonds are operative. It is only the ionic bond which is unrepresented in the structures of the elements; in the nature of things this bond cannot arise in a structure consisting of a single atomic species.

THE STRUCTURES OF THE REPRESENTATIVE ELEMENTS

The inert gases

7.02. We have already discussed the structures of these elements in § 6.02, where we saw that they all possess one or other of the two characteristic close-packed structures. These structures reflect the distinctive geometrical properties of the van der Waals bond, and the physical properties of the inert gases indicate the weakness of this bond.

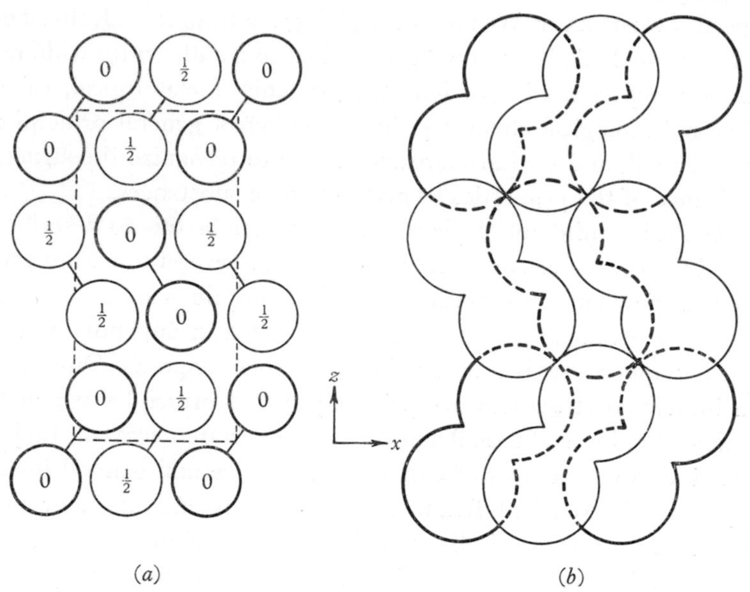

Fig. 7.01. Plan of the unit cell of the orthorhombic structure of iodine projected on a plane perpendicular to the y axis. In (a) atomic centres are shown and the heights of the atoms are indicated in units of b. In (b) the atoms are assigned their correct van der Waals radii to emphasize the packing.

Elements of group 7: the halogens

7.03. The halogens are typical univalent elements, and all occur in the solid state as diatomic molecules, X–X, in which the interatomic binding is due to covalent forces. The arrangement of these molecules in the structure of iodine is shown in fig. 7.01, from which the compact disposition of the molecules, bound to one another by van der Waals forces, can be seen. The centres of the molecules are located at the corners and the centres of the faces of the orthorhombic unit cell, and this arrangement may be compared with that of the spherical atoms in the cubic close-packed structure of a metal: the structure may in fact be regarded as a distortion of cubic close packing arising from the unsymmetrical shape of the molecules concerned.

Elements of group 6: the chalcogens

7.04. The elements of group 6 are divalent, and immediately several structural possibilities arise: the two bonds from any one atom may both be directed to a single atom to give a diatomic molecule, as in O_2, or alternatively they may be directed to two different atoms to give polyatomic molecules in the form of closed rings or indefinitely extended chains. All these possibilities are in fact found among the crystal structures of these elements.

Oxygen and sulphur

7.05. In solid oxygen diatomic molecules O_2 occur. In sulphur, however, in the orthorhombic form, S_8 molecules having the puckered ring configuration shown in fig. 7.02 are found. The S–S distance is

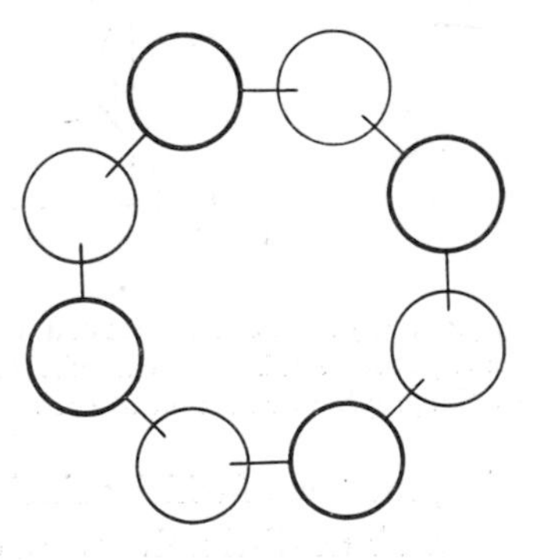

Fig. 7.02. The molecule S_8 in the crystal structure of orthorhombic sulphur.

2·04 Å, corresponding to the covalent radius of the sulphur atom, and the puckered form of the ring arises from the characteristic spatial distribution of the two covalent bonds from each atom at an angle of 108°. The discrete and compact S_8 molecules are relatively loosely bound to one another in the crystal by van der Waals forces, and retain their identity in the melt and in the vapour near the boiling point. At higher temperatures they break down into S_2 molecules.

Selenium and tellurium

7.06. Similar puckered rings are found in the structure of one form of selenium, but this element is polymorphous and more commonly occurs with the alternative crystal structure shown in fig. 7.03. In this structure each atom is again bound to two others by covalent bonds, but now the molecules take the form of indefinitely extended helical

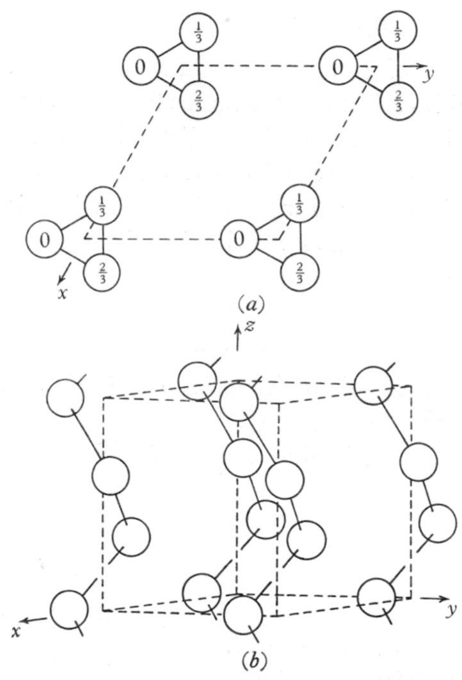

Fig. 7.03. (*a*) Plan of the unit cell of the hexagonal structure of selenium projected on a plane perpendicular to the *z* axis. The heights of the atoms are indicated in units of *c*. (*b*) Clinographic projection of the same unit cell.

chains, the axes of which coincide with the vertical edges of the hexagonal unit cell. The distinctive form of these molecules once more arises from the characteristic mutual disposition of the two bonds from each atom at an angle of about 104°. The Se–Se distance within the chain is 2·36 Å but that between atoms in adjacent chains is 3·49 Å, corresponding to the weaker van der Waals forces which hold the chains together. Even so, this distance is considerably less than that corresponding to the van der Waals radius of the selenium atom, and it is probable that the inter-chain bonding possesses some metallic character, as would be expected from the semi-metallic nature of the element. The mutual disposition of the individual chains reflects the undirected nature of the interchain bonding, for it will be seen that each is symmetrically surrounded by six neighbours and that the chains are close-packed in the structure like rods in a bundle. The structure of selenium is our first example of a molecular structure in which the molecules have infinite extension in one dimension.

Tellurium has the same structure as selenium, and it is probable that similar chain molecules exist in plastic sulphur.

Elements of group 5

Nitrogen and phosphorus

7.07. The elements of group 5 are trivalent. In nitrogen the three bonds from any one atom are all directed to a single other atom to form the very stable molecule N_2, and in the crystal structure of the solid

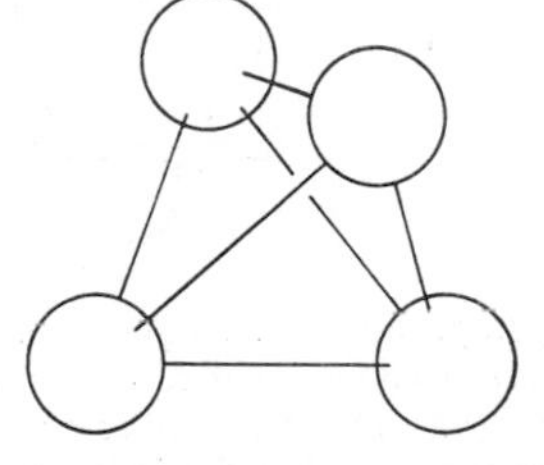

Fig. 7.04. The structure of the molecule P_4.

these molecules are held together by van der Waals forces. In all the other elements of the group the three bonds from any one atom are directed to three different atoms. The simplest possible arrangement is found in the phosphorus molecule, P_4 (fig. 7.04), in which the atoms are located at the corners of a regular tetrahedron and in which each atom is covalently bonded to the other three. This molecule is found in the vapour and in solution, and probably also occurs in the solid in yellow phosphorus. The orbitals involved in the formation of the molecule are not known with certainty. The simplest explanation in terms of p^3 orbitals is untenable because it would involve interbond angles of about 90° (compared with the 60° angle actually observed) and would suggest that an analogous molecule N_4 should be found in nitrogen. A more likely explanation is that hybridized pd^2 bonds are involved. These give a theoretical interbond angle of 66°, not very different from that observed, and, being unavailable in nitrogen, would account for the non-existence of an N_4 molecule.

Arsenic, antimony and bismuth

7.08. Arsenic, antimony and bismuth are all isomorphous, and in the structure of these elements each atom is covalently bound to three others to form an indefinitely extended puckered sheet of the type shown in fig. 7.05. Each such sheet may be regarded as a two-dimensional molecule of infinite extent analogous to the one-dimensional molecules in the structure of selenium, and the whole structure is formed by the superposition of these sheets to give a rhombohedral arrangement.

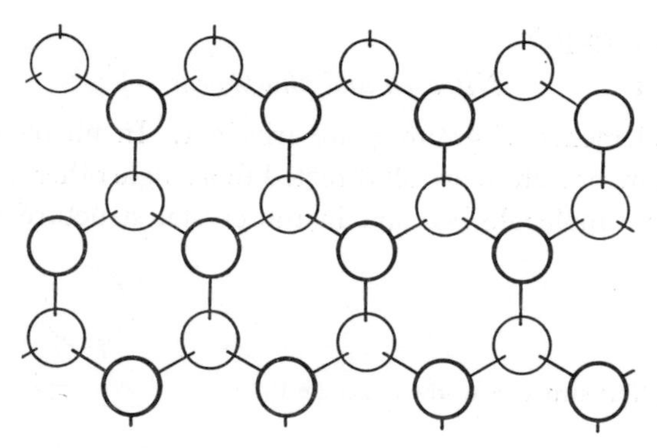

Fig. 7.05. Plan of a single sheet of the rhombohedral structure of arsenic projected on a plane perpendicular to the principal axis. The atoms are at two different heights, as shown, and the structure as a whole is formed by the superposition of identical sheets in such a way that the lower atoms of each sheet fall vertically above the 'holes' in the sheet beneath.

Each atom therefore has three neighbours in its own sheet at a distance d_1 and, in addition, three more remote neighbours in the adjacent sheet at a greater distance d_2. These distances in the three structures are as follows:

	d_1 (Å)	d_2 (Å)	$d_2 - d_1$ (Å)
As	2·51	3·15	0·64
Sb	2·87	3·37	0·50
Bi	3·10	3·47	0·37

Within the molecular sheets the interbond angles are almost exactly 90°, suggesting that simple p^3 orbitals are involved in the covalent bonding. The forces between the sheets are primarily van der Waals bonds, but, as in the selenium structure, there is no doubt that these forces also have an appreciable component of metallic character which increases with increasing atomic number, as is shown by the progressive diminution in the difference between d_2 and d_1. In spite of the obviously metallic character of antimony and bismuth, however, it is important to notice the marked difference between their structure of relatively low co-ordination and the highly co-ordinated structures of the typically true metals. In the melt the influence of the covalent bonding is lost, and liquid antimony and bismuth have a nearly close-packed arrangement of atoms the density of which is actually greater than that of the solid. It is, of course, this property of expansion on solidification which accounts for the importance of these metals in typecasting.

Elements of group 4

Carbon as diamond

7.09. The elements of group 4 are quadrivalent, and for the first time the possibility arises of a structure which owes its coherence in three dimensions to covalent bonds alone. We have already discussed the diamond structure as a typical example of such an arrangement (§4.12). In this structure each carbon atom is tetrahedrally co-ordinated by four neighbours to which it is bound by covalent sp^3 hybrid bonds, and the whole crystal constitutes a single giant molecule. The interatomic distance of 1·54 Å is characteristic of the C–C single bond. Diamond is an excellent insulator, a property which arises from the fact that the Brillouin zone contains four valency electrons per atom and is completely filled, corresponding to the state of affairs represented in fig. 5.17*e*.

Carbon as graphite

7.10. Carbon is dimorphous and occurs naturally not only as diamond but also as graphite, with the quite different structural arrangement shown in fig. 7.06. In this structure the carbon atoms lie in a series of parallel sheets, in each of which they are arranged at the corners of a set of plane regular hexagons. The sheets are so superposed that only one-half of the atoms in each sheet lie vertically above and below atoms in the adjacent sheets; the carbon atoms are thus of two kinds in respect of their environment and the repeat distance along the c axis of the cell is twice the distance between neighbouring sheets. Within each sheet the C–C distance is 1·42 Å, but the closest distance of approach between successive sheets is 3·35 Å. This distance is too large to represent any but a van der Waals bond, and the structure must accordingly be regarded as a molecular one in which each sheet constitutes an infinite two-dimensional molecule, the forces between the sheets being purely residual in nature. The characteristic mechanical properties of graphite, and its striking difference from diamond, arise immediately from the weakness of this bonding.

The bonding between the carbon atoms in the individual sheets in graphite can be discussed in terms of either the valence-bond or the molecular-orbital treatment. On the basis of the former theory the

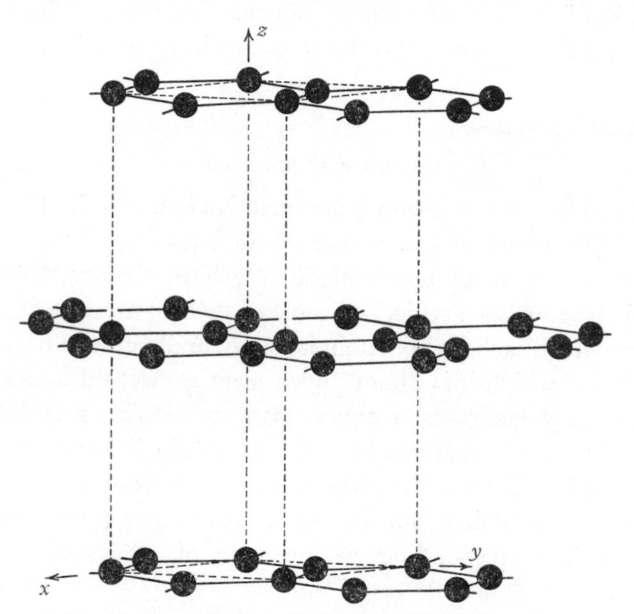

Fig. 7.06. Clinographic projection of the hexagonal structure of graphite. The unit cell is indicated by broken lines.

sheets are to be regarded as consisting of an array of fused benzene rings in which the equivalence of the atoms and the regular hexagonal form of the rings arises from resonance between many possible configurations, such as

```
      /     \      /     \
  =C         C=C          C—
    \       /    \       //
     C=C           C—C
    /     \      //     \
  =C         C—C          C—
    \      //    \      //
     C—C           C—C
   //    \       //    \
```

In this way each bond effectively acquires one-third double-bond character, just as in benzene (§§4.14 and 4.23) each bond acquires one-half double-bond character. This is consistent with the observed C–C separation of 1·42 Å, a distance intermediate between those found in diamond (1.54 Å) and benzene (1·39 Å).

The molecular-orbital treatment similarly emphasizes the relationship between graphite and benzene, for it pictures the bonding in the graphite sheets as due to σ bonds arising from sp^2 hybridization of the carbon atom. The three sp^2 hybrid bonds, as we have seen (§4.08), are coplanar and mutually inclined at 120°, but they account for only three of the four electrons available per atom and leave a p orbital standing perpendicular to their plane. Overlap between these p orbitals of neighbouring atoms will give rise to weaker π bonds, so that the binding between each pair of carbon atoms is due to one σ and one π bond and is intermediate in character between a single and a double bond. The electron density associated with the π bonds will take the form of delocalized ring-shaped figures above and below the rings of carbon atoms, just as in benzene itself (§4.28). It is the mobility of these π electrons, restricted in their movement to translations parallel to the layers, which accounts for the pronounced anisotropy in the electrical conductivity of graphite: parallel to the sheets the conductivity is some 10^5 times greater than in a perpendicular direction.

Silicon and germanium

7.11. The graphite structure is found only in carbon, but the diamond arrangement is found also in a number of other elements of group 4. Silicon and germanium have this structure but differ from diamond in being intrinsic semi-conductors, and indeed find extensive technical applications on this account. The semi-conducting properties arise from the fact that the Brillouin zone, although full as in diamond, is separated from a zone of higher energy by only a narrow energy gap, corresponding to the condition represented in fig. 5.17*f*.

Tin

7.12. Tin is dimorphous, and in the grey form, stable at low temperatures, has the diamond structure with an interatomic distance of 2·80 Å. White tin, however, the common form stable at ordinary temperatures, has a more complex structure in which each atom is irregularly co-ordinated by six neighbours, the shortest Sn–Sn distance being 3·02 Å. The marked difference between this interatomic distance and that in grey tin is discussed below (§7.13; see also §5.27).

Lead

7.13. In lead there is no evidence of covalent binding, and the structure is the cubic close-packed arrangement characteristic of a true metal. The effective radius of the lead atom, however, is large compared with that of other true metals of high atomic number, and it is probable that this is to be ascribed to incomplete ionization arising from the stability of the lone pair of 6*s* electrons (see §3.01). Similarly, incomplete ionization in tin accounts for the large atomic radius in the white form of this element, but the fact that the radius is markedly different in grey tin emphasizes that the state of ionization may vary from one structure to another. This point is of particular importance in connexion with the structures of alloys; whereas the lead atom, for example, is incompletely ionized in the structure of the element it may, nevertheless, occur in the fully ionized state, with a different radius, in intermetallic systems.

Elements of group 3

7.14. With the exception of boron, all the elements of group 3 are metallic, but aluminium and thallium alone of these elements have the strictly close-packed structures characteristic of a true metal. The structures of boron and gallium are complex and will not be discussed in detail here. That of indium is tetragonal with atoms at the corners and face centres of the tetragonal unit cell. The axial ratio c/a is 1·08, so the structure is pseudo-cubic and may be regarded as a slightly distorted form of cubic close packing. Thallium is dimorphous, with both cubic body-centred and hexagonal close-packed structures. Aluminium is cubic close packed.

Indium and thallium resemble white tin and lead in having atomic radii large compared with those of true metals of similar atomic number, and the explanation is again to be found in incomplete ionization arising from the stability of the lone pair of 5*s* and 6*s* electrons, respectively. As a result of this stability the elements behave in many respects as if they belonged to group 1, just as lead has many of the properties of a group 2 element.

Elements of groups 2 and 1

7.15. The representative elements of groups 2 and 1 comprise magnesium, the alkaline earths and the alkali metals. All are typically metallic and all possess one or other of the structures characteristic of the true metals (§§5.03–5.05).

7.16. The above review of the crystal structures of the representative elements serves to show that we find among these elements examples of a wide range of structure types. Broadly speaking we may say that the elements of the later groups and of the earlier periods have typically molecular structures, but that, as we proceed in a given period towards the earlier groups, or in a given group towards the later periods, the structures become more and more metallic in character. Even so, the structures of some of the obviously metallic elements (e.g. bismuth, antimony and tin) still show the influence of covalent bonding, and the structures characteristic of the truly metallic elements are found only in the alkali and alkaline earth metals and in aluminium, thallium and lead.

THE STRUCTURES OF THE TRANSITION, LANTHANIDE AND ACTINIDE ELEMENTS

7.17. The elements of the transition, lanthanide and actinide series are all metallic, and the great majority of them possess one or more of the three crystal structures characteristic of the true metals. The transition series is of the greatest importance in chemical theory and also includes many elements which find wide application in metallurgical technology.

The transition elements

7.18. The transition metals are characterized by the fact that not only the *s* and *p* electrons of the outermost shell but also the *d* electrons of the penultimate shell are available for bonding. As a result, the theory of the interatomic binding in these elements is particularly complex and certainly cannot be regarded as having reached a state of finality. Some of the considerations involved have already been outlined in chapter 5, but here, where we are concerned primarily with structure, it would be quite inappropriate to attempt a detailed review of the great volume of work in this field, especially as many such reviews are readily available (see appendix 1). It is sufficient for our purpose to emphasize that crystal structure alone is far from a complete characterization of the properties of the elements concerned, and that metals with the same crystal structure may, nevertheless, differ profoundly in their physical properties.

Although the majority of the transition elements possess structures characteristic of the true metals there are certain exceptions which call for mention.

Zinc, cadmium and mercury

7.19. These metals have been described as transition elements, but from a structural point of view they display in many respects the characteristics associated with the series of representative elements discussed above. We have seen in this series a progressive transition from molecular to truly metallic structures, an element in the *n*th group showing a tendency to assume a structure in which each atom is covalently bound to $(8-n)$ close neighbours. We would thus expect the elements of group 2 to have structures in which each atom is 6-co-ordinated, and in mercury we find just such a structure in its simplest

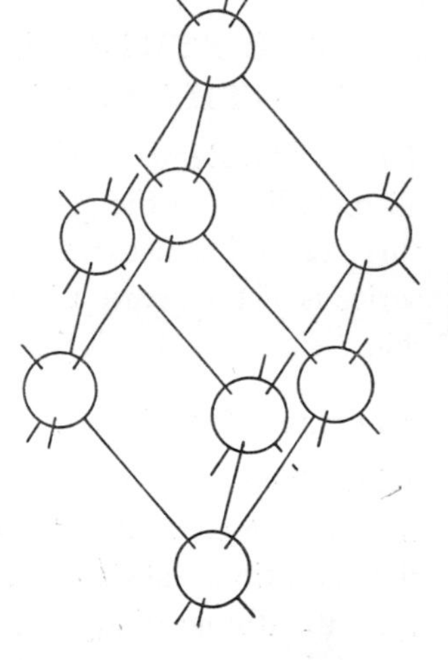

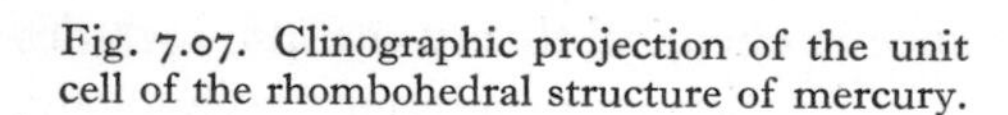

Fig. 7.07. Clinographic projection of the unit cell of the rhombohedral structure of mercury.

form: the atoms are arranged at the corners of a rhombohedral unit cell so that each has six equidistant neighbours (fig. 7.07).

In zinc and cadmium the resemblance to the structure of a true metal is more marked, and the structures of these elements can be most easily described as a distorted form of hexagonal close packing in which the individual close-packed sheets are more widely spaced than in the structure of the true metals. In consequence the axial ratio c/a is greater (1·856 in zinc and 1·886 in cadmium) than the value 1·633 characteristic of the ideally close-packed arrangement, and each atom, although co-ordinated by twelve neighbours, has six of these neighbours, in its own sheet, more closely bonded than the rest. It will be noted that in this series of group 2 elements zinc displays most closely, and mercury least closely, the structural characteristics of a true metal, so that the progressive tendency towards more metallic properties with increasing atomic number, which we have observed in the later groups, is here reversed.

The application of the $(8-n)$ rule to elements preceding group 4 implies the availability of $(8-n)$ electrons per atom for covalent bond formation and is to this extent artificial unless a mechanism for the provision of these electrons can be proposed. A possible mechanism in the case of zinc, based on the valence-bond treatment of metal theory, has already been outlined in §5.28, but it is difficult to feel satisfied that this is more than an *ad hoc* explanation designed to explain the observed crystal structure of the element: if the structure of zinc were unknown there would be few grounds for treating it as other than a simple divalent element.

Manganese

7.20. Manganese is of interest on account of its polymorphism and the complexity of some of its structures. The structure of δ-Mn, stable at the melting point, is body-centred cubic (A_2) and that of γ-Mn, stable between 1100 and 1137 °C, is cubic close-packed (A_1), both structures characteristic of the true metals. The structures of β- and α-Mn, however, the forms stable at lower temperatures, are far more complex. In these structures the atoms are of more than one kind, the co-ordination is irregular and the shortest interatomic distances are considerably smaller than those corresponding to the atomic radii observed in δ- and γ-Mn. The Brillouin zone pattern for the β-Mn structure has been calculated and gives rise to an $N(E)$ curve with a peak at 1·41 electrons per atom and a complete zone containing 1·62 electrons per atom, suggesting that the structure consists of atoms in both the univalent and divalent states. There is confirmatory evidence for this from a number of alloy systems which have the same geometrical disposition of atoms as β-Mn and in which this structural arrangement occurs at a composition corresponding to an electron:atom ratio of 1·5. In α-Mn, too, it is probable that the manganese atoms exist in more than one valency state.

Cobalt

7.21. Cobalt is dimorphous and crystallizes with both the hexagonal and cubic close-packed structures. We have already seen that energetically the difference between these two structures is slight, and that geometrically they are very similar, differing only in the mutual disposition of alternate individually close-packed layers. In hexagonal close packing these layers are identical, and the repeat sequence can be represented as:

$$...ABABAB... \quad \text{or} \quad ...BCBCBC... \quad \text{or} \quad CACACA...,$$

whereas in cubic close packing alternate layers are different and the sequence is

$$...ABCABC....$$

In cobalt, by suitable heat treatment, it is possible to obtain a close-packed structure which conforms to neither of these arrangements but in which the sequence of layers is, say,

$$...ABABAB\mathbf{ABC}BCBCBC....$$

Here it will be seen that the majority of layers are arranged as in hexagonal close packing but that the regularity of the arrangement is disturbed by a group of three layers (denoted by bold symbols) which are disposed as in cubic close packing. Such a disturbance is termed a *stacking fault*, and in cobalt these stacking faults occur at random and on the average about once in every ten layers.

7.22. The structure of cobalt is the first example we have encountered of a *defect structure*, a class of structures in which the arrangement of the atoms departs from the perfect geometrical regularity of an ideal structure. In such cases it is no longer possible to identify a unit cell which is strictly representative of the structure as a whole and which, by indefinite repetition, builds up the complete crystal; all we can do is to choose a cell on a statistical basis and associate with each atomic site a certain probability that it will in fact be occupied. Defect structures are by no means uncommon, and we shall find other examples of such structures among those to be described later.

The lanthanide and actinide elements

7.23. These elements are characterized by the fact that the differentiating electrons lie deeply buried in the extranuclear structure in the third highest energy shell (the N shell in the lanthanides and the O shell in the actinides); in consequence the elements in each series are closely similar in their chemical properties. The lanthanide metals are of very limited technical importance, and that of the actinide elements is primarily concerned with the part they play in nuclear reactions. We shall not, therefore, discuss their structures in detail here although certain points merit brief mention.

The known structures of the lanthanide and actinide metals are indicated in table 5.01, from which it will be seen that the structures characteristic of the true metals, and particularly the hexagonal close-packed arrangement, are common. Polymorphism, however, is of frequent occurrence among these elements, and plutonium, for example, crystallizes in no fewer than six modifications—the A_1 and A_2 structures indicated and four others of greater complexity. Praseodymium, neodymium and samarium are of interest in that they possess close-packed structures in which the sequence of layers is

$$...A\ B\ A\ C\ A\ B\ A\ C....$$

The structure thus resembles that of cobalt in being a mixture of hexagonal and cubic close packing, but differs from it in that the sequence of layers is regular (repeating every fourth layer) instead of random. The arrangement is accordingly not a defect structure and can be referred to an ideal hexagonal unit cell of which the repeat distance in the principal direction (and therefore the c/a ratio) is twice that of the simple A_3 structure.

7.24. This completes our discussion of the structures of the elements. We shall have occasion to refer again to those of the metals when we come to discuss the structures of intermetallic systems. Since, however, some of these structures closely resemble those of simple chemical compounds it is convenient to defer our consideration of alloy systems until after these other structures have been described.

8 THE STRUCTURES OF SOME SIMPLE COMPOUNDS

INTRODUCTION

8.01. In this chapter we shall discuss the structures of a number of simple compounds of composition $A_m X_z$, and also of some complex oxides and sulphides of composition $A_m B_n X_z$. We shall find among these compounds representatives of ionic, covalent and molecular structures.

AX STRUCTURES

Halides

8.02. The structures of a number of *AX* halides are shown in table 8.01 and are discussed in the following paragraphs.

Table 8.01. *The structures of some AX halides*

Caesium chloride structure (C.N. 8:8)			Sodium chloride structure (C.N. 6:6)				Zincblende or wurtzite structure (C.N. 4:4)			
—	—	—	LiF	LiCl	LiBr	LiI	—	—	—	—
—	—	—	NaF	NaCl	NaBr	NaI	—	—	—	—
—	—	—	KF	KCl	KBr	KI	—	—	—	—
RbCl*	RbBr*	RbI*	RbF	RbCl	RbBr	RbI	—	—	—	—
CsCl	CsBr	CsI	CsF	CsCl*	—	—	—	—	—	—
—	—	—	—	—	—	—	CuF	CuCl	CuBr	CuI
—	—	—	AgF	AgCl	AgBr	—	—	—	—	AgI
NH_4Cl	NH_4Br	NH_4I*	—	NH_4Cl*	NH_4Br*	NH_4I	NH_4F	—	—	—

* These structures are not found at normal temperatures and pressures.

Alkali halides

8.03. We have already considered in detail the structures of the alkali halides, all of which crystallize with either the sodium chloride or the caesium chloride arrangement (§§3.04 and 3.05). All of these compounds are essentially ionic, and the degree of ionic character depends on the difference in electronegativity of the atoms concerned; it is thus a maximum in caesium fluoride and a minimum in lithium iodide. As we have seen, the radius ratio r^+/r^- is the primary factor in determining whether a given halide possesses the sodium chloride or the caesium chloride structure. It is not, however, the only factor, for some halides (e.g. KF, RbCl and RbBr) do not have the structure to be expected on purely geometrical grounds and others exhibit at high temperatures and pressures structures different from those found under normal conditions: at high pressures RbCl, RbBr and RbI have the caesium chloride structure and at high temperatures CsCl has that of sodium chloride.

Silver and cuprous halides

8.04. The ionic radii of Cu^{+} and Ag^{+} (0·96 and 1·26 Å, respectively) are intermediate between those of Na^{+} and K^{+}, and on the basis of purely geometrical considerations all the halides of these metals would be expected to resemble the alkali halides in their structures. In fact, however, AgI and the four cuprous halides have the zincblende structure (§3.06), and only AgF, AgCl and AgBr have the expected sodium chloride arrangement; and of these AgCl and AgBr are insoluble in water and unstable to light. In AgI the interatomic distance is actually less than in AgBr, and in this salt, and in all the cuprous halides, the binding is primarily covalent. In cuprous iodide, for example, the four bonds from each copper and iodine atom can be represented thus:

$$\rightarrow \underset{\uparrow}{\overset{|}{\mathrm{Cu}}} \leftarrow \qquad \leftarrow \underset{\downarrow}{\overset{|}{\mathrm{I}}} \rightarrow$$

In this way the copper atom (with electronic configuration 2, 8, 18, $4s^1$) effectively acquires seven electrons to give the krypton configuration 2, 8, 18, $4s^2$, $4p^6$, and its four bonds may be regarded as sp^3 hybrids (of one $4s$ and three $4p$ orbitals) with the normal tetrahedral configuration. In terms of the concept of formal charges the atoms must be represented as Cu^{3-} and I^{3+}.

8.05. The structure of silver iodide is of interest also in another respect. This compound is trimorphous. The γ-form, stable at room temperature, has the zincblende structure described above, but 137 °C this changes to a β form with the closely related wurtzite structure (§4.13). In these two arrangements the sites occupied by the iodine atoms are disposed as in cubic and hexagonal close packing, respectively, so that the structures may be regarded as a close-packed array of iodine atoms with the silver atoms in the tetrahedral interstices. In α-AgI, stable above 146 °C, a less tightly packed cubic body-centred arrangement of iodine atoms obtains, but now the silver atoms have no fixed positions and are able to wander freely in a fluid state throughout the structure. At this temperature the electrical conductivity rises abruptly and the framework of silver atoms may be regarded as 'melting' into itself; the final breakdown of the iodine framework takes place only when the true melting point of 555 °C is reached. In its α-form silver iodide is another example of a defect structure.

Ammonium halides

8.06. In the majority of its salts the ammonium ion NH_4^+ behaves as a spherical ion of radius about 1·48 Å. This radius is very close to that of the rubidium ion, and the corresponding ammonium and rubidium salts are therefore often isostructural. Thus NH_4Cl, NH_4Br and NH_4I all have the sodium chloride structure at a sufficiently high temperature, although at lower temperatures they also exhibit the caesium chloride arrangement. Ammonium fluoride, however, crystallizes with the wurtzite structure with an (NH_4)–F distance of 2·66 Å, a distance appreciably less than that to be expected on the basis of the radii of the ions (1·48 + 1·36 = 2·84 Å). The reason for this abnormal behaviour of ammonium fluoride is important and will be discussed later (§ 12.04).

The occurrence of the NH_4^+ ion in the highly symmetrical sodium chloride and caesium chloride structures is seemingly inconsistent with its tetrahedral configuration and can be explained only on the assumption that the ion effectively acquires spherical symmetry by free rotation under the influence of the energy of thermal agitation. We shall encounter many other examples of structures in which ions or molecules are in free rotation, either at all temperatures or above a certain transition temperature; all are examples of yet a further type of defect structure.

Hydrides and hydroxides

8.07. The majority of the structures of compounds containing hydrogen are so distinctive in their properties that they are best discussed as a class apart (see chapter 12). These remarks, however, do not apply to the salt-like hydrides or to the hydroxides of the more electropositive metals, many of which form typically ionic structures resembling the corresponding halides in their properties.

The hydrides *A*H of all the alkali metals are known; they all form stable, colourless crystals of relatively high melting point and all have the sodium chloride structure in which the hydrogen occurs as the negative ion H^- of radius 1·54 Å. This radius is intermediate between the radii of the ions F^- and Cl^-, so that the structural resemblance between hydrides and halides is readily understandable.

In the limited number of *A*OH hydroxides which have as yet been studied the OH^- ion behaves as a spherical entity of radius 1·53 Å (again intermediate in size between the F^- and Cl^- ions), and some of the crystal structures are analogous to those of the corresponding halides. Thus KOH (at high temperatures) has the sodium chloride structure.

Oxides and sulphides

8.08. The oxides and sulphides of composition *A*O and *A*S show a far wider range of structural types than the *AX* halides. Some of these compounds are primarily ionic but in others the bonding is predominantly covalent; there are, moreover, some important differences between oxides and sulphides to which we refer later (§8.60). The structures we shall consider are summarized in table 8.02, but there are also others which we shall not discuss.

Table 8.02. *The structures of some AX oxides and sulphides*

Sodium chloride structure (C.N. 6:6)		Zincblende structure (C.N. 4:4)		Wurtzite structure (C.N. 4:4)		PdO and related structures (C.N. 4:4)		NiAs structure (C.N. 6:6)
—	—	BeO	BeS	—	—	—	—	—
MgO	MgS	—	—	—	—	—	—	—
CaO	CaS	—	—	—	—	—	—	—
SrO	SrS	—	—	—	—	—	—	—
BaO	BaS	—	—	—	—	—	—	—
VO	—	—	—	—	—	—	—	VS
MnO	MnS	—	—	—	MnS	—	—	—
FeO	—	—	—	—	—	—	—	FeS
CoO	—	—	—	—	—	—	—	CoS
NiO	—	—	—	—	—	—	—	NiS
—	—	—	—	—	—	PdO	PdS	—
—	—	—	—	—	—	PtO	PtS	—
—	—	—	—	—	—	CuO	—	—
—	—	—	ZnS	ZnO	ZnS	—	—	—
—	—	—	CdS	—	CdS	—	—	—
—	—	—	HgS	—	—	—	—	—
—	PbS	—	—	—	—	—	—	—

Oxides

8.09. The oxides of all the alkaline earth metals except beryllium, and also of some of the transition metals, have the typically ionic sodium chloride structure, an arrangement consistent with the radius ratio of the ions concerned and with the large differences between their electronegativities. Beryllium oxide, BeO, has the zincblende structure. This, too, is consistent with the picture of ionic bonding, on account of the very small radius of the Be^{2+} ion, but it is, nevertheless, probable that in this structure the bonds have an appreciable degree of covalent character. If, however, they are treated as purely ionic, BeO is one of the few examples of an ionic AX compound in which the radius ratio is sufficiently small to give the 4:4 co-ordinated zincblende structure.

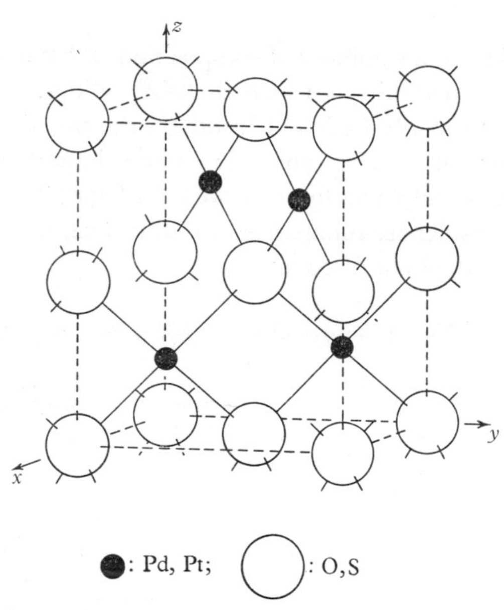

Fig. 8.01. Clinographic projection of the unit cell of the tetragonal structure of PdO, PtO, PdS and PtS.

8.10. The influence of covalent bonding in the oxides is shown more clearly in the structures of ZnO, PdO, PtO and CuO. Zinc oxide has the wurtzite structure, although the sodium chloride arrangement would be expected if the bonding was ionic on account of the relatively large radius of the ion Zn^{2+}. The structure of PdO and PtO is shown in fig. 8.01 and will be seen to be quite different from any we have as yet encountered in ionic crystals. Each oxygen atom is tetrahedrally co-ordinated by metal atoms, but the metal atoms themselves are surrounded by four oxygen atoms disposed in a plane almost at the corners of a square. The structure of CuO is a slightly distorted variant of the same arrangement. This distribution of bonds is an elegant example of the stereochemistry of the metals concerned, all of which (in the divalent state) can form four planar dsp^2 hybrid bonds. Thus in CuO, for example, the four bonds from each copper and oxygen atom can be represented as

$$\begin{array}{c}\downarrow\\ \text{—Cu—}\\ \uparrow\end{array} \quad \text{and} \quad \begin{array}{c}\uparrow\\ \text{—O—}\\ \downarrow\end{array}$$

In this way the copper atom acquires a share in six extra electrons to achieve the configuration 2, 8, 18, $4s^2 4p^5$, and it is hybridization of one $3d$, one $4s$ and two $4p$ orbitals which forms the dsp^2 bond. In PdO the corresponding bond structure gives rise to the configuration 2, 8, 18, 18, $5s^2 5p^4$. One d, one s and two p orbitals are again available for dsp^2 hybridization, but it is possible that the single unpaired electron in the $4p$ orbital in copper accounts for the structural difference between CuO and PdO, and for the fact that there are no isomorphous compounds of copper and palladium.

The distinction between these planar bonds in divalent copper and the tetrahedral sp^3 bonds in univalent copper will be noted.

Sulphides

8.11. The sulphides of the alkaline earth metals are isostructural with the corresponding oxides (see table 8.02) and, except BeS, are all essentially ionic structures. The sulphides MnS and PbS also have the sodium chloride arrangement. The sulphides of V, Fe, Co and Ni, however, are different in structure from the oxides, showing that the tendency towards covalent bonding is more pronounced in sulphides than in oxides, as is to be expected from the relative electronegativities of oxygen and sulphur. The sulphides of Pd and Pt have structures closely resembling those of the corresponding oxides, and again reveal the characteristic planar distribution of the four dsp^2 bonds in these elements.

The nickel arsenide structure

8.12. The sulphides of V, Fe, Co and Ni have a structure not represented among the oxides. This arrangement, found also in NiAs after which it is named, is illustrated in fig. 8.02. Each A atom is co-ordinated by six X neighbours at the corners of a distorted octahedron but the six A neighbours of an X atom are disposed at the corners of a trigonal prism. The co-ordination, however, is not as simple as this description would imply, for each A atom has also two other atoms of the same kind (those vertically above and below it) only slightly more remote than its X neighbours. The bonding is obscure and may vary from one compound to another, but it is probable that the A–X bonds are primarily covalent and that metallic bonds operate in addition between the A atoms in a vertical direction. This is consistent with the pronounced anisotropy of the structure: the thermal expansion of nickel arsenide itself is eight times greater perpendicular to the principal axis than along this direction.

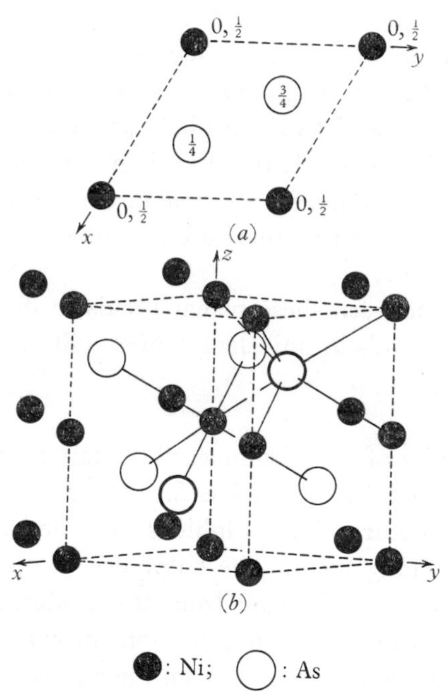

Fig. 8.02. (*a*) Plan of the unit cell of the hexagonal structure of nickel arsenide, NiAs, projected on a plane perpendicular to the *z* axis. (*b*) Clinographic projection of the same structure. The two As atoms represented by heavy circles are those within the unit cell; the others lie outside the cell but have been added to show the co-ordination about the Ni atom at 0, 0, ½.

The nickel arsenide structure occurs not only in the sulphides but also in many other compounds containing a transition metal and one of the elements Sn, Pb, As, Sb, Bi, Se and Te. Many of these systems are essentially intermetallic in their properties and will be discussed further in the chapter devoted to alloys. Here, however, it is interesting to note that as the system becomes more metallic so the bonding in the vertical direction becomes stronger. Thus in FeS the Fe–S and Fe–Fe distances differ by 0·44 Å, whereas in CoSb, owing to compression of the structure in the z direction, this difference has shrunk to 0·02 Å.

Nitrides

8.13. A large number of elements form nitrides of composition AN. Those of the group 3 metals Al, Ga and In have the wurtzite structure and are normal covalent compounds since the number of electrons available is just that required to form four tetrahedral bonds about each atom. Formally the structures could alternatively be regarded as ionic, containing the ions A^{3+} and N^{3-}, and on this picture, too, the co-ordination would be fourfold on account of the small radius ratio r^+/r^-. Such a viewpoint is, however, hardly tenable, for the difference in electronegativity (1·5 in the case of AlN) is insufficient to give more than a limited degree of ionic character to the bond.

8.14. The nitrides (and phosphides) of many of the transition metals form crystals with the sodium chloride structure. This description, however, must not be interpreted as implying that they are ionic in character, for in fact they display many of the properties of intermetallic systems. For this reason a discussion of these nitrides is deferred to chapter 13.

Boron nitride

8.15. Boron nitride, BN, is unique in having a structure resembling that of graphite (§7.10): it can be described as derived from the graphite structure by replacing the carbon atoms by boron and nitrogen atoms alternately. The planar distribution of the three bonds about the boron atom arises, as we have seen (§4.08), from sp^2 hydridization, and that about the nitrogen atom, which is to be contrasted with the pyramidal distribution of bonds in ammonia, must arise from the same cause. Nitrogen in its ground state has the configuration $1s^2, 2s^2 2p_x^1 2p_y^1 2p_z^1$, but if the two $2s$ electrons are uncoupled and one is promoted to, say, the p_z orbital we obtain the configuration $1s^2$, $2s^1 2p_x^1 2p_y^1 2p_z^2$ with the unpaired $2s$, $2p_x$ and $2p_y$ orbitals available for sp^2 hybridization. The bonding in boron nitride is thus similar to that in graphite, but with this difference: in graphite the carbon atoms, after sp^2 hybridization, have a p_z orbital containing one electron standing perpendicular to the plane of the rings, and overlap between these produces a delocalized π bond to which the electrical conductivity may be ascribed; in boron nitride the corresponding p_z orbitals are vacant in the boron atom and are occupied by two electrons in that of nitrogen. Overlap to form π bonds is therefore no longer possible.

Carbides and silicides

8.16. Many carbides and silicides of composition AX are formed by transition metals. These carbides and silicides are characterized by very high melting points, extreme hardness, optical opacity and relatively high electrical conductivity. Many of them have the sodium chloride structure but they are not ionic compounds; rather do they resemble the corresponding nitrides and phosphides in simulating alloy systems in many of their properties. For this reason they will be discussed later.

Other carbides, notably those of the more electropositive metals, are quite different in their properties, being colourless, transparent insulators resembling inorganic salts rather than metal systems. These carbides are ionic; some typical structures will be described in the sections of this chapter devoted to AX_2 and A_mX_z compounds.

Silicon carbide

8.17. Silicon carbide (carborundum, SiC) is of especial interest on account of its rich polymorphism, no fewer than six structures being known. As is to be expected, each carbon and silicon atom is tetrahedrally co-ordinated by four atoms of the other kind, and two of the forms of carborundum have the zincblende and wurtzite structures. The close relationship between these two structures has already been discussed (§4.13), and is emphasized by the many AX compounds (including ZnS itself) in which both are found. It is illustrated in fig. 8.03, where the cubic zincblende structure has been drawn with one of the cube diagonals vertical and parallel to the principal axis of the wurtzite structure. When viewed in this way it will be seen that both structures can be visualized as formed by the superposition of a series of puckered sheets of atoms, but that in zincblende successive sheets are identical (albeit translated) whereas in wurtzite they differ and are related by a rotation through 180° about the principal axis. In the two structures the sequence of sheets can therefore be symbolized as

$$\ldots A\ A\ A\ A \ldots \quad \text{and} \quad \ldots A\ B\ A\ B \ldots,$$

respectively. The remaining four forms of carborundum are formed by the superposition of these same two puckered sheets but in more complex sequences, as for example,

...***A*** *A B B* ***A*** *A B B*... or ...***A*** *A A B B B* ***A*** *A A B B B*...,

where the span of the repeat unit is indicated by bold symbols. In one form the structure repeats only after 33 layers, and the *c* dimension of the hexagonal unit cell is 82·9 Å. The relationship between these complex structures of carborundum and the simpler zincblende and wurtzite arrangements may be compared with that between the structure of praseodymium and those of the simple close-packed metals (§7.23).

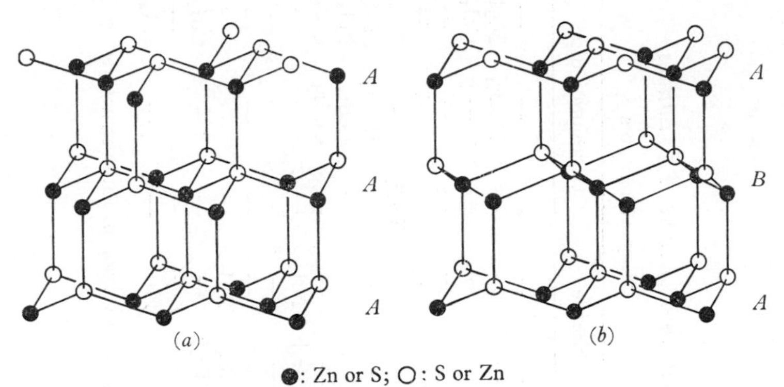

●: Zn or S; ○: S or Zn

Fig. 8.03. Clinographic projections of the structures of (*a*) zincblende and (*b*) wurtzite showing the sequence of layers in the two structures.

AX_2 STRUCTURES

8.18. The different structures adopted by AX_2 compounds are far more numerous and show a far greater diversity of type than those found in *AX* compounds. Here we shall discuss only a limited number of structures of common occurrence, but it must be borne in mind that there are in addition many other compounds in which the structural arrangement is different.

Halides

8.19. The structures of the AX_2 halides which we shall discuss are shown in table 8.03.

Ionic halides

8.20. A considerable number of AX_2 halides are essentially ionic, and have structures determined by the relative sizes of the ions concerned.

Table 8.03. *The structures of some AX_2 halides*

Symmetrical structures						Layer structures									
Fluorite		Rutile			β-Cristobalite	Cadmium chloride			Cadmium iodide			HgI_2	Chain structures		Isolated molecules
—	—	—	—	—	BeF_2	—	—	—	—	—	—	—	—	—	—
—	—	MgF_2	—	—	—	$MgCl_2$	—	—	—	$MgBr_2$	MgI_2	—	—	—	—
CaF_2	—	—	$CaCl_2$*	$CaBr_2$*	—	—	—	—	—	—	CaI_2	—	—	—	—
SrF_2	$SrCl_2$	—	—	—	—	—	—	—	—	—	—	—	—	—	—
BaF_2	$BaCl_2$	—	—	—	—	—	—	—	—	—	—	—	—	—	—
—	—	—	—	—	—	—	—	—	$TiCl_2$	—	TiI_2	—	—	—	—
—	—	MnF_2	—	—	—	$MnCl_2$	—	—	—	$MnBr_2$	MnI_2	—	—	—	—
—	—	FeF_2	—	—	—	$FeCl_2$	—	—	—	$FeBr_2$	FeI_2	—	—	—	—
—	—	CoF_2	—	—	—	$CoCl_2$	—	—	—	$CoBr_2$	CoI_2	—	—	—	—
—	—	NiF_2	—	—	—	$NiCl_2$	$NiBr_2$†	NiI_2	—	—	—	—	—	—	—
—	—	PdF_2	—	—	—	—	—	—	—	—	—	—	$PdCl_2$	—	—
CuF_2	—	—	—	—	—	—	—	—	—	—	—	—	$CuCl_2$	$CuBr_2$	—
—	—	ZnF_2	—	—	—	$ZnCl_2$	$ZnBr_2$	ZnI_2	—	—	—	—	—	—	—
CdF_2	—	—	—	—	—	$CdCl_2$	$CdBr_2$†	—	—	—	CdI_2	—	—	—	—
HgF_2	—	—	—	—	—	—	—	—	—	—	—	HgI_2	—	—	$HgCl_2$
PbF_2	—	—	—	—	—	—	—	PbI_2	—	—	PbI_2	—	—	—	—

* These compounds have a somewhat distorted version of rutile structure. † See §8.25.

Since the anions are generally the larger ions it is the co-ordination of these about the cations which determines the structural arrangement, and this co-ordination may be 8-, 6- or 4-fold, as in the structures of caesium chloride, sodium chloride and zincblende, respectively. The co-ordination about the anion, however, will clearly be only one-half of that about the cation, so that possible co-ordinations and the corresponding radius ratio conditions are as follows:

Co-ordination	8:4	6:3	4:2
r^+/r^-	> 0·7	0·7–0·3	< 0·3

Structures corresponding to all these possibilities are known.

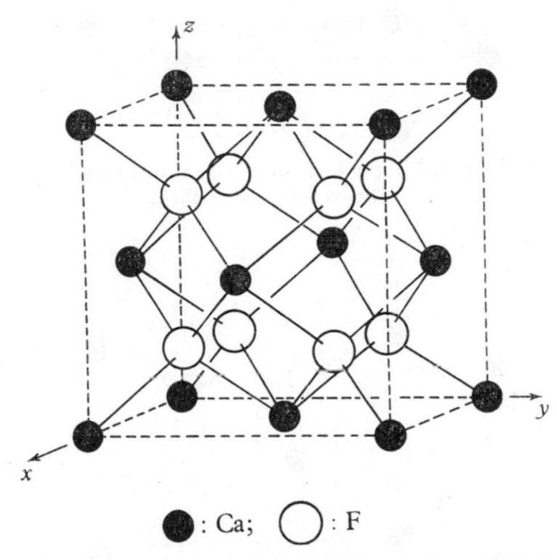

Fig. 8.04. Clinographic projection of the unit cell of the cubic structure of fluorite, CaF_2.

The fluorite structure

8.21. A co-ordination of 8:4 is found in the structure of fluorite, CaF_2 (fig. 8.04). Here the calcium ions are arranged at the corners and face centres of a cubic unit cell and the fluorine ions are at the centres of the eight cubelets into which the cell may be divided. Each calcium ion is therefore co-ordinated by eight fluorine neighbours at the corners of a cube while the calcium neighbours of a fluorine ion are four in number, disposed at the corners of a regular tetrahedron. This is the only structure in which 8:4 co-ordination is found.

The rutile structure

8.22. A co-ordination of 6:3 occurs in several AX_2 structures, of which the commonest is the tetragonal rutile structure (fig. 8.05), named after one of the mineral forms of TiO_2. In this structure each A atom is

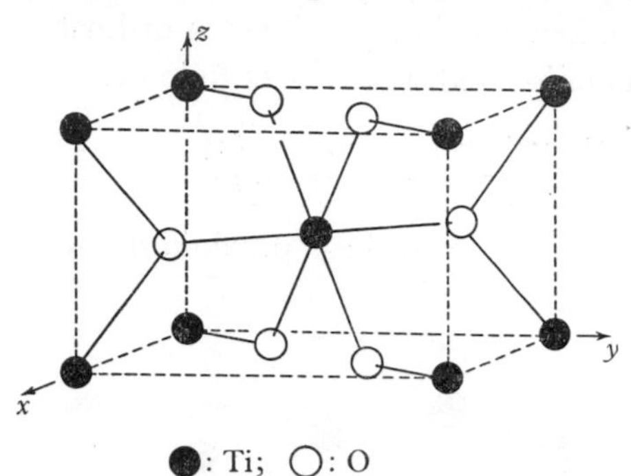

Fig. 8.05. Clinographic projection of the unit cell of the tetragonal structure of rutile, TiO_2.

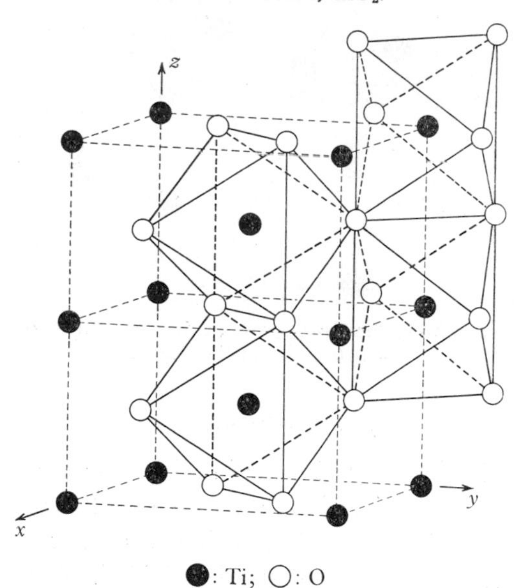

Fig. 8.06. Clinographic projection of the tetragonal structure of rutile, TiO_2, showing the co-ordinating octahedra of anions round the cations and the way in which these octahedra are linked in bands by sharing horizontal edges.

surrounded by six X neighbours at the corners of a slightly distorted regular octahedron, while the three A atoms co-ordinating each X atom lie in a plane at the corners of a nearly equilateral triangle. If several unit cells of this structure are considered it will be seen that each of the co-ordinating octahedra of X atoms shares its two horizontal edges with adjacent octahedra. The octahedra are thus linked in bands, which run vertically through the structure and pass through the centre and corners of the unit cell, as shown in fig. 8.06.

The β-cristobalite structure

8.23. A co-ordination of 4:2 is found among AX_2 halides only in BeF_2, which has the (idealized) β-cristobalite structure, named after one of the forms of SiO_2. The cubic unit cell of this structure is shown in

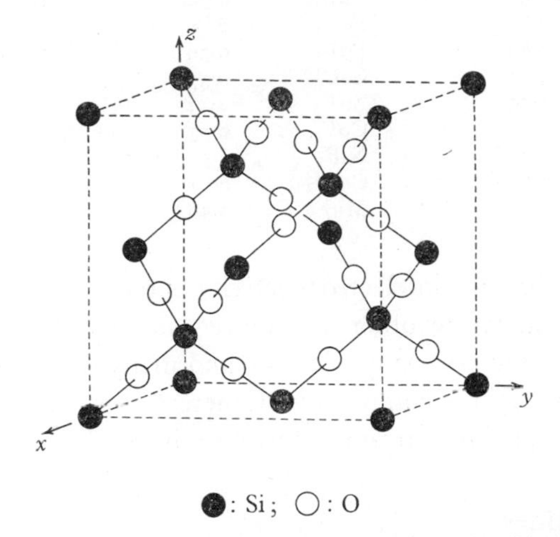

●: Si; ○: O

Fig. 8.07. Clinographic projection of the unit cell of the (idealized) cubic β-cristobalite structure.

fig. 8.07, and it will be seen that the structure can be most simply described as an arrangement of A atoms occupying the positions of the carbon atoms in diamond (or of the zinc and sulphur atoms in zinc-blende), with an X atom midway between each pair of A atoms. Every A atom is therefore surrounded by four X neighbours at the corners of a regular tetrahedron, while every X atom is co-ordinated by only two neighbours arranged diametrically opposite to one another.

8.24. The three structures just described are found in a number of AX_2 halides in which the bonding is primarily ionic, and it will be seen from table 8.04 that the great majority of the compounds quoted have the structure to be expected on the basis of the radius ratio of the ions concerned. Broadly speaking, we may say that these symmetrical structures occur in those halides in which the A and X atoms differ widely in electronegativity: thus they are common in fluorides, rare in chlorides and bromides (where they are formed only by the strongly electropositive alkaline earth metals), and unknown in iodides.

Table 8.04. *The radius ratios of some AX_2 halides*

Fluorite structure $r^+/r^- > 0{\cdot}7$		Rutile structure $0{\cdot}7$–$0{\cdot}3$		β-Cristobalite structure $< 0{\cdot}3$	
BaF_2	0·99	MnF_2	0·59	BeF_2	0·23
PbF_2	0·88	FeF_2	0·59		
SrF_2	0·83	PdF_2	0·59		
HgF_2	0·81	$CaCl_2$	0·55		
$BaCl_2$	0·75	ZnF_2	0·54		
CaF_2	0·73	CoF_2	0·53		
CdF_2	0·71	NiF_2	0·51		
$SrCl_2$	0·63	$CaBr_2$	0·51		
		MgF_2	0·48		

When the difference in electronegativity is too small to give rise to a typically ionic structure other and more complex structural arrangements result. There are many of these, and many of them are found in only a limited number of compounds. Others, however, are of sufficiently common occurrence to warrant description here.

Molecular halides

The cadmium chloride and cadmium iodide structures

8.25. Many AX_2 halides, particularly those of the transition metals, show one or other of the closely related cadmium chloride and cadmium iodide structures. Both of these structures are formed by the superposition of a series of composite layers, each of which consists of a sheet of cadmium atoms sandwiched between two sheets of atoms of the halogen. The arrangement of one such layer is shown in fig. 8.08, and it will be seen that a characteristic feature of the structure is the asymmetry of the co-ordination: the cadmium atoms are symmetrically surrounded by six halogen atoms at the corners of an octahedron, whereas the three cadmium neighbours of each halogen atom all lie to one side of it. In cadmium iodide the structure as a whole is built up by the superposition of such layers in identical orientation, and the structure can therefore be described in terms of the very simple hexagonal unit cell shown in fig. 8.09. It will be noted that if the cadmium atoms are ignored the distribution of the iodine atoms alone is that found in hexagonal close packing.

The cadmium chloride structure differs from that of cadmium iodide only in the geometrical disposition of successive layers; these layers are now so arranged that the halogen atoms are in cubic close packing. This close relationship between the two structures is emphasized by the behaviour of $CdBr_2$ and $NiBr_2$. These compounds both have the

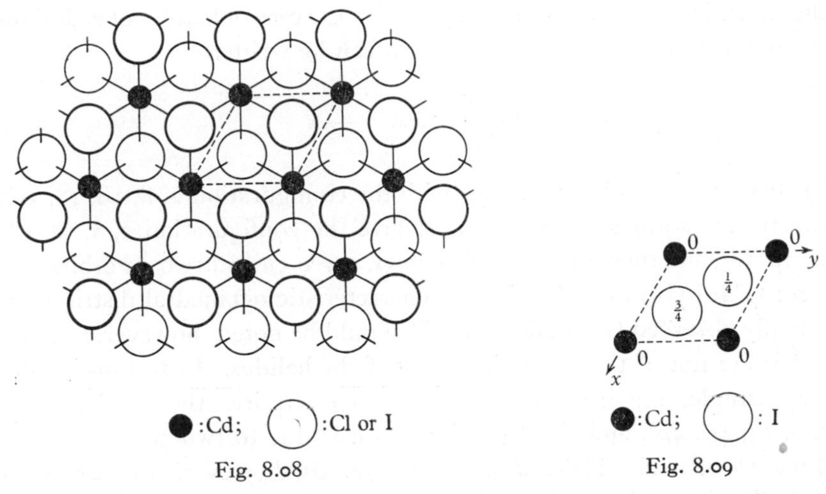

Fig. 8.08 Fig. 8.09

Fig. 8.08. Plan of a single layer of the structures of cadmium chloride, $CdCl_2$, and cadmium iodide, CdI_2, projected on a plane perpendicular to the z azis. The unit cell of cadmium iodide is indicated.

Fig. 8.09. Plan of the unit cell of the hexagonal structure of cadmium iodide, CdI_2, projected on a plane perpendicular to the z axis.

cadmium chloride structure, but in addition they possess defect structures derived from that of cadmium chloride by stacking successive layers in irregular sequence. The arrangement of the bromine atoms alone is therefore sometimes as in cubic and sometimes as in hexagonal close-packing; in this respect the structure resembles that of cobalt (§7.21).

8.26. It is clear that the bonding in the cadmium chloride and cadmium iodide structures cannot be purely ionic, for adjacent atoms in neighbouring layers are of the same kind and the forces between them can be only of the van der Waals type. Indeed, this is also evident from the large interlayer spacing in these structures; in cadmium chloride, for example, the Cd–Cl distance within the layers is 2·74 Å whereas the Cl–Cl distance between adjacent layers is 3·68 Å. Moreover, in their physical properties the structures reveal the weakness of this interlayer bonding, for they generally display excellent cleavage parallel to the layers and a marked anisotropy in thermal expansion. The structures can, therefore, be regarded as molecular arrangements in which each sheet constitutes a single molecule of infinite extent in two dimensions, as in the structures of arsenic and graphite, and in which the bonding in the sheets is due to covalent links. Thus in the case of halides of cadmium the bond system round the two atoms can be written

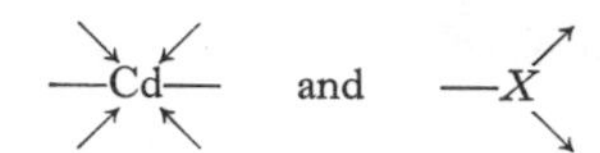

In this way the cadmium atom (with the configuration 2, 8, 18, 18, $5s^2$) effectively acquires ten electrons to give the configuration 2, 8, 18, 18, $5s^2 5p^6 5d^4$, and there are available one $5s$, three $5p$ and two $5d$ orbitals to form sp^3d^2 hybrid bonds with the characteristic octahedral distribution actually observed (see table 4.01). It should be noted, however, that the bonds are not of the same kind in all of the halides. In ferrous iodide, for example, the iron atom (2, 8, 14, $4s^2$) acquires the configuration 2, 8, 18, $4s^2$, $4p^6$, and the bond formation is due to two $3d$, one $4s$ and three $4p$ orbitals. These d^2sp^3 hybrid bonds, however, are also octahedrally disposed.

Although we have just treated the cadmium chloride and cadmium iodide structures as covalent it nevertheless seems probable that the *A*–*X* bonds within the sheets still retain an appreciable degree of ionic character and that their true state is better described as a resonance between covalent and ionic bonding in which the influence of the former predominates. The reason for this view is that the difference in electronegativity of the *A* and *X* atoms appears still to have an influence in determining which of these closely related structures is formed, that of cadmium chloride being the more ionic of the two. Thus it will be seen from table 8.03 that the cadmium chloride structure is found in many chlorides but that the cadmium iodide arrangement is favoured by the corresponding bromides and iodides. In fact in some cases (e.g. the halides of Mg, Ca, Mn, Cd, etc.) it is possible to trace a progressive transition from a typically ionic structure through those of cadmium chloride and cadmium iodide as we pass from the fluoride to the iodide and as the difference in electronegativity progressively diminishes. The view that the bonding is at least partially ionic in some of the halides is also supported by magnetic evidence.

The mercuric iodide structure

8.27. Another structure of the layer type is found in HgI_2. Here each mercury atom is co-ordinated by only four halogen neighbours, arranged at the corners of a regular tetrahedron, and each of these halogen atoms

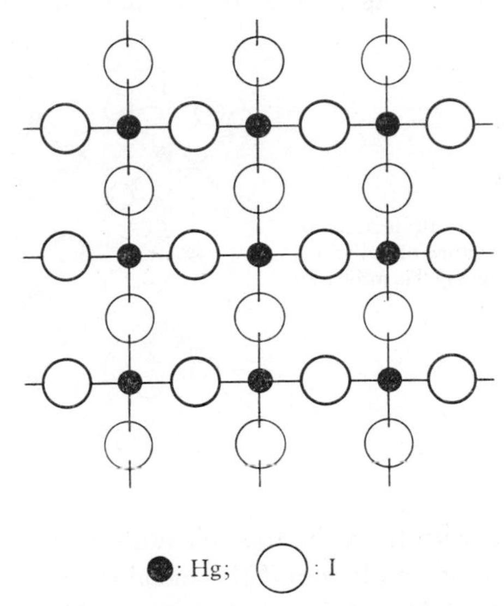

Fig. 8.10. Plan of a single layer of the structure of mercuric iodide, HgI_2.

is bound to only two mercury neighbours to give a sheet of the form shown in fig. 8.10. The structure as a whole is made up by the superposition of such sheets. The bond structure round the two atoms may be represented as

$$-\underset{\uparrow}{\overset{\downarrow}{\text{Hg}}}- \quad \text{and} \quad \leftarrow\text{I}-$$

Each mercury atom (with the configuration 2, 8, 18, 32, 18, $6s^2$) in this way acquires six electrons to give the configuration 2, 8, 8, 32, 18, $6s^2 6p^6$, and one $6s$ and three $6p$ orbitals are available to form the tetrahedral sp^3 hybrid bonds actually observed.*

* Structures of the $CdCl_2$, CdI_2 and HgI_2 types are often described as 'layer lattices'. The phrase is unfortunate, for the word 'lattice' has a specific meaning in crystallography which is not here applicable. 'Layer structures' is a better term.

The palladous chloride and related structures

8.28. Yet another essentially molecular arrangement is found in the structures of $PdCl_2$, $CuCl_2$ and $CuBr_2$. Although these structures differ in detail, in all of them the metal atom is co-ordinated by four halogen atoms at the corners of a square, and these squares are joined by opposite edges to form indefinitely extended planar chains. These chains are disposed in the structure with their axes parallel (as shown for $PdCl_2$ in fig. 8.11) and constitute infinite one-dimensional molecules which

Fig. 8.11. Clinographic projection of the orthorhombic structure of palladous chloride, $PdCl_2$. The unit cell is indicated by broken lines.

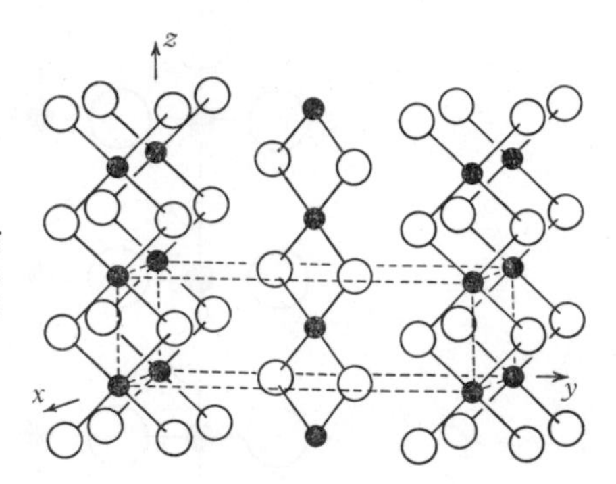

○ : Pd; ◯ :Cl

may be compared with those in the selenium structure (§ 7.06). The planar distribution of bonds round the metal atoms is the same as in the structures of the oxides of the same elements (§ 8.10), and represents the characteristic spatial disposition of dsp^2 hybrid bonds. Thus in $PdCl_2$, say, the bond structure within the chain may be represented

Cl Cl
Pd Pd Pd
Cl Cl

The mercuric chloride structure

8.29. The structure of $HgCl_2$ is of interest in that it consists of discrete Cl–Hg–Cl molecules, in contrast to the infinite two-dimensional molecules found in HgI_2 (§ 8.27). These linear molecules are arranged in the crystal in a manner similar to that of the I_2 molecules in iodine (§ 7.03), and are bound to one another only by van der Waals bonds. In this compound the mercury atom acquires the configuration 2, 8, 18, 32,

18, $6s^2 6p^2$, so that one $6s$ and one $6p$ orbital are available to form the two linearly directed sp hybrid bonds. It is thus possible to formulate bond configurations which account satisfactorily for each of the very different structures of $HgCl_2$ and HgI_2, but it is less easy to understand just exactly why they should be so different, consisting in the former case of discrete molecules and in the latter of molecular sheets. In fact isolated HgI_2 molecules do, indeed, exist in the vapour, and it is interesting to note that there is an appreciable difference in Hg–I distance as between crystal and vapour, corresponding to the difference in bond character. This distance is 2·78 Å for the sp^3 bonds in the crystal and 2·61 Å for the stronger sp bonds in the molecules in the vapour. The structure of $HgBr_2$ is intermediate in character between those of CdI_2 and $HgCl_2$: it is a layer structure in which each mercury atom is co-ordinated by a distorted octahedron of bromine neighbours, the distortion being such as to bring two of these neighbours much closer than the other four.

Hydrides and hydroxides

8.30. Ionic hydrides of composition AH_2 are formed only by the alkaline earth metals; they contain the H^- ion, are salt-like in character and may be compared with AH halides of the alkali metals. The remaining AH_2 hydrides are quite different in their properties, in respect of which they most closely resemble metal systems. They are therefore more conveniently discussed later (§§ 13·37–13·39).

We have already seen that in a limited number of AOH hydroxides the OH group behaves as a negative ion of radius 1·53 Å, intermediate between the radii of the F^- and Cl^- ions, and that the structures of these hydroxides are analogous to those of the corresponding halides. The same is true of certain of the $A(OH)_2$ hydroxides. None of these hydroxides has any of the symmetrical structures characteristic of truly ionic bonding, but those of Mg, Ca, Mn, Fe, Co, Ni and Cd have the cadmium iodide structure. Some other hydroxides, however, have structures quite different from those of the halides. These structures, and the reason for their abnormal properties, are discussed in §§ 12.10–12.12.

Oxides and sulphides

8.31. The oxides and sulphides of composition AO_2 and AS_2 resemble the AO and AS compounds in showing a wide variety of structural types. Some of those which we shall consider are summarized in table 8.05. It will be seen from this table that oxides and the corresponding sulphides are not in general isostructural. We revert to this point later (§ 8.60).

Table 8.05. *The structures of some AX_2 oxides and sulphides*

Fluorite structure (C.N. 8:4)	Rutile structure (C.N. 6:3)	Silica structures (C.N. 4:2)	Cadmium iodide structure	Molybdenum sulphide structure	Chain structure
ZrO_2	GeO_2	SiO_2	TiS_2	MoS_2	SiS_2
HfO_2	SnO_2	GeO_2	ZrS_2	WS_2	
PoO_2	PbO_2		SnS_2		
CeO_2	TiO_2		TaS_2		
PrO_2	VO_2		PtS_2		
ThO_2	NbO_2				
PaO_2	TaO_2				
UO_2	CrO_2				
NpO_2	MoO_2				
PuO_2	WO_2				
AmO_2	MnO_2				
	IrO_2				

Oxides

8.32. On account of the strongly electronegative character of oxygen many oxides AO_2 are primarily ionic, and have symmetrical structures of the fluorite, rutile and β-cristobalite types, already described (§§ 8.21–8.23), composed of A^{4+} and O^{2-} ions. The particular structure which obtains is determined by geometrical considerations, as will be seen from the radius ratios quoted for some of these oxides in table 8.06. Layer structures of the cadmium chloride and cadmium iodide types, which are common in AX_2 compounds containing less electronegative anions, are not found among oxides.

Table 8.06. *The radius ratios of some AX_2 oxides*

Fluorite structure $r^+/r^- > 0{\cdot}7$		Rutile structure $0{\cdot}7$–$0{\cdot}3$		Silica structures $< 0{\cdot}3$	
CeO_2	0·72	PbO_2	0·60	GeO_2	0·38
ThO_2	0·68	SnO_2	0·51	SiO_2	0·29
PrO_2	0·66	TiO_2	0·49		
PaO_2	0·65	WO_2	0·47		
UO_2	0·64	OsO_2	0·46		
NpO_2	0·63	IrO_2	0·46		
PuO_2	0·62	RuO_2	0·45		
AmO_2	0·61	VO_2	0·43		
ZrO_2	0·57	CrO_2	0·40		
HfO_2	0·56	MnO_2	0·39		
		GeO_2	0·38		

The silica structures

8.33. The structure of silica, SiO_2, is of particular interest and importance because of its relationship to the silicate minerals to be discussed later. Silica is trimorphous and occurs naturally as three forms stable in the temperature ranges indicated:

Quartz	Tridymite	Cristobalite
(below 870 °C)	(870–1470 °C)	(above 1470 °C)

In addition, there are α and β modifications of each of these forms, differing only slightly in structure. The idealized structure of β-cristobalite has already been described (§8.23 and fig. 8.07); the actual structure differs from this in that the oxygen atoms are somewhat displaced from the straight lines joining pairs of silicon atoms so that the two bonds to each oxygen atom are no longer collinear. The structure of tridymite is also 4:2 co-ordinated and can be regarded as related to that of wurtzite in exactly the same way as the structure of cristobalite is related to that of zincblende: if the zinc and sulphur atoms in wurtzite are replaced by silicon atoms, and if oxygen atoms are introduced midway between each pair of these atoms, the (idealized) structure of tridymite results. In quartz 4:2 co-ordination is again found, but now the co-ordinating tetrahedra about the silicon atoms are yet differently arranged to give a denser structure which is discussed in detail in §9.15.

The nature of the Si–O bond in silica calls for brief discussion. The difference in electronegativity between silicon and oxygen is 1·7, and this value corresponds to a bond which is about 40 per cent ionic. It is therefore not possible to regard silica in its various forms as a purely ionic compound, and the Si–O bond must be treated as possessing a considerable degree of covalent character. This view is supported by the fact that the oxygen atoms in all the structures are displaced from the line joining silicon atoms, an arrangement clearly energetically impossible in a truly ionic structure and showing a tendency towards the characteristic spatial disposition of covalent bonds.

Peroxide and superoxide structures

8.34. The structures of certain AO_2 compounds not shown in table 8.05 also merit brief consideration. The peroxides CaO_2, etc., of the alkaline earth metals, although ionic, are quite different from the oxides so far discussed in that in them the anions are the diatomic units O_2^{2-};

similarly, in the superoxides KO_2, RbO_2 and CsO_2 diatomic ions O_2^- are found. All these oxides are therefore more closely related to AX than to AX_2 compounds. The structure of SrO_2 is shown in fig. 8.12 and it will be seen that it may be regarded as the sodium chloride arrangement with Sr^{2+} and linear O_2^{2-} ions in place of Na^+ and Cl^-. The O_2^{2-} ions are all arranged with their axes parallel to one edge of the unit cell, the symmetry of which is accordingly lowered from cubic to tetragonal. The other peroxides and superoxides have similar structures.

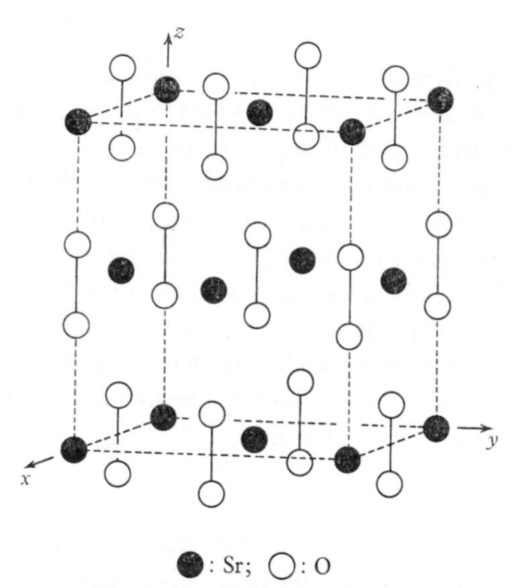

Fig. 8.12. Clinographic projection of the unit cell of the tetragonal structure of strontium peroxide, SrO_2.

Carbon dioxide

8.35. The structure of carbon dioxide, CO_2, is purely molecular. Discrete, linear O–C–O molecules are arranged with the carbon atoms at the corners and face centres of a cubic unit cell and with the axes of the molecules parallel to the cube diagonals (fig. 8.13). This is an elegant example of a typically molecular structure, for it will be seen that the centres of the molecules are disposed as in cubic close packing, and that the arrangement of the molecules themselves is as compact as is consistent with their aspherical shape. The interatomic distance between the oxygen atoms of adjacent molecules is about 3·2 Å, corresponding to the weak van der Waals forces which hold them together, and the C–O distance within each molecule is 1·16 Å. This latter distance is significantly less than would be expected on the basis of the simple structure O═C═O and probably arises from resonance between this structure and the forms

$$\overset{\oplus}{O}\equiv C\rightarrow\overset{\ominus}{O} \quad \text{and} \quad \overset{\ominus}{O}\leftarrow C\equiv\overset{\oplus}{O}$$

Although these two forms are polar they will, of course, contribute equally to the resonance hybrid so that the resultant moment will be

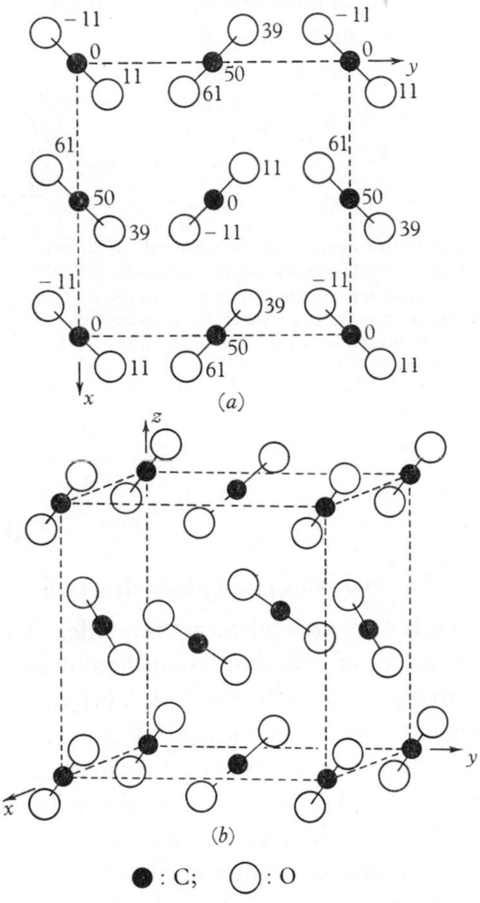

● : C; ○ : O

Fig. 8.13. (*a*) Plan of the unit cell of the cubic structure of carbon dioxide, CO_2, projected on a plane perpendicular to the *z* axis. The heights of the atoms are expressed in units of *c*/100. (*b*) Clinographic projection of the same structure.

zero, as is in fact observed. We have here yet another example of a structure in which a precision determination of bond length throws light on the nature of the interatomic binding.

Sulphides

8.36. Sulphur is less electronegative than oxygen, and in consequence no sulphide of composition AS_2 crystallizes with any of the typically ionic structures commonly found among the oxides AO_2. A number of sulphides have the cadmium iodide layer structure, but many others, particularly those of the transition metals, have structures unrepresented among the compounds so far considered. A few of these are of sufficiently common occurrence to warrant discussion.

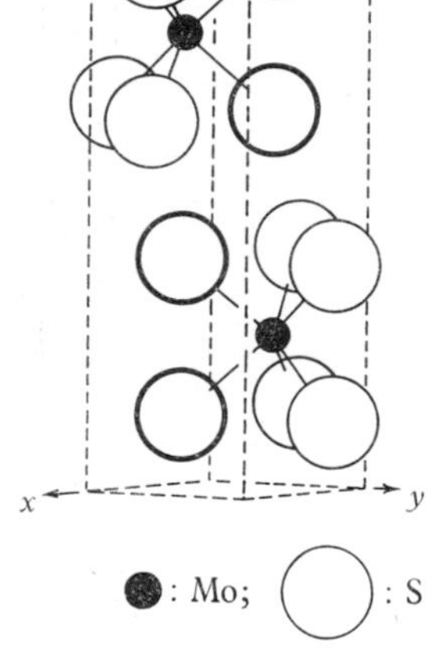

Fig. 8.14. Clinographic projection of the unit cell of the hexagonal structure of molybdenum sulphide, MoS_2. The four S atoms represented by heavy circles are those within the unit cell; the others lie outside the cell but have been added to show the co-ordination about the Mo atoms.

The molybdenum sulphide structure

8.37. The structure of molybdenum sulphide, MoS_2 (fig. 8.14), resembles the structures of cadmium chloride and cadmium iodide in that it is a layer arrangement in which each layer consists of a sheet of A atoms sandwiched between two sheets of X atoms. The co-ordination about the A atoms, however, is no longer octahedral, the six X neighbours being arranged at the corners of a trigonal prism (cf. the co-ordination of Ni about As in nickel arsenide, § 8.12). The layers are superposed in such a way that alternate layers are identical, and are held together only by weak van der Waals bonds. The structure is thus another example of a molecular arrangement in which each layer constitutes an infinite two-dimensional molecule. The bond distribution about the two atoms may be written as

$$\rightarrow \text{Mo} \leftarrow \quad \text{and} \quad \leftarrow \text{S}$$

In this way the molybdenum atom of configuration 2, 8, 18, 13, $5s^1$ effectively acquires eight electrons to give the configuration 2, 8, 18, 18, $5s^2\,5p^2$. The orbitals involved in bond formation are four $4d$, one $5s$ and one $5p$, and hybridization of these gives d^4sp bonds with the characteristic trigonal prismatic disposition actually observed (see table 4.01).

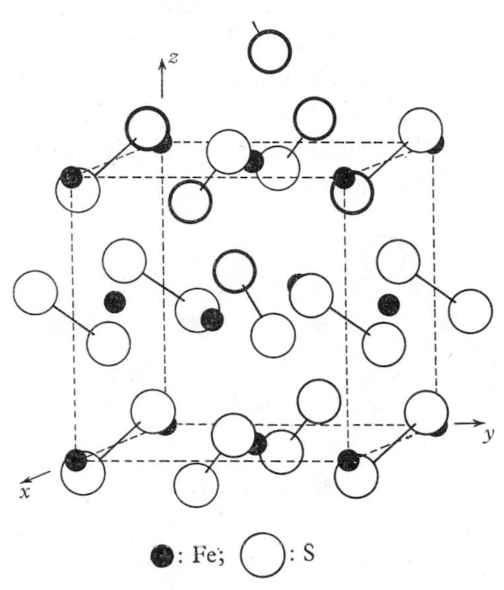

Fig. 8.15. Clinographic projection of the unit cell of the cubic structure of pyrites, FeS_2; some S atoms outside the unit cell are shown. Six of the S atoms are represented by heavy circles to emphasize the octahedral co-ordination about the Fe atom at the centre of the top face.

The pyrites and marcasite structures

8.38. In the pyrites and marcasite structures of FeS_2 discrete S_2 groups are found. Geometrically the pyrites structure (fig. 8.15) is closely related to that of sodium chloride, from which it may be regarded as being derived by replacing Na by Fe and Cl by S_2 groups. This description, however, conceals the nature of the co-ordination, for the S–S distance within the S_2 group is such that each iron atom is surrounded by six sulphur atoms at the corners of a nearly regular octahedron, while each sulphur atom is bound to one other sulphur atom and to three iron atoms. The bond distribution about the atoms can thus be written

—Fe— and —S—S— (with arrows indicating the additional bonds)

The octahedral bonds about the iron atom are d^2sp^3 hybrids formed by two $3d$, one $4s$ and three $4p$ orbitals (see table 4.01), and the bonding throughout the structure is wholly covalent. The crystal may thus be regarded as constituting a single giant molecule, just as a crystal of diamond is a single molecule of carbon. It is interesting to note that the $[FeS_6]$ octahedra in the structure share faces, and that the shared edges bounding these faces are actually longer than the unshared edges. This is additional evidence for the covalent character of the binding; in ionic crystals co-ordinating polyhedra of anions rarely share faces, and when they do so the shared edges are usually shortened owing to the mutual repulsion of the cations.

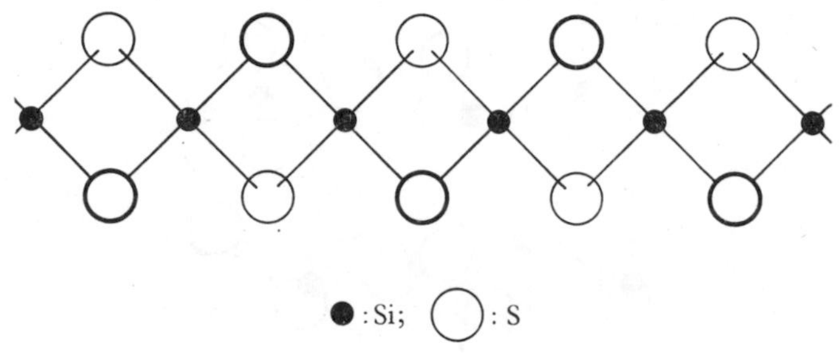

Fig. 8.16. The structure of the chain molecule of SiS_2. Each Si atom is tetrahedrally co-ordinated by S, and the tetrahedra are linked into an infinite chain by sharing opposite edges.

Silicon disulphide

8.39. The structure of SiS_2 is of interest by virtue of its comparison with that of SiO_2. As in SiO_2, the silicon atoms are tetrahedrally co-ordinated, but, owing to the less ionic character of the bonding, the tetrahedra in SiS_2 are able to approach more closely and so share opposite edges to form infinite molecular chains of the form shown in fig. 8.16. In all the forms of silica the co-ordinating tetrahedra share only corners. The chains in the SiS_2 structure are packed together in much the same manner as the chains in selenium (§7.06), so that each is symmetrically surrounded by six others.

Carbides

8.40. The great majority of carbides of composition AC_2 resemble intermetallic systems in many of their properties, and will be discussed later (§§ 13.37–13.39). Some, however, principally the compounds of the more electropositive metals, are strikingly different in their properties and form typically ionic crystal structures. Thus the carbides CaC_2, SrC_2, BaC_2, LaC_2, CeC_2, PrC_2 and NdC_2 all have tetragonal structures analogous to that of SrO_2 (fig. 8.12) with the O_2^{2-} ions replaced by C_2^{2-} and with $c/a > 1$. The structure of ThC_2 (fig. 8.17) is closely related, and again may be compared with that of sodium chloride; but now the axes of the C–C ion lie perpendicular to the principal axis, instead of parallel to it, and give a tetragonal cell with $c/a < 1$.

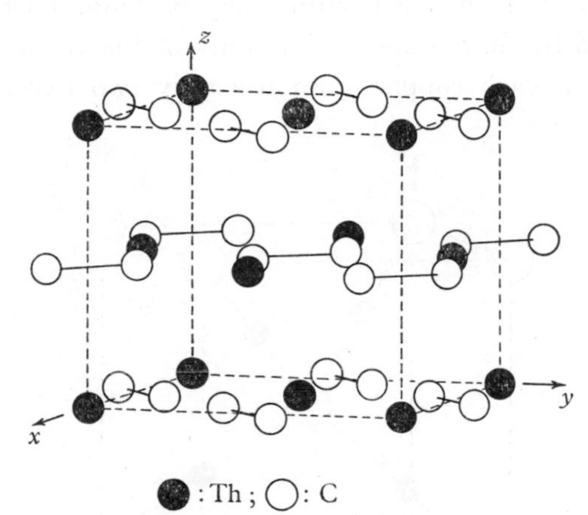

Fig. 8.17. Clinographic projection of the unit cell of the tetragonal structure of thorium carbide, ThC_2.

A_2X STRUCTURES

8.41. Many structural types are represented among A_2X compounds. We shall describe only two, which are of common occurrence among metallic oxides and sulphides. There are, of course, also numerous molecular structures of this composition. Among these the structure of ice is of particular importance and is discussed separately in § 12.05.

The anti-fluorite structure

8.42. Many of the oxides, sulphides, selenides and tellurides of the alkali metals (e.g. Li_2O, Li_2S, etc.) have the so-called anti-fluorite structure, i.e. a fluorite structure in which the positions of the anions and cations are interchanged. Most of these can be regarded as essentially ionic compounds. The co-ordination is 4:8.

The cuprite structure

8.43. The mineral cuprite, Cu_2O, and the corresponding silver oxide, Ag_2O, have the very simple cubic structure shown in fig. 8.18, with 2:4 co-ordination. The oxygen atoms are located at the corners and centre of the unit cell, and the copper atoms occupy the centres of four of the eight cubelets into which the cell may be divided. The arrangement of the copper atoms alone is thus as in cubic close packing. Each oxygen atom is co-ordinated by four copper neighbours at the corners of a regular tetrahedron, but each copper atom has only two oxygen neighbours

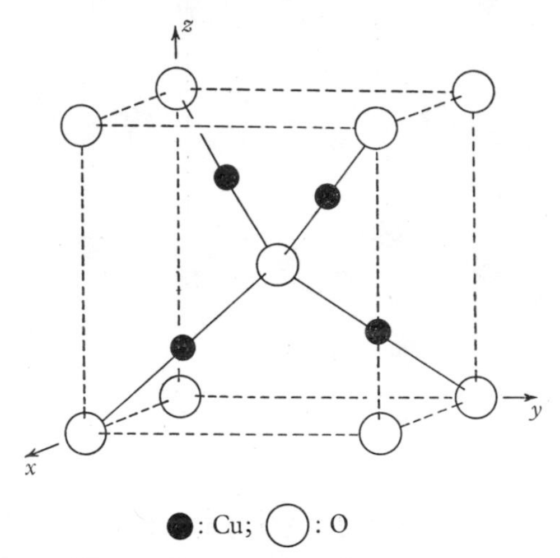

Fig. 8.18. Clinographic projection of the unit cell of the cubic structure of cuprite, Cu_2O.

symmetrically disposed about it on a straight line. This structure is unique among inorganic compounds in that it consists of two identical interpenetrating frameworks which are not directly bonded together: starting from any one atom it is possible to reach half, but only half, of the remainder by travelling along Cu–O bonds.

The bond structure in Cu_2O may be formulated thus

$$—\mathrm{Cu}\leftarrow \qquad \begin{array}{c}\uparrow\\ —\mathrm{O}—\\ \downarrow\end{array}$$

This arrangement gives the copper atom the electronic configuration 2, 8, 18, $4s^2 4p^2$, so that the orbitals involved in the bonds are one $4s$ and one $4p$; hybridization of these gives sp bonds with the characteristic linear configuration actually observed (see table 4.01). This is the fourth simple copper compound whose structure we have considered, the others being CuI (§8.04), CuO (§8.10) and $CuCl_2$ (§8.28), and it is interesting to compare the stereochemistry of univalent and divalent copper in these structures. In both CuO and $CuCl_2$ the Cu^{II} atom forms four coplanar dsp^2 bonds, in CuI the Cu^I atom again forms four bonds but now they are sp^3 bonds tetrahedrally disposed, while finally in Cu_2O the Cu^I atom forms two sp bonds arranged in line.

A_mX_z STRUCTURES

8.44. Among compounds of this type few typically ionic crystals are found, and the structures are generally of far greater diversity and complexity than those of the relatively simple compounds so far considered. With increasing valency of the cation the tendency towards covalent binding becomes more and more pronounced, and layer or molecular structures are increasingly common. Thus among AX_3 compounds only the fluorides and a few oxides have symmetrical structures, while many of the other halides, hydroxides, oxides and sulphides have layer structures differing from each other in detail but all displaying the general features already described as characteristic of such structures. Probably no AX_z compounds with $z > 3$ have ionic structures, and in the majority of them purely molecular arrangements are found.

The aluminium fluoride structure

8.45. One symmetrical structure of common occurrence among AX_3 compounds is that of aluminium fluoride, AlF_3, shown (in a somewhat idealized form) in fig. 8.19. Aluminium ions are situated at the corners of the cubic unit cell, with fluorine ions at the mid-points of the cell edges. The co-ordination is thus 6:2, and each Al^{3+} ion is surrounded by a regular octahedron of F^- ions. These octahedra are linked together by sharing only corners. This structure (or a slightly deformed variant of it) is found in the fluorides AlF_3, ScF_3, FeF_3, CoF_3, RhF_3 and PdF_3 and in the oxides CrO_3, WO_3 and ReO_3.

The corundum and haematite structures

8.46. The oxides Al_2O_3 and Fe_2O_3 are both polymorphous, and exist in several forms. The α form of these oxides (corundum and haematite), and also the oxides Cr_2O_3, Ti_2O_3, V_2O_3, α-Ga_2O_3 and Rh_2O_3, have a structure which may be described as a hexagonal close-packed array of oxygen atoms with metal atoms in two-thirds of the octahedrally co-ordinated interstices. Each metal atom is thus co-ordinated by six oxygen atoms, each of which in its turn has four metal atom neighbours. The co-ordination is 6:4. The γ form of Al_2O_3 and Fe_2O_3 has a quite different structure which is described below (§ 8.57).

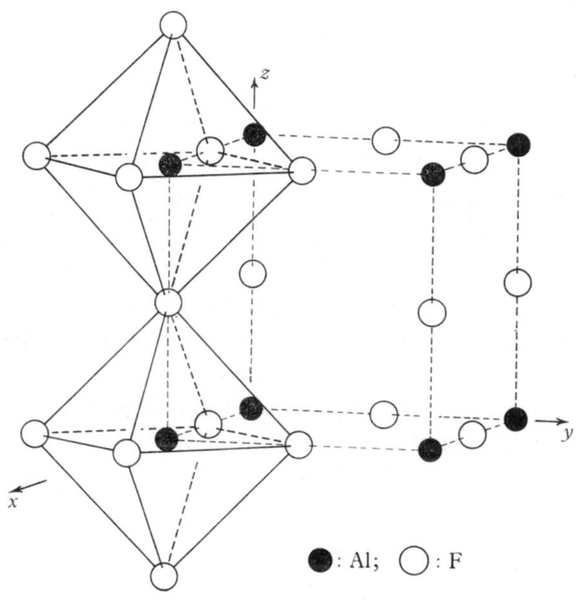

Fig. 8.19. Clinographic projection of the unit cell of the (idealized) cubic structure of aluminium fluoride, AlF_3. Two co-ordinating $[AlF_6]$ octahedra are indicated to show their linkage by sharing corners.

$A_mB_nX_z$ STRUCTURES

8.47. The composition $A_mB_nX_z$ clearly embraces a vast number of compounds, including, among others, the salts of the inorganic acids. At this point, however, it is convenient to confine our discussion for the most part to those compounds in which A and B both represent electropositive elements while X is a non-metal, usually oxygen, sulphur or a halogen. We shall therefore be concerned here with the complex oxides, sulphides and halides of two metals, all of which can be regarded as primarily ionic compounds with structures containing atoms A, B and X in the ionized state. Naturally the oxides are by far the most important and it is they which have been most extensively studied. Many of these oxides prove to be of particular crystallographic significance on account of a number of interesting structural features which they display, while in recent years some have also assumed great practical importance by virtue of their ferroelectric properties. A discussion at some length is therefore warranted.

ABX_3 structures

8.48. Two structures are of common occurrence among ABX_3 compounds and are usually named after the minerals perovskite, $CaTiO_3$, and ilmenite, $FeTiO_3$.

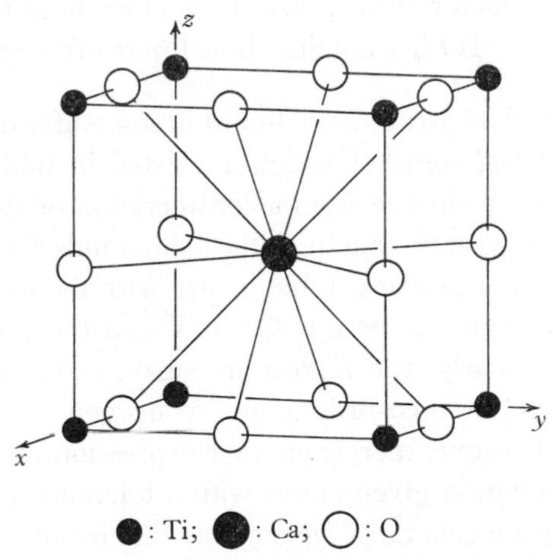

Fig. 8.20. Clinographic projection of the unit cell of the (idealized) cubic structure of perovskite, $CaTiO_3$.

The perovskite structure

8.49. The perovskite structure in its idealized form is cubic and has the atomic arrangement shown in fig. 8.20, with one formula unit of $CaTiO_3$ in the unit cell. Titanium atoms occupy the corners of the cell, a calcium atom is at its centre and the oxygen atoms lie at the mid-points of its edges. Each Ca is thus 12- and each Ti 6-co-ordinated by oxygen neighbours while each oxygen is linked to four Ca and two Ti atoms. As is to be expected, it is the larger metal atom (or ion), calcium in this case, which occupies the position of higher co-ordination. It will be noted that geometrically the structure can be regarded as a close-packed array of (O + Ca) atoms, with Ti atoms occupying some of the octahedrally co-ordinated interstices. Its close relationship to the structure of AlF_3 (see fig. 8.19) will also be observed: if the Ca atoms are omitted those of Ti and O occupy the positions of Al and F, respectively, in the latter structure.

It is clear that in such a symmetrical structure a simple relationship must exist between the radii of the component ions. Ideally this relationship is

$$r_A + r_X = \sqrt{2}\,(r_B + r_X),$$

where A is the larger cation, but it is found that in practice the structure appears whenever the condition

$$r_A + r_X = t\sqrt{2}\,(r_B + r_X)$$

holds. Here t is a 'factor of tolerance' which may lie within the approximate limits 0·7–1·0. If t lies outside these limits other structures obtain.

8.50. The perovskite structure is found in some fifty or more complex oxides and fluorides, some of which are listed in table 8.07. Several points of importance emerge from a consideration of this list.

In the first place we note that in all the compounds the A ions are large (e.g. K, Ca, Sr, Ba) and comparable in size with the oxygen or fluorine ion, as is to be expected since the A and X ions together form a close-packed array. Similarly, the B ions are small, since they must have a radius appropriate to 6-co-ordination by oxygen or fluorine. These conditions are, of course, merely another expression of the fact that the radii satisfy the relation given above with a tolerance factor within the range quoted. Quite generally, we may say that for oxides and fluorides the radii of the A and B ions must lie within the ranges 1·0–1·4 Å and 0·45–0·75 Å, respectively.

Table 8.07. *Some compounds with the perovskite structure*

Not all of these compounds have the ideal structure of fig. 8.20 (see text)

$NaNbO_3$	$CaTiO_3$	$CaSnO_3$	$BaPrO_3$	$YAlO_3$	$KMgF_3$
$KNbO_3$	$SrTiO_3$	$SrSnO_3$	$SrHfO_3$	$LaAlO_3$	$PbMgF_3$
$NaWO_3$	$BaTiO_3$	$BaSnO_3$	$BaHfO_3$	$LaCrO_3$	$KNiF_3$
	$CdTiO_3$	$CaCeO_3$	$BaThO_3$	$LaMnO_3$	$KZnF_3$
	$PbTiO_3$	$SrCeO_3$		$LaFeO_3$	
	$CaZrO_3$	$BaCeO_3$			
	$SrZrO_3$	$CdCeO_3$			
	$BaZrO_3$	$PbCeO_3$			
	$PbZrO_3$				

A second point of interest is that among oxides the perovskite structure is not exclusively restricted to those compounds in which the A and B ions are divalent and quadrivalent, respectively, as is shown by the fact that $KNbO_3$ and $LaAlO_3$ have this structure. It thus appears that the valencies of the individual cations in the structure are of only secondary importance, and that any pair of ions can occur provided that they have radii appropriate to the co-ordination and an aggregate valency of 6 to confer electrical neutrality on the structure as a whole; among the oxides listed in table 8.07 are compounds with pairs of cations of valencies 1 and 5, 2 and 4, and 3 and 3. This point is made even more clearly by the fact that the perovskite structure is also found in a number of oxides (not shown in table 8.07) in which the *A* and/or *B* sites are not all occupied by atoms of the same kind. Thus $(K_{\frac{1}{2}}La_{\frac{1}{2}})TiO_3$ has the perovskite structure with the *A* ions replaced by equal numbers of ions of K and La, while in $Sr(Ga_{\frac{1}{2}}Nb_{\frac{1}{2}})O_3$ the *B* ions are replaced by equal numbers of ions of Ga and Nb. In $(Ba_{\frac{1}{2}}K_{\frac{1}{2}})(Ti_{\frac{1}{2}}Nb_{\frac{1}{2}})O_3$ the same structure is again found, with (Ba+K) in place of *A* and (Ti+Nb) in place of *B*. It is, of course, clear that such arrangements must constitute defect structures, for a single unit cell will typify the structure as a whole only if we consider the *A* and/or *B* sites to be statistically occupied by equal number of the substituent ions.

A still more extreme example shows that the perovskite structure can even occur with some of the *A* sites unoccupied. Sodium tungsten bronze has the ideal composition $NaWO_3$, with the perovskite structure, but this compound shows very variable composition and colour, and is better represented by the formula Na_xWO_3 with $1 > x > 0$. In the sodium-poor varieties the structure remains essentially unaltered but some of the sites normally occupied by sodium are vacant. To preserve neutrality one tungsten ion is converted from W^{5+} to W^{6+} for every site so unoccupied, and this change in ionization gives rise to the characteristic alteration in colour and explains its association with the sodium content. In the extreme case, when no sodium is present, we have WO_3, the structure of which is closely related to that of AlF_3. We have already shown how this structure, in its turn, is related to that of perovskite.

A third point to be noted from table 8.07 is that among the compounds with the perovskite structure are many 'titanates', 'niobates', 'stannates', etc., which would normally be regarded as inorganic salts. Structurally, however, there is no justification for this view. We shall later find that in the true salts of inorganic acids finite complex anions have a discrete existence in the crystal structure: in calcium carbonate, for example, anions CO_3^{2-} are clearly recognizable and the structure as a whole is built up of these anions and of Ca^{2+} cations arranged in a manner very similar to that of the ions in sodium chloride. In calcium 'titanate' on the other hand, each titanium ion is co-ordinated symmetrically by six oxygen neighbours and no TiO_3^{2-} complex ion can be discerned. Thus, in spite of the resemblance between the empirical formulae $CaCO_3$ and $CaTiO_3$, the compounds are in structure entirely distinct, and while the former is a salt the latter should more properly be regarded as a complex oxide.

The final point which we would make about the perovskite structure is that the 'ideal' highly symmetrical cubic structure so far described is found in only a limited number of the compounds given in table 8.07. At high temperatures, or when the tolerance factor is very close to unity, this simple structure does indeed often occur, but in many compounds the actual structure is a pseudosymmetric variant of the ideal arrangement, derived from it by small displacements of the atoms. In some cases these displacements result in a slight distortion of the unit cell, the symmetry of which is accordingly reduced, and in others the deformation is such that adjacent cells are no longer precisely identical so that the true unit cell comprises more than one of the smaller ideal units. The number of these pseudosymmetric structures is too great to describe in detail here, but it is important to stress that in many cases the degree of departure from the ideal arrangement is only very slight. Thus $BaTiO_3$ has a tetragonal unit cell with axial ratio $c/a = 1{\cdot}01$ derived from the cubic cell by an extension parallel to one of the cube edges of only 1 per cent. Even so, these departures from the ideal structure are of profound importance, for it is to them that the ferroelectric properties of many of these oxides must be ascribed. Ferroelectricity is not compatible with the high symmetry of the ideal structure, and it is only in those members of the perovskite family which have structures of lower symmetry that the property can occur.

The ilmenite structure

8.51. When the A ion in an ABX_3 compound is too small to form the perovskite structure, say when its radius is less than about 1·0 Å, the alternative ilmenite arrangement sometimes occurs. This structure is closely related to that of corundum and haematite (§8.46) and may be described as an hexagonal close-packed array of X ions (usually oxygen ions) with A and B ions each occupying one-third of the octahedrally co-ordinated interstices. Thus both A and B ions are now 6-co-ordinated by anion neighbours.

8.52. A few oxides with the ilmenite structure are listed in table 8.08. It will be seen from this list that, with the exception of $CdTiO_3$, all are oxides in which the A ion is considerably smaller than the corresponding ion in the perovskite structure. The dimorphism of $CdTiO_3$, which at high temperatures has the perovskite structure, is readily explained by the fact that the radius of the Cd^{2+} ion (0·97 Å) is very close to the critical value for transition between the perovskite and ilmenite arrangements. Again it will be noted that the list includes a number of 'titanates' which should more properly be described as complex oxides.

Table 8.08. *Some compounds with the ilmenite structure*

$MgTiO_3$	$FeTiO_3$	$CdTiO_3$*	α-Al_2O_3	α-Fe_2O_3
$MnTiO_3$	$CoTiO_3$	$LiNbO_3$	Ti_2O_3	Rh_2O_3
	$NiTiO_3$		V_2O_3	Ga_2O_3

* Low temperature form.

All the oxides listed in table 8.08 are isomorphous, and solid solution between them is common. Thus $MgTiO_3$ and $FeTiO_3$ form solid solutions of composition $Mg_xFe_{(1-x)}TiO_3$ in which the Mg and Fe ions occupy the A sites at random, and $FeTiO_3$ and α-Fe_2O_3 at high temperatures form a complete range of solid solution between the two extreme compositions. In this latter case, as the proportion of α-Fe_2O_3 increases the Fe^{2+} and Ti^{4+} ions in $FeTiO_3$ are progressively replaced by Fe^{3+} ions until ultimately these occupy both A and B positions.

AB_2X_4 structures

8.53. The only AB_2X_4 structure which we shall discuss is that of spinel, $MgAl_2O_4$.

The spinel structure

8.54. The spinel structure is found among a very large number of oxides AB_2X_4 and also in a limited number of sulphides, selenides, fluorides and cyanides of the same composition, some of which are recorded in table 8.09. The cubic unit cell of this structure is shown in fig. 8.21 and contains 32 X ions. Each A ion is tetrahedrally co-ordinated by four and each B ion is octahedrally co-ordinated by six X neighbours, and each X ion is bound to one A and to three B ions. The co-ordination may therefore be summarized thus:

$$A-4X, \quad B-6X, \quad A-X-3B.$$

It will be realized from fig. 8.21 that if the X ions alone are considered the positions of these ions could be described in terms of a cubic sub-cell of volume only one-eighth that of the true cell, with the X ions at its corners and face centres. In other words, the structure is one in which the X ions are arranged as in cubic close packing, with the A and B ions in the tetrahedrally and octahedrally co-ordinated interstices, respectively.

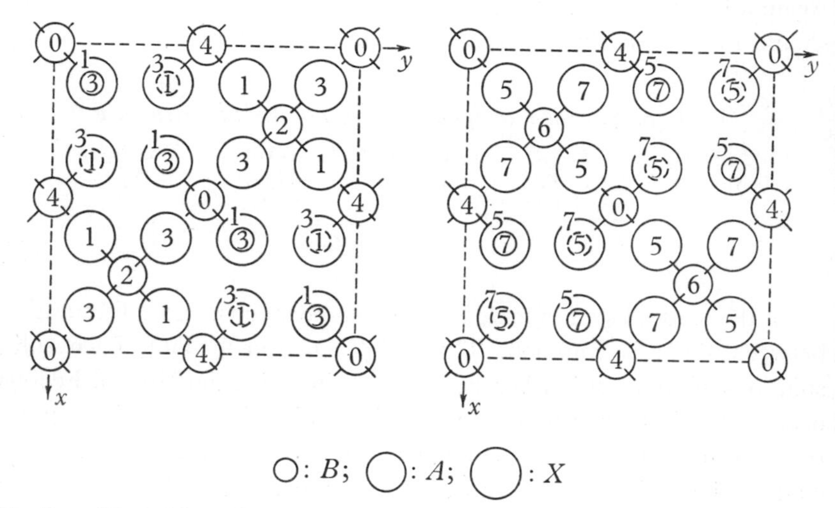

Fig. 8.21. Plan of the unit cell of the cubic structure of spinel, AB_2X_4, projected on a plane perpendicular to the z axis. The heights of the atoms are indicated in units of $\frac{1}{8}c$. For clarity the lower and upper halves of the cell are shown separately, and only the co-ordination about the A ions is indicated.

In the majority of the oxides with the spinel structure A is a divalent and B a trivalent ion, but this condition is not essential, and it is found (as in the perovskite structure) that the total cation charge is the most significant factor and that other cation combinations which yield a neutral structure are admissible.

8.55. The structure so far described is the 'normal' spinel structure and is the arrangement found in many AB_2O_4 oxides. Some oxides of this composition, however, have an alternative 'inversed' structure, which may be best described in terms of a specific example, such as $MgFe_2O_4$. In this oxide the pattern of sites in the structure and the distribution of the oxygen ions is exactly as in the normal spinels. The arrangement of the cations, however, is different: the 4-co-ordinated

Table 8.09. *Some compounds with the spinel structure*

$BeLi_2F_4$	$MgCr_2O_4$*	$MgFe_2O_4$†	$FeNi_2O_4$	$MgGa_2O_4$†
$MoNa_2F_4$	$MnCr_2O_4$	$TiFe_2O_4$†	$GeNi_2O_4$	$ZnGa_2O_4$
WNa_2O_4	$FeCr_2O_4$	$MnFe_2O_4$	$FeNi_2S_4$	$CaGa_2O_4$*
$ZnK_2(CN)_4$	$CoCr_2O_4$	$FeFe_2O_4$†	$NiNi_2S_4$	$MgIn_2O_4$†
$CdK_2(CN)_4$	$NiCr_2O_4$*	Fe_2O_3	$MgRh_2O_4$*	$CaIn_2O_4$
$HgK_2(CN)_4$	$ZnCr_2O_4$*	$CoFe_2O_4$	$ZnRh_2O_4$*	$MnIn_2O_4$
$TiMg_2O_4$†	$CdCr_2O_4$*	$NiFe_2O_4$	$TiZn_2O_4$†	$FeIn_2O_4$†
VMg_2O_4†	$MnCr_2S_4$	$CuFe_2O_4$†	$SnZn_2O_4$†	$CoIn_2O_4$†
$SnMg_2O_4$	$FeCr_2S_4$	$ZnFe_2O_4$*	$MgAl_2O_4$	$NiIn_2O_4$†
MgV_2O_4*	$CoCr_2S_4$	$CdFe_2O_4$*	$SrAl_2O_4$	$CdIn_2O_4$
FeV_2O_4	$CdCr_2S_4$	$AlFe_2O_4$	$CrAl_2O_4$	$HgIn_2O_4$
ZnV_2O_4*	$HgCr_2S_4$*	$PbFe_2O_4$	$MoAl_2O_4$	
	$ZnCr_2Se_4$	$MgCo_2O_4$	$FeAl_2O_4$	
	$CdCr_2Se_4$*	$TiCo_2O_4$	$CoAl_2O_4$	
	$TiMn_2O_4$	$CoCo_2O_4$	$NiAl_2O_4$*	
		$CuCo_2O_4$	$CuAl_2O_4$	
		$ZnCo_2O_4$	$ZnAl_2O_4$*	
		$SnCo_2O_4$†	Al_2O_3	
		$CoCo_2S_4$	$ZnAl_2S_4$	
		$CuCo_2S_4$		

* Normal structure. † Inversed structure.

sites are occupied, not by the Mg^{2+} ions, but by one-half of the Fe^{3+} ions, while the rest of these ions, together with all the Mg^{2+} ions, are distributed at random over the sites of 6-co-ordination. If we wish to emphasize this distinction we may write the formula of the inversed structure as $Fe(MgFe)O_4$. The structure type of the compounds listed in table 8.09 is indicated, where known.

8.56. The factors which determine the appearance of one or other of the two spinel structures are by no means clear. It might be expected that the 4-co-ordinated sites would always be occupied by the smaller cations, so that spinels with $r_A < r_B$ would have the normal and those with $r_A > r_B$ the inversed structure. This, however, is far from being the case, and if anything the reverse is more nearly true. Thus there appears to be a tendency for tri- and quadrivalent ions to prefer 6-co-ordinated sites (except for the ions Fe^{3+}, In^{3+} and Ga^{3+}, which prefer 4-co-ordination) and for the ions Zn^{2+} and Cd^{2+} to show a special preference for 4-co-ordination. We can therefore arrange ions in a series

Zn^{2+}, Cd^{2+}
Fe^{3+}, In^{3+}, Ga^{3+}
Other divalent ions
Other trivalent ions
Quadrivalent ions

such that of the two metal ions in any given spinel the one standing higher in the series will tend to occupy the 4-co-ordinated sites. Thus we should expect the majority of $A^{2+}B_2^{3+}O_4$ spinels (all except those containing Fe^{3+}, In^{3+} or Ga^{3+} ions without Zn^{2+} or Cd^{2+} ions) to have the normal structure and all $A^{4+}B_2^{2+}O_4$ spinels to have the inversed arrangement, as indeed appears to be the case where the structure type is known. As specific examples of the application of this principle we may contrast the structures of the following pairs of compounds

Normal structure	Inversed structure
MgV_2O_4	VMg_2O_4
$MgCr_2O_4$	$MgFe_2O_4$
$ZnFe_2O_4$	$CuFe_2O_4$
$NiAl_2O_4$	$NiIn_2O_4$

8.57. The inclusion of magnetite, Fe_3O_4, in table 8.09 is of interest. This oxide has the inversed spinel structure, and its relationship to the other spinels is made clear by writing the formula as $Fe^{3+}(Fe^{2+}Fe^{3+})O_4$. It is probable that the semi-conducting properties of magnetite can be attributed to an interchange of electrons between the Fe^{2+} and Fe^{3+} ions in the *B* sites, and that semi-conductivity in other spinels is associated, as here, with the presence of ions of different valencies in crystallographically equivalent positions.

Of even greater interest is the relationship of the spinels to Al_2O_3 and Fe_2O_3, for it has long been known that many spinels can take up indefinite quantities of these oxides in solid solution. As normally prepared, Fe_2O_3 exists in the α form with the structure already described (§8.46), but, by the careful oxidation of Fe_3O_4, a γ form, with an entirely different structure, is obtained. The cubic unit cell of this new structure is of about the same size as that of the spinel structure of Fe_3O_4, and the arrangement first proposed was one in which the iron ions occupied the same sites as in the spinel with the necessary four additional oxygen ions per unit cell accommodated in interstices in the structure. It was, however, difficult to understand how such large ions as those of oxygen could be introduced into a structure already so closely packed, and it is now recognized that the actual structure must be regarded, not as a spinel structure with an excess of oxygen, but rather as a spinel structure with a cation deficiency. The structure of γ-Fe_2O_3 (and of γ-Al_2O_3) is one in which 32 oxygen ions per unit cell are arranged exactly as in spinel, with the corresponding number of iron or aluminium ions, namely $21\frac{1}{3}$, distributed at random over the 24 sites normally occupied by the cations. On the average there are therefore $2\frac{2}{3}$ vacant cation sites per unit cell.

In the light of this structure the conversion of Fe_3O_4 into γ-Fe_2O_3 by oxidation is readily understood. The unit cell of Fe_3O_4 contains 8 Fe^{2+}, 16 Fe^{3+} and 32 O^{2-} ions. As oxidation proceeds the oxygen ions are undisturbed but the Fe^{2+} ions are replaced by two-thirds their number of Fe^{3+} ions. Finally, when all eight Fe^{2+} ions have been thus replaced, an extra $5\frac{1}{3}$ Fe^{3+} ions have been introduced and $2\frac{2}{3}$ sites are left vacant. In a similar way we can understand the ready solid solution of Al_2O_3 in the spinels. Starting, say, with $MgAl_2O_4$, the aluminium content may be gradually increased by the substitution of Al^{3+} for Mg^{2+}. In other solid solutions which we shall discuss later electrical neutrality is preserved in such a substitution by another simultaneous substitution elsewhere in the structure, but here it is achieved simply by the appearance of vacant sites in positions which would normally be occupied by cations. When the substitution has been completed just $2\frac{2}{3}$ such sites appear per unit cell and the γ-Al_2O_3 structure results.

SOME CHEMICAL CONSIDERATIONS

8.58. In the present chapter we have described the structures of a considerable number of relatively simple compounds, mostly the halides, oxides and sulphides of the metallic elements. It has been convenient to present these structures collectively in this way because many of the principles of structural architecture are illustrated as well by simple structures as by those which are more complex. Some of these principles are discussed more fully in the next chapter, but at this point it is desirable that we should summarize certain of the more important conclusions of the present chapter by a brief review in which the chemical rather than the geometrical significance of the structures is considered.

8.59. In considering the halides, oxides and sulphides of the metallic elements it is important to bear in mind from the start that almost always the non-metallic atoms or ions are far larger than those of the metals. For this reason the greater part of the volume of the structure is occupied by non-metallic atoms, and in many cases, as we have seen, the structure may be described as a close-packed framework of these atoms with those of the metallic element disposed in its interstices. Often this framework is in itself so stable that it is a matter of relatively little moment which particular metal atoms are present and which particular interstices they occupy. Accordingly, solid solution and variable composition are common. Examples of this among complex oxides have already been given, but one further example of an even simpler compound may be quoted.

The oxide FeO has already been described as a simple ionic compound with the sodium chloride structure, in which, of course, the oxygen ions are arranged as in cubic close packing. It is found, however, that the composition rarely corresponds to the ideal formula, and that much more commonly a considerable excess of oxygen is present, corresponding, say, to an approximate composition $FeO_{1\cdot10}$. As in the relationship of γ-Fe_2O_3 to Fe_3O_4, it is impossible to countenance the presence of excess oxygen in a structure already close packed, and a more accurate description is in terms of a deficiency of iron, say $Fe_{0\cdot9}O$. The pattern of oxygen sites is as in the ideal structure, but as Fe^{2+} ions are replaced by Fe^{3+} vacant cation sites appear in the structure. Since these sites are randomly arranged their exact number, and accordingly the exact composition, are matters of no significance. The relevance of this to the chemical laws of fixed composition and simple proportions is apparent, and is discussed more fully in the next chapter (§9.35).

8.60. One further question which may be briefly discussed at this point is the relationship of oxides to sulphides. Although oxygen and sulphur are closely related chemically there are, nevertheless, important differences which are reflected in the corresponding structures. In the first place, oxygen is more electronegative than sulphur, and simple ionic structures are therefore far more common in oxides than in sulphides. Thus simple symmetrical ionic structures often occur in oxides of composition AO and AO_2 but in sulphides are found only in the AS compounds of the most electropositive metals; symmetrical AS_2 structures are unknown (see tables 8.02 and 8.05). On the other hand, the sulphides show a range of more complex structures not represented among oxides, and often the sulphide structure is surprisingly complex when that of the corresponding oxide is simple, as, for example, in the pairs of compounds FeO and FeS, MoO_2 and MoS_2, SiO_2 and SiS_2, and others. Furthermore, many common sulphides have no counterpart among oxides, and vice versa. Thus FeS_2 is of widespread occurrence as the minerals pyrites and marcasite but FeO_2 does not exist; PbO_2 is a common oxide whereas PbS_2 is unknown.

At the other extreme, many sulphides show a marked resemblance to intermetallic systems, displaying variation in composition to a much more marked degree than the oxides and having the lustre and electrical conductivity characteristic of such systems. The selenides and tellurides show the same distinctive properties to an even more pronounced degree.

9 SOME STRUCTURAL PRINCIPLES

INTRODUCTION

9.01. In the preceding two chapters we have described the crystal structures of a considerable number of elements and simple compounds, and in the course of these discussions many of the principles determining the architecture of the solid state have implicitly emerged. It is now desirable, however, that we should consider some of these principles in a more explicit way, and this we do in the present chapter. Although there are as yet many structures still to be considered, we shall find that most of the ideas we wish to discuss can be well illustrated in terms of the simple structures already described, and that other more complex structures merely provide further illustrations of the same general principles.

STRUCTURES AS CO-ORDINATING POLYHEDRA

9.02. We have repeatedly seen that it is often convenient to describe the co-ordination about a given atom in a crystal structure in terms of a polyhedron at the corners of which the co-ordinating atoms are disposed; the structure as a whole can then be described by defining the way in which these polyhedra are packed together. The utility of this concept is restricted primarily to the simpler and more symmetrical structures, but among them it is a concept of considerable value for it often enables us to emphasize important features of the structural arrangement without giving a detailed description of the individual atomic positions. Indeed, even when these atomic positions are known the significance of the structure can rarely be appreciated until the form of the co-ordinating polyhedra has been described.

In essentially ionic crystals it is normally the co-ordinating polyhedra of anions about cations with which we are concerned since the anions are almost always the largest ions present and it is the co-ordination of them about the cations, rather than that of the cations about the anions, which is determined by the radius ratio; the co-ordination about the anions is governed by the number of cations available. In the structure of sodium chloride, for example, both Na^+ and Cl^- ions are admittedly octahedrally co-ordinated by ions of the other kind but it is more fruitful to picture the structure in terms of the co-ordinating octahedra of Cl^- about Na^+. It is this co-ordination which is limited to 6 by the radius ratio, whereas there is room around the Cl^- for a much larger number of Na^+ neighbours; it is only considerations of electrical neutrality which limit this number also to 6. The structure as a whole can then be envisaged as an assembly of $[NaCl_6]$ octahedra so disposed that every edge is common to two such octahedra (fig. 3.02). A second purely practical reason for considering primarily the co-ordination round the cations lies in the fact that in more complex ionic compounds (such as the complex oxides) two or more cations often occur, whereas more than one anion is rarely found. Each cation is therefore regularly co-ordinated by anions of only one kind whereas the anions are often irregularly linked to several different cations, as, for example, in the perovskite and spinel structures.

9.03. In simple covalent crystals the concept of co-ordinating polyhedra is no less important. In them, however, the form of the polyhedron about any atom is determined, not by geometrical considerations of atomic size, but by the covalency of that atom and the characteristic spatial distribution of the bonds which it forms. Thus in CuI the tetrahedral arrangement of iodine atoms about copper reflects the disposition of the four sp^3 hybrid bonds of the Cu^I atom, whereas in CuO the dsp^2 bonds of the Cu^{II} atom lie in a plane and the co-ordinating 'polyhedron' is accordingly a square. In such cases it is necessary to consider the form of the polyhedra about both types of atom, for they are no longer interdependent as in ionic crystals: in both CuI and CuO the non-metal atoms are tetrahedrally co-ordinated, and the completely different structures arise from the difference in the co-ordination about the copper atoms.

Electrostatic bond strength

9.04. The picture of simple ionic structures as consisting of polyhedra of anions round the cations enables a quantitative measure to be assigned to each bond. If we consider a structure in which a cation A, carrying a charge $+ze$, is co-ordinated by n anions X, then each A–X bond is said to have an *electrostatic bond strength* of z/n. Thus in sodium chloride each Na–Cl bond is of strength 1/6, in caesium chloride each Cs–Cl bond is of strength 1/8 and in fluorite each Ca–F bond is of strength $2/8 = \frac{1}{4}$; in the perovskite structure of $CaTiO_3$, where each Ca^{2+} ion is 12-

co-ordinated and each Ti^{4+} ion 6-co-ordinated, the strengths of the Ca–O and Ti–O bonds are $2/12 = \frac{1}{6}$ and $4/6 = \frac{2}{3}$, respectively. It must not, of course, be assumed that the electrostatic bond strength is a measure of the force between the ions, for this will depend not only on the charge but also on the interionic distance and is, for example, greater in the caesium chloride structure of CsCl (Cs–Cl = 3·56 Å) than in the sodium chloride structure of RbI (Rb–I = 3·66 Å). Nevertheless, the concept is one of considerable value, as we shall shortly see.

Pauling's rules for ionic structures

9.05. Certain general principles underlying the structures of ionic crystals, some of which we have already explicitly discussed and others of which have appeared implicitly in the structures we have considered, have been summarized by Pauling and codified in the form of five rules. Although these rules are for the most part the expression in formal terms of principles earlier recognized by Goldschmidt, their enunciation at this point serves as a convenient summary of these principles and affords an opportunity of recapitulating some of the most important structural features of a number of ionic structures described in the preceding chapter. It must be clearly understood that the rules apply only to structures in which all the bonds are predominantly ionic, and that they cease to hold in structures in which the bonds have any considerable degree of covalent or metallic character. Conversely, if the rules are found not to hold, then we have indirect evidence that the structure under consideration is not ionic.

9.06. THE FIRST RULE

A co-ordinated polyhedron of anions is formed about each cation, the cation–anion distance being determined by the radius sum and the co-ordination number of the cation by the radius ratio.

This rule is the expression of the fundamental principle of ionic structure building, namely, that in these structures the ions may be treated to a first approximation as rigid spheres in contact, each of characteristic and constant radii, and that the way in which these spherical ions are packed together is determined by their relative sizes.

9.07. THE SECOND RULE (the Electrostatic Valency Principle)

In a stable co-ordinated structure the total strength of the valency bonds which reach an anion from all the neighbouring cations is equal to the charge of the anion.

This rule is an expression of the tendency of any structure to assume a configuration of minimum potential energy in which the ions strive as far as possible to surround themselves by neighbours of opposite sign so that electrical charges are neutralized locally. If, as is often the case, the co-ordination round the cations in a structure can be predicted, the rule frequently enables that round the anions to be determined. Thus, to take the perovskite structure of $CaTiO_3$ once more as an example, we have in this structure Ca–O and Ti–O bonds of electrostatic valency strengths $\frac{1}{6}$ and $\frac{2}{3}$, respectively. If we now consider the oxygen ion it is clear that its total valency of 2 is satisfied if it in its turn is bound to four Ca^{2+} and to two Ti^{4+} ions.* The $[CaO_{12}]$ and $[TiO_6]$ co-ordinating polyhedra must therefore be so linked that each corner is common to four of the former and to two of the latter; this information in itself is sufficient to define the idealized structural arrangement uniquely. The second rule is not always rigidly obeyed, and it is clear, for example, that it cannot apply in detail in the defect spinel structures with vacant cation sites. Nevertheless, large departures from the rule are not to be expected and have never been observed.

9.08. THE THIRD RULE

The existence of edges, and particularly of faces, common to two anion polyhedra in a co-ordinated structure decreases its stability; this effect is large for cations with high valency and small co-ordination number, and is especially large when the radius ratio approaches the lower limit of stability of the polyhedron.

This rule arises immediately from the fact that an edge or a face common to two anion polyhedra necessitates the close approach of two cations, and a corresponding increase in the potential energy of the system as compared with the state in which only corners are shared and the cations are as far apart as possible. It is readily seen that the effect will be the more marked the higher the cation charge and the lower the co-ordination, as may be illustrated by a comparison of the caesium chloride, sodium chloride and zincblende structures. In caesium chloride each cation is surrounded by anions at the corners of a cube, and these cubes share faces; in sodium chloride the anions lie at the corners of octahedra, which share edges; and, finally, in the zincblende structure the anions are disposed at the corners of tetrahedra and these

* Other combinations of Ca–O and Ti–O bonds are, of course, mathematically possible but are structurally highly improbable.

tetrahedra share only vertices. When polyhedra are linked by common edges there is a tendency for these shared edges to be shorter than any which are unshared, the separation of the positive ions being in this way somewhat increased. Thus in the rutile structure of TiO_2 (§8.22) each titanium ion is octahedrally co-ordinated by oxygen but the co-ordinating octahedra are not perfectly regular because the two shared edges are slightly shorter than the remaining ten. There are, in addition, two other naturally occurring forms of TiO_2, namely, brookite and anatase. In both of these the co-ordination about the titanium ions is again octahedral, as would be expected from the radius ratio, but the structures differ in the way in which the octahedra are united: in brookite three and in anatase four of the twelve edges are shared. We would thus expect rutile to be the most stable and anatase the least stable of the three forms.

The third rule is also of considerable value in a converse sense in that an apparent violation of the rule is strong evidence that the structure under consideration is not ionic. Thus we have already argued that the pyrites structure of FeS_2 (§8.38) is covalent because in it the [FeS_6] octahedra are so linked that shared edges are actually longer than those unshared. Similar arguments apply to SiS_2 (§8.39). In all the forms of silica, which are at least partially ionic, the [SiO_4] tetrahedra share only corners, owing to the strong mutual repulsion between the Si^{4+} ions. In SiS_2, however, the corresponding tetrahedra share opposite edges. Such an arrangement is most unlikely in an ionic structure and we accordingly conclude that the Si–S bond is predominantly covalent.

9.09. The fourth rule

In a crystal containing different cations those of high valency and small co-ordination number tend not to share polyhedron elements with each other.

This rule is essentially a corollary to the third rule, and follows from similar arguments. Examples will be encountered later, but at this stage we may note that in perovskite the [CaO_{12}] polyhedra share edges whereas the [TiO_6] octahedra, in which the co-ordination is lower and the cations are more highly charged, have only corners in common.

9.10. The fifth rule

The number of essentially different kinds of constituent in a crystal tends to be small.

This rule implies that as far as possible the environment of all chemically similar atoms in a structure will be similar. Thus, even if the second rule admits several types of co-ordination about the anions, only one of these types of co-ordination may be expected to obtain and to be common to all anions. To illustrate this rule we may consider the structure of garnet. This mineral, the structure of which will be considered later (§ 11.12), in one form has the composition $Ca_3Al_2Si_3O_{12}$, and for our present purposes may be treated as an ionic crystal. The Ca^{2+}, Al^{3+} and Si^{4+} ions are 8-, 6- and 4-co-ordinated, respectively, by O^{2-} ions, so that the electrostatic bond strengths of the several bonds are as follows:

$$\text{Ca–O}: 2/8 = \tfrac{1}{4}; \quad \text{Al–O}: 3/6 = \tfrac{1}{2}; \quad \text{Si–O}: 4/4 = 1.$$

Let us now consider an oxygen ion. It is clear that its bond strength of 2 could be satisfied by a number of alternative combinations of bonds, but if we make the assumption that every oxygen ion has the same environment (and is therefore linked to cations of all three types) then it follows that only one of these arrangements is possible, namely, that in which each O^{2-} ion is linked to one Si^{4+}, to one Al^{3+} and to two Ca^{2+} ions:

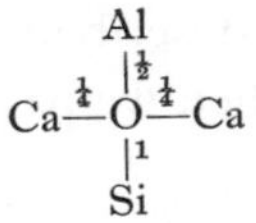

$$\begin{array}{ccccc} & & \text{Al} & & \\ & & |\tfrac{1}{2} & & \\ \text{Ca} & \overset{\frac{1}{4}}{—} & \text{O} & \overset{\frac{1}{4}}{—} & \text{Ca} \\ & & |1 & & \\ & & \text{Si} & & \end{array}$$

This arrangement is in fact found in garnet, but it should be noted that although all the O^{2-} ions are similarly co-ordinated they do not occupy geometrically equivalent positions. Nor is there any implication in the rule that they should do so; the rule requires only that the oxygen ions should be similarly co-ordinated, not that they should be structurally indistinguishable.

It must be admitted that this fifth rule is of somewhat limited value, and that in many structures the co-ordination round ions of a given type is of more than one kind. We have chosen as an example a silicate in which all the oxygen ions are similarly co-ordinated, but we shall find later that this is true of only a limited number of such compounds and that in the majority of silicates the oxygen ions are not all alike in respect of their environment. The same is also true of many other ionic structures.

CRYSTAL STRUCTURE AND MORPHOLOGY

9.11. Before the discovery of X-ray diffraction, crystals could be classified only on the basis of morphology, and in terms of their symmetry were assigned to one or other of the thirty-two classes of the seven systems. Such a basis for classification inevitably suffers from the limitation that it is restricted to substances which can exist as well formed crystals or whose symmetry can be determined by physical means, whereas, as we have seen, there are many substances which are crystalline in the widest sense but which rarely, if ever, can be grown as sizeable single crystals suitable for morphological study. The investigation of the internal structure of a crystal naturally provides material for a classification both more precise and more detailed than that based on external form alone, but, even so, many of the classical concepts, such as polymorphism, isomorphism and solid solution, which originally had a purely morphological foundation, are still of the utmost importance. It is therefore of interest to review some of these concepts in the light of the more recent work on structure analysis, and to interpret their significance in terms of the more detailed features of the structures involved. It is convenient to make such a review at this point in terms of structures already described in order that the results of our discussion may be available when we pass on to consider more complex structures later.

Polymorphism

9.12. A substance is said to be polymorphous when it is capable of existing in two or more forms with different crystal structures. We have already encountered numerous instances of this phenomenon as, for example, carbon, selenium, some of the metallic elements, zinc sulphide, ferric oxide, silica, and many others. In some of these examples one form alone is found under a given set of physical conditions and a reversible transition between forms is brought about by a change in these conditions, in which case the forms are said to be *enantiotropic*. Thus iron has the cubic close-packed structure between the temperatures 906 and 1401 °C, and the cubic body-centred structure at temperatures outside this range. Even so, the rate at which the transition takes place may vary between wide limits; at the one extreme it may be virtually instantaneous and at the other it may be so slow that a form is capable of indefinite existence under conditions in which it is, strictly speaking, only metastable. Carbon, for example, has survived as a mineral throughout geological times as both graphite and diamond. When the transition between forms is irreversible the forms are said to be *monotropic*.

The relationships between the polymorphous forms of different substances are so varied that any systematic classification is difficult. Nevertheless, it is convenient to recognize certain fairly well defined types, and we shall classify polymorphic changes under the following heads:

(*a*) Changes in which the immediate co-ordination of the atoms is not significantly altered.

(*b*) Changes in which a change in immediate co-ordination occurs.

(*c*) Changes involving a transition between 'ideal' and defect structures.

(*d*) Changes in which a change in bond type takes place.

We shall consider these classes separately, but it must be recognized that many changes are intermediate in character between one class and another, while others are such that they can be treated equally well as falling under more than one heading. Examples of these various types of polymorphic change are summarized in table 9.01, in which, for convenience of reference, some substances to be considered in later chapters have been included.

Table 9.01. *The classification of polymorphic changes*

(*a*) Changes without significant change in immediate co-ordination

Reconstructive

Ni, Se, zincblende ↔ wurtzite, quartz ↔ tridymite ↔ cristobalite, rutile ↔ brookite ↔ anatase

Displacive

α-quartz ↔ β-quartz, perovskites, resorcinol

(*b*) Changes with change in immediate co-ordination

RbCl, aragonite ↔ calcite, $CaTiO_3$, α-Fe ↔ γ-Fe

(*c*) Changes between 'ideal' and defect structures

HgCl, CH_4, NaCN, NH_4NO_3, AgI, Ag_2HgI_4

(*d*) Changes involving change in bond type

White Sn ↔ grey Sn, diamond ↔ graphite

Polymorphic changes without significant change in immediate co-ordination

9.13. Some substances already described which show polymorphic changes of this type are nickel, zinc sulphide, silica and titanium dioxide. Nickel exists in both cubic and hexagonal close-packed forms. We have already considered the close relationship between these arrangements and it is not surprising that energetically they differ little and that dimorphism should occur. Rather is it surprising that this type of dimorphism is not more common among the close-packed metals and that cobalt alone should show the defect structure (§7.21) in which cubic and hexagonal close-packing are combined. In the two forms of zinc sulphide (zincblende and wurtzite) the immediate environment of zinc and sulphur atoms is identical and the differences between the structures appear only when next-nearest neighbours are considered. Similarly, in all three forms of silica (quartz, tridymite and cristobalite) the silicon atoms are tetrahedrally co-ordinated by oxygen, and in the three forms of TiO_2 (rutile, brookite and anatase) the co-ordination about titanium is octahedral.

A number of polymorphous molecular structures also fall into this class. In the two structures of selenium already described (§7.06) the immediate environment of any one selenium atom is the same in both the infinite helical chains and the finite Se_8 rings, but in respect of the more remote environment the structures are entirely different.

9.14. Although the several forms of a given compound may differ little structurally and energetically, so that the formation of one or another when the crystal grows may be readily understood, it does not necessarily follow that ready interconversion of these forms in the solid state will take place. The process may involve the rupture and re-formation of interatomic bonds, and the activation energy of this operation may be too large to permit its easy accomplishment. Thus the structure of zincblende can be converted into that of wurtzite only by breaking some of the interatomic bonds, and the same is true of the interconversion of all the other structures given as examples in the previous section. In this sense these polymorphic changes may be described as *reconstructive* in that a reconstruction of the bond pattern is involved in the change. Such changes often take place only slowly or with reluctance, and show marked hysteresis if reversible.

9.15. On the other hand there also exist changes of a different type which may be described as *displacive* in that they involve only a small displacement of atoms without any rearrangement of the bond structure. An example of such a change is afforded by those perovskites which have the ideal structure at high temperatures but which transform to a slightly distorted form of this arrangement on cooling. Another example is quartz. We have explained (§ 8.33) that this is the variety of silica stable below 870 °C but that it exists in two slightly different forms. The α or low-temperature form, stable below 573 °C, has a rhombohedral structure with the silicon atoms arranged in the manner represented schematically in fig. 9.01 *a*. In the β or high-temperature form, stable

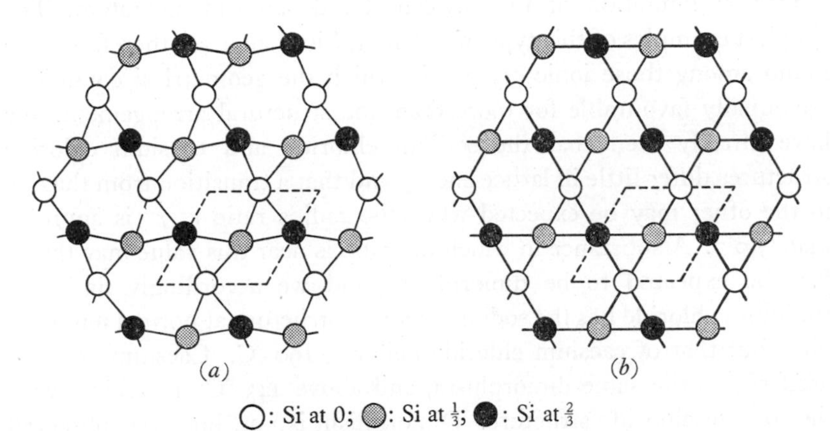

Fig. 9.01. (*a*) Plan of the rhombohedral structure of α-quartz, SiO_2, projected on a plane perpendicular to the principal axis. (*b*) Plan of the hexagonal structure of β-quartz projected on a plane perpendicular to the *z* axis. In both diagrams only silicon atoms are shown; the oxygen atoms are tetrahedrally disposed about those of silicon in such a way that all corners of the [SiO_4] tetrahedra are shared.

from 573 to 870 °C, these atoms are disposed more symmetrically to give the arrangement of hexagonal symmetry shown in fig. 9.01 *b*. In both forms the silicon atoms are, of course, tetrahedrally co-ordinated by oxygen, and it will be readily seen that the transformation from the one to the other can be effected by a small displacement of the silicon atoms without any disturbance of the Si–O bonds. Displacive changes of this kind usually take place readily and reversibly, and it is interesting in this connexion to contrast the rapid interconversion of the α and β forms of quartz at 573 °C with the very slow rate at which the quartz, tridymite and cristobalite structures transform into one another; although tridymite and cristobalite are metastable at normal temperatures both occur as minerals.

Displacive changes also relate the polymorphous forms of many organic compounds. Often in these forms the individual molecules are identical, and the structures differ only in the way in which these molecules are packed together under the influence of van der Waals forces.

Polymorphic changes with change in immediate co-ordination

9.16. In covalent structures changes in immediate co-ordination are not usually possible, owing to the characteristic spatial distribution and numerical limitation of the covalent bonds around any atom. The simplest examples of this type of polymorphic change are therefore to be found among those ionic crystals in which the geometrical conditions are equally favourable for more than one structural arrangement. We have already seen that the sodium chloride and caesium chloride structures differ little in lattice energy and that a transition from the one to the other may be expected when the radius ratio r^+/r^- is approximately 0·7. A substance in which the ratio is near this value may therefore be expected to be dimorphous, and we accordingly find that rubidium chloride has the sodium chloride structure at normal temperatures but that of caesium chloride below −190 °C. Caesium chloride itself shows the same dimorphism, and above 445 °C crystallizes with the sodium chloride structure. Dimorphism is not, however, observed in the alkali halides with small cations.

A further example of a compound in which the geometrical conditions are favourable for dimorphism is calcium carbonate. This compound occurs naturally as both calcite and aragonite, two distinct minerals with quite different structures (see §§ 10.10 and 10.11). The calcite structure is common to a considerable number of carbonates (and nitrates) in which the radius of the cation is less than about 1·0 Å, but when the radius exceeds this value the aragonite structure is found. Calcium carbonate itself is dimorphous because the radius of the calcium ion (0·99 Å) is close to this critical value, but dimorphism is not found in those carbonates in which the cation is either appreciably smaller or appreciably larger. The closely analogous case of the dimorphism of $CdTiO_3$ has already been described (§ 8.52).

Another example of polymorphic change involving a change in co-ordination is that between the cubic body-centred and cubic close-packed structures of α- and γ-iron. Although a change in co-ordination takes place, it should nevertheless be noted that the transformation can be achieved by a purely displacive mechanism, and accordingly takes place readily. In fig. 9.02 four unit cells of the cubic body-centred α-iron structure are shown, but in addition a tetragonal all-face-centred unit cell is indicated. If this tetragonal cell is now imagined to shrink laterally until it becomes a cube each iron atom acquires four additional neighbours and the cubic face-centred cell of γ-iron is obtained.* On the other hand the transition from the cubic body-centred to the hexagonal close-packed structure cannot be achieved by a simple displacement of atoms, and in this case a reconstructive process is involved.

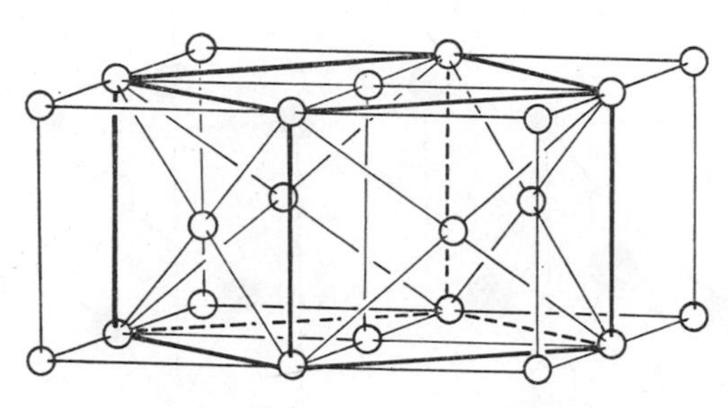

Fig. 9.02. Four unit cells of the cubic body-centred structure of α-iron showing the all-face-centred tetragonal unit cell in terms of which the structure may alternatively be described.

Polymorphic changes between 'ideal' and defect structures

9.17. The commonest polymorphic changes of this type are those associated with the onset of free rotation in a crystal structure. Many simple molecular compounds (HCl, HBr and CH_4 are examples) show a transition with rising temperature from a complex structure to a simple close-packed arrangement in which the molecules effectively acquire spherical symmetry by free rotation. Similar effects are displayed by a number of ionic crystals containing complex ions of unsymmetrical shape. Thus NaCN, KCN and RbCN have complex structures at low temperatures but transform at higher temperatures to the sodium chloride structure in which the CN^- ions behave as spherical entities. In some cases rotation may take place only about one axis, so that the molecule or group acquires cylindrical rather than spherical symmetry, and in other cases rotations about different axes may be excited successively at different temperatures. An extreme example of this is ammonium nitrate, in which both cation and anion are capable of

* In the same way, the caesium chloride and sodium chloride structures are related by a simple displacive mechanism. If the caesium chloride structure be suitably distended in a direction parallel to one of the cube diagonals each ion 'loses' two of its neighbours (those along the direction of distortion) while the remaining six assume positions at the corners of a regular octahedron.

rotation. No fewer than six different crystalline modifications, stable at different temperatures, are known. The high-temperature form, stable above 125 °C, has the caesium chloride structure in which both NH_4^+ and NO_3^- groups behave as spherical ions. The other forms, stable at lower temperatures, have more complex structures in which some or all of the rotational degrees of freedom of one or both ions are inhibited.

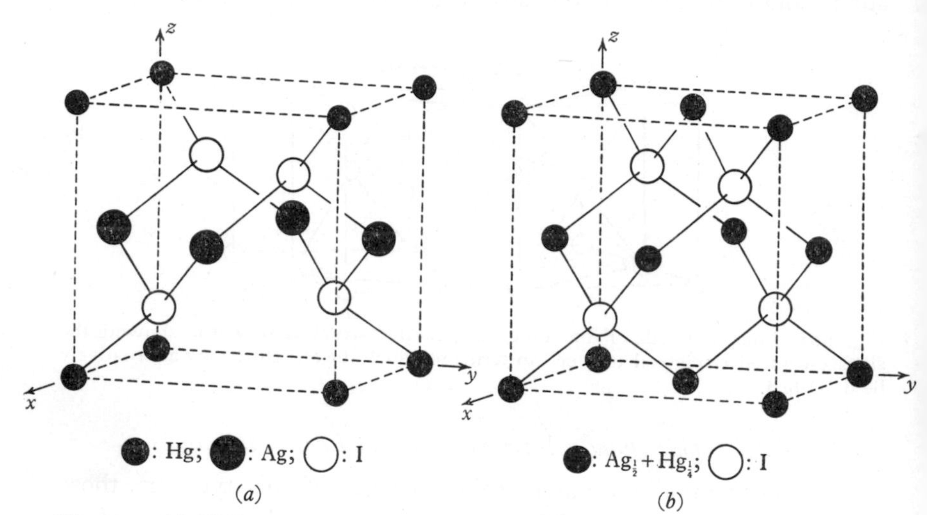

Fig. 9.03. (*a*) Clinographic projection of the unit cell of the tetragonal (pseudo-cubic) structure of the low-temperature β form of Ag_2HgI_4. (*b*) Clinographic projection of the unit cell of the cubic structure of the high-temperature α form.

9.18. A second type of polymorphic change involving a defect structure is that already discussed in the case of silver iodide (§8.05). Between 137 and 146 °C the β form of this compound has the wurtzite structure, but in the α form, stable from 146 °C to the melting point, the silver atoms are freely mobile in a cubic body-centred framework of iodine. A somewhat analogous effect is observed in Ag_2HgI_4. At ordinary temperatures the β form of this iodide has the tetragonal (but pseudo-cubic) structure shown in fig. 9.03*a*. The iodine atoms are arranged as in cubic close packing, in some of the tetrahedral interstices of which the Hg and Ag atoms are disposed in ordered array, the Hg atoms at the corners of the cell and those of Ag at the mid-points of its vertical faces. Above 51 °C, however, the α form has a strictly cubic structure (fig. 9.03*b*). In this the arrangement of the iodine atoms is as before, but now the Hg and Ag atoms are distributed in a disordered array over *all* the tetrahedrally co-ordinated interstices. There are, however, only three atoms available to occupy the four such sites in each unit cell, so that each of these sites must be regarded as statistically occupied by $[Ag_{\frac{1}{2}}+Hg_{\frac{1}{4}}]$. When viewed in this way the high-temperature structure may be regarded very simply as derived from that of zincblende by replacing S and Zn atoms by I and $[Ag_{\frac{1}{2}}+Hg_{\frac{1}{4}}]$, respectively. A marked increase in electrical conductivity takes place on transition from the β to the α form, and this argues that in the latter structure the Hg and Ag atoms can move readily between the sites which they occupy.

Although the polymorphic transformations between the α and β forms of AgI and of Ag_2HgI_4 are in many respects analogous there is the significant difference that in the former the pattern of iodine sites changes in the transformation whereas in the latter it does not. In consequence the transformation of AgI takes place abruptly at a characteristic temperature whereas that of Ag_2HgI occurs over an extended temperature range. This point is of considerable importance, for it implies that the transformation between the ordered structure of the β form and the disordered array of the α form takes place gradually and that intermediate conditions of partial order can exist. This type of *order–disorder transformation* is of especial importance and common occurrence in intermetallic systems and will be discussed in more detail when these systems are described (§§13.11–13.15).

Polymorphic changes involving change in bond type

9.19. Changes of this type are not common, and are necessarily confined to those structures in which alternative types of interatomic bonding force can operate. One example is the transformation between white and grey tin (§7.12), the former having metallic properties while the latter is a covalent crystal. Another is the transformation between the purely covalent structure of diamond and the molecular arrangement in graphite, in which the sheets of carbon atoms are held together only weakly by van der Waals forces (§§7.09, 7.10). Polymorphic changes of this kind usually take place only slowly and with difficulty. Thus white tin can exist indefinitely in a supercooled condition and graphite cannot, unfortunately, be readily converted into diamond.

9.20. We conclude our discussion of polymorphism with some brief general remarks. The physical factors most commonly involved in bringing about an enantiotropic polymorphic transformation are changes of temperature and of pressure. As far as the effect of pressure is concerned we generally find, as is to be expected, that an increase in pressure promotes a transition towards structures of higher density and higher co-ordination, as is shown by the simple examples given in table 9.02. The effect of temperature change is more complex, but if we confine our attention to transformations involving a change in immediate co-ordination we can say that the form stable at the higher temperature tends to be the structure of lower co-ordination and (if a change of symmetry takes place) of higher symmetry (see table 9.03). In simple ionic crystals a rise in temperature thus tends to bring about a structural transition in the same direction as would arise from the replacement of the cation by one of smaller radius, and a rise in pressure tends to raise the transition temperature.

A very full account of the structural aspects of polymorphism, including a discussion of the thermodynamics of the transformation processes, will be found in the work by Winkler cited in appendix 1.

Table 9.02. *The effect of pressure on structure*

	Low-pressure form		High-pressure form	
Substance	Structure type	Co-ordination	Structure type	Co-ordination
RbCl, RbBr, RbI	Sodium chloride	6:6	Caesium chloride	8:8
Cs, Fe	Cubic body centred	8	Cubic close packed	12
GeO_2	Quartz	4:2	Rutile	6:3

Table 9.03. *The effect of temperature on structure*

	Low-temperature form		Transition temperature (°C)	High-temperature form	
Substance	Structure type	Co-ordination		Structure type	Co-ordination
CsCl	Caesium chloride	8:8	445	Sodium chloride	6:6
RbCl	Caesium chloride	8:8	−190	Sodium chloride	6:6
Ti, Zr, Tl	Hexagonal close packed	12	—	Cubic body centred	8
$CaCO_3$	Aragonite	9*	—	Calcite	6*
KNO_3	Aragonite	9*	128	Calcite	6*

* This is the number of oxygen atoms co-ordinating the cation.

Isomorphism

9.21. Substances are said to be isomorphous when a close but inexactly defined crystallographic relationship exists between them. The concept was originally introduced in 1819 by Mitscherlich, who observed that certain compounds, closely related chemically, also formed crystals of the same or very similar habit, and who indeed assumed a close chemical analogy to be essential for analogy of crystal form. It soon became clear, however, that such a condition was not, in fact, necessary, and various supplementary tests for detecting isomorphism were proposed. The ability of two substances to form solid solutions and orientated overgrowths, or to precipitate each other from supersaturated solution, were applied as criteria, but such tests proved no more discriminating than the simple condition of crystallographic resemblance, and still revealed as isomorphous chemically unrelated bodies. Thus the silicates $CaAl_2Si_2O_8$ and $NaAlSi_3O_8$ form a continuous series of solid solution, and, again, sodium nitrate is isomorphous with calcite, $CaCO_3$, by which it is precipitated from supersaturated solution and on which it will form orientated overgrowths. On the other hand many substances now known to be structurally very closely related as, for example, sodium chloride and caesium fluoride, show no such tendencies.

Any attempt to define isomorphism on a purely morphological basis encounters the inevitable difficulty that the morphological description of a crystal in terms only of its external form is insufficiently discerning. It is, of course, necessary that isomorphous crystals should belong to the same class and, in all systems but the cubic, that the crystallographic parameters—the axial ratios and interaxial angles—should correspond closely. Such correspondences may, however, be fortuitous, and with a crystal of simple habit it may even be difficult to decide to which of the several classes of a given system the crystal belongs. In the cubic system there is the additional difficulty that all crystals of a given class are morphologically identical, so that on morphological grounds cubic crystals with quite different structures, such as a cubic close-packed metal, sodium chloride and fluorite, must be regarded as isomorphous.

9.22. When we turn to consider the internal structure of a crystal we can adopt a definition of isomorphism which is both more practical and more precise by regarding crystals as isomorphous only if they are composed of similar structural units similarly arranged. Such a definition of isomorphism is purely geometrical and does not demand any resemblance of chemical properties. It associates together as isomorphous substances those which have the same, or nearly the same, crystal structure, even though they are chemically quite unrelated and show no tendency to form mixed crystals or orientated overgrowths. Thus in our sense NaCl, PbS, and ScN, all with the sodium chloride structure, are isomorphous, whereas NaCl and CaF_2, of the same crystal class but with different structures, are not.

On the basis of this structural definition the conditions to be satisfied if two substances are to be isomorphous are that they should have analogous chemical formulae, that they should be composed of atoms or ions of comparable relative sizes and that the interatomic bonding forces should be of the same kind, since these are the factors which determine the appearance of a given structure type. Even though some analogy of chemical formula is demanded it must, however, be recognized that a wide latitude is admissible because the possibility of replacing a single atom or ion by a radical or complex group must also be entertained. Thus NaCl, NH_4I and KCN are isomorphous, and other more extreme examples of isomorphous crystals of seemingly very different chemical composition will be encountered later.

We proceed now to consider certain other relationships between crystal structures which are closely associated with isomorphism.

Anti-isomorphism

9.23. Two substances are said to be anti-isomorphous when their crystal structure are geometrically identical but with the positions of corresponding atoms or ions interchanged. We have already quoted Li_2O and Li_2S (§8.42), with the anti-fluorite structure, as examples. Anti-isomorphism is not common, and among simple ionic crystals is found only among those with the anti-fluorite structure. The reason for this is readily understood if we consider a hypothetical compound A_2X with the anti-rutile structure. The octahedral co-ordination of O^{2-} about Ti^{4+} in rutile demands that the radius ratio r^+/r^- should lie within the range 0·7–0·3, so that for an anti-rutile structure to be stable the ratio r^-/r^+ would have to fall between these same limits. There are, however, no pairs of ions whose radii meet this condition. On the other hand the condition for the appearance of the fluorite structure is $r^+/r^- > 0{\cdot}7$, and the corresponding condition for the anti-fluorite structure, namely $r^-/r^+ > 0{\cdot}7$, is easily satisfied.

Polymeric isomorphism

9.24. This term is used to describe a structural relationship, closely akin to isomorphism, in which the correspondence between the compounds considered is revealed in a comparison not between single unit cells of the structures but between equivalent groups of several unit cells. The isomorphism of zincblende, ZnS, and the mineral chalcopyrite,

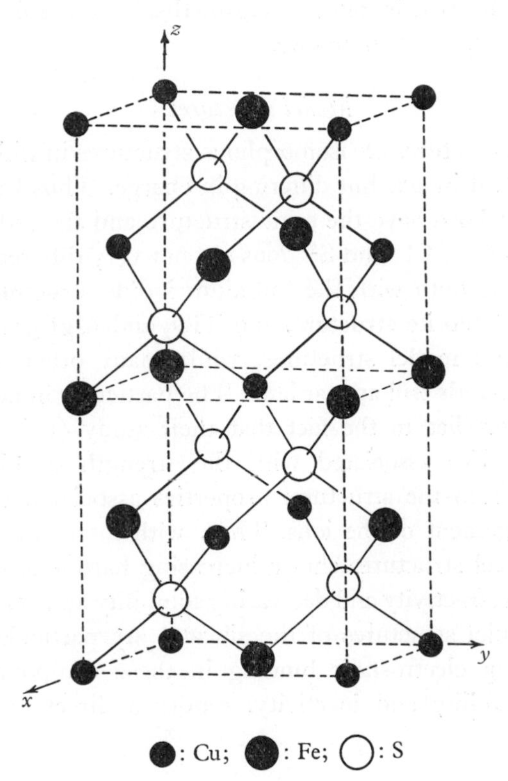

Fig. 9.04. Clinographic projection of the unit cell of the tetragonal structure of $CuFeS_2$.

$CuFeS_2$, is of this type. In the tetragonal structure of chalcopyrite (fig. 9.04) each Cu and Fe atom is tetrahedrally co-ordinated by four S atoms, and each S atom is tetrahedrally co-ordinated by two Cu and two Fe atoms. If the distinction between Cu and Fe is ignored the structure is seen to be identical with that of zincblende, the unit cell consisting of two zincblende cells superimposed. When this distinction is taken into account, however, the atoms at the corners of the sub-cell are no longer identical, and the larger cell becomes the true repeat unit. A closely analogous relationship exists between the structures of rutile, TiO_2, and $FeNb_2O_6$, the latter being identical with that of rutile if the Fe and Nb atoms are not distinguished. When this distinction is recognized the true unit cell is three times as large and corresponds to three cells of rutile superimposed.

Owing to the close relationship of cell dimensions, polymeric isomorphs often form orientated overgrowths. Solid solution between them, however, appears to be rare.

Model structures

9.25. Model structures are isomorphous structures in which ions occur closely equivalent in size but differing in charge. Thus CaS is a model of NaCl is that both have the same structure and the radii of the Na^+ and Ca^{2+} and of the Cl^- and S^{2-} ions are not very different. Similarly, $ZrSe_2$ and CdI_2, both with the cadmium iodide structure, ThO_2 and CaF_2, with the fluorite structure, and TiO_2 and MgF_2, with the rutile arrangement, are model structures, while many others among more complex compounds will appear later. The particular importance of the model structures lies in the fact that their study enables the purely physical properties associated with the strength of binding to be distinguished from the structural properties associated with the geometrical arrangement of the ions. Thus, with increasing ionic charge a series of model structures shows increasing hardness, melting point and molecular refractivity and decreasing solubility and general chemical reactivity. Model structures of the silicates are particularly valuable, since the strong electrostatic binding in these compounds, and the resulting insolubility and inactivity, render a direct chemical study difficult.

DEFECT STRUCTURES

9.26. In an 'ideal' crystal structure every unit cell is identical, and every crystallographically equivalent site in the unit cell is identically occupied.* Thus a single unit cell is representative of the crystal as a whole, and if the content of one unit cell is determined the structure of

* Crystallographically equivalent sites are those related by the operation of the symmetry elements of the space group to which the crystal belongs. If one such site is occupied by an atom of a given kind the symmetry requires that all the sites should be so occupied.

the whole crystal is known. In defect structures this is not so; it is now no longer possible to describe the structure in terms of a typical unit cell conforming strictly to the demands of symmetry, and at best all we can do is to identify a unit cell which is statistically representative. Defect structures are by no means uncommon, and in many cases possess important physical properties which can be directly attributed to the defects. Although we have already encountered a number of examples in passing, the interest of these structures justifies a more systematic review at this point.

Defect structures are of many different types and do not readily lend themselves to classification. We may, however, recognize two broadly distinct classes, namely:

(*a*) Structures in which some of the atoms are mobile and do not occupy definite sites.

(*b*) Structures in which all atoms occupy definite sites but in which the distribution of some of the atoms among these sites is statistical. Examples of these types of defect structure are summarized in table 9.04 and are discussed in detail below.

There are, in addition, quite other types of crystal defect which are best distinguished by being described as structural 'faults'; these we shall discuss separately (§§9.36–9.45).

Table 9.04. *The classification of defect structures*

(*a*) Structures with mobile atoms

Free translation

α-AgI, Ag_2HgI_4

Free rotation

NH_4Cl, NH_4Br, NH_4I, H_2, CH_4, HCl, HBr, NH_4NO_3, NaCN, KCN, RbCN, CsCN

(*b*) Structures with statistical distribution of atoms

Fully occupied sites

Co, $KLaTi_2O_6$, $MgFe_2O_4$, $KNdTi_2O_6$, Sr_2GaNbO_6, Sr_2CrTaO_6, $(BaK)(TiNb)O_6$

Partially occupied sites

Bronze, Fe_2O_3, Fe_xO

Structures with mobile atoms

9.27. Defect structures of this type may be further subdivided according as to whether the mobile atoms are free to move without restraint or are restricted in their motion to rotation about a fixed point.

Free translation

9.28. We have already given the high-temperature form of silver iodide, α-AgI, as an example of a structure in which the silver atoms are freely mobile (§8.05). In this structure the iodine atoms are disposed in a cubic body-centred arrangement and those of silver are presumably in the tetrahedrally co-ordinated interstices. There are, however, twelve such interstices in the unit cell, and the ability of the two silver atoms per cell to move freely between them is reflected in the high electrical conductivity of α-AgI compared with that of the β and γ forms. A similar state of affairs exists in the high-temperature form of Ag_2HgI_4 (§9.18). Here the three atoms 2Ag + Hg in the unit cell are statistically distributed over four equivalent sites, and, again, the high conductivity is to be attributed to free mobility between these sites.

Free rotation

9.29. Molecules or atom groups are in free rotation in the structures of many crystals, and in our discussion of polymorphism (§9.17) we have already given a number of examples of structures in which this phenomenon is observed. A group in free rotation about one or more axes effectively acquires higher intrinsic symmetry, but it should be noted that symmetry evidence alone is not unequivocal proof of the existence of free rotation: the apparent symmetry could be due to a distribution of fixed groups in random orientation. Thus it is by no means certain that the NH_4 groups in NH_4Cl, NH_4Br and NH_4I are in free rotation at very low temperatures in spite of the high symmetry of their structures. Nevertheless, there is in many structures ample supplementary evidence from other physical properties that free rotation does, in fact, take place.

From the physical point of view free rotation is merely an extreme case of the normal energy of thermal agitation. In all solids at temperatures above the absolute zero the molecules or groups of atoms possess energy of both translational and rotational vibration about the mean equilibrium position, and the amplitude of the rotational vibrations depends upon the temperature and the way in which the potential energy of the molecule varies with its orientation. If the potential energy in the equilibrium position is less than that in other possible orientations by an amount appreciably greater than kT per molecule, rotation will be confined to small oscillations; but if the energy difference is never as large as kT, the kinetic energy of the molecule will be sufficient to bring about complete rotation. We see from this simple picture that free rotation will be most common with symmetrically shaped molecules or groups of small moment of inertia in loosely bound structures at high temperatures, and that markedly unsymmetrical groups may have rotations about different axes excited successively at different temperatures. We also see that rotation is what may be described as a *co-operative phenomenon*, for if we consider a structure in which the molecules are in rotational vibration it is clear that the restraining forces acting on any one molecule are themselves largely determined by the amplitude of the rotations of its neighbours, and vanish when these rotations are sufficiently intense. Thus free rotation may be expected to set in rather abruptly over a quite narrow range of temperature.

From what has been said above it is clear that free rotation will be most common in molecular structures which contain small molecules, and indeed a quantitative treatment reveals that it may be expected in many such structures below their melting points. In hydrogen, free rotation of the H_2 molecule occurs at the absolute zero, and the crystal structure down to the lowest temperature at which it has been observed is hexagonal close packed. In methane, rotation sets in at about 20 °K, and the transition is accompanied by an anomalous increase in specific heat between 18 and 22·8 °K. Above the latter temperature the structure is cubic close packed. Similar transitions take place in HCl and HBr, the high-temperature forms of which have close-packed structures while at low temperatures more complex structures of lower symmetry obtain. Free rotation is naturally less common in molecular structures containing large molecules, but it is found in certain long-chain compounds, to be discussed later, in which rotation about the axis of the chain takes place (§ 14.22).

Rotation in inorganic compounds is necessarily confined to those structures in which discrete atomic groups are to be found; among the structures so far discussed the ions NH_4^+, CN^- and OH^- are examples. There are, however, many other inorganic structures of greater complexity, notably the salts of the common oxy-acids, which contain complex anions and in which rotation of these anions about one or more axes is often observed. These structures are discussed in chapter 10.

Structures with statistical distribution of atoms

9.30. Defect structures of this type may be conveniently further subdivided into two classes according as to whether crystallographically equivalent sites are or are not fully occupied.

Fully occupied equivalent sites

9.31. We have already discussed a number of structures of this type. In cobalt (§7.21), in one of its forms, successive layers of atoms are arranged sometimes as in hexagonal and sometimes as in cubic close packing. In the perovskite structure of $KLaTi_2O_6$ (§8.50) the K^+ and La^{3+} ions are distributed at random in equal numbers over crystallographically equivalent sites, and the same is true of the Mg^{2+} and one-half of the Fe^{3+} ions in the inversed spinel structure of $MgFe_2O_4$ (§8.55). There are, however, many other examples of such structures, particularly among intermetallic systems, and some of these will be described later. One further example, however, may conveniently be discussed here.

The compounds $LiFeO_2$ and Li_2TiO_3 both have the sodium chloride structure with a random distribution of the metal ions in the cation sites. These two compounds form solid solutions not only between themselves but also with MgO (which also has the sodium chloride structure), and the flexibility of this system is clearly due to the close similarity of the radii of the Li^+, Mg^{2+}, Fe^{3+} and Ti^{4+} ions. The structures are also of interest from the point of view of Pauling's electrostatic valency rule, for in the defect structure of, say, $LiFeO_2$ it is clear that electrical charges cannot be balanced locally in detail: in the neighbourhood of the Fe^{3+} ions there must be an excess, and in the neighbourhood of the Li^+ ions a deficiency, of positive charge.

Solid solutions, which provide another example of this and other types of defect structure, are of such importance and of such common occurrence that their properties are considered in more detail separately (§§9.33, 9.34).

Partially occupied equivalent sites

9.32. Examples of structures in which crystallographically equivalent sites are only partially occupied have also already been given. In the spinel structure of γ-Fe_2O_3 (§8.57) $21\frac{1}{3}$ iron atoms per unit cell are distributed at random over 24 equivalent positions, and in the perovskite structure of sodium bronze (§8.50) and in the sodium chloride structure of Fe_xO (§8.59) some of the cation sites are vacant. Such structures often show variable composition, for once the possibility of vacancies in the structure is accepted no particular significance is to be attached to the exact number of sites which are in fact unoccupied. In an ionic crystal all that is necessary is that electrical neutrality should be preserved, and in sodium bronze and in Fe_xO this is achieved by a change in the valency state of an appropriate number of tungsten or iron atoms. In the particular case of the bronze, any composition between $NaWO_3$ and WO_3 can be achieved.

Defect structures of this type are also commonly found in intermetallic systems and in solid solutions.

Solid solution

9.33. Many pairs of elements or chemical compounds are capable of crystallizing together in the form of a homogeneous solid solution. Thus from solutions containing KCl and KBr it is possible to grow homogeneous crystals of any composition between the two extremes, and these crystals all have the sodium chloride structure with the cation sites occupied by K^+ and the anion sites occupied by Cl^- and Br^- distributed at random. They are therefore to be regarded as defect structures.

The ability of two substances to form solid solutions is determined primarily by geometrical rather than chemical considerations. If the two structures are the same, and if the radii of the substituent atoms differ by not more than about 15 per cent of that of the smaller, a wide range of solid solution may be expected at room temperature. At higher temperatures a somewhat greater degree of tolerance is permitted, but, speaking generally, if the difference in radius exceeds this limit only restricted solid solution occurs. In the example just quoted the radii of the Cl^- and Br^- ions (1.81 and 1·95 Å, respectively) differ by only about 8·5 per cent and extensive solid solution is to be expected, but in other pairs of alkali halides in which the difference in radius is proportionately greater the extent of solid solution is correspondingly reduced. Thus KCl and KI are mutually soluble over only a limited range of composition, NaCl and KCl form solid solutions only at high temperatures, and LiCl and KCl are mutually insoluble under all conditions. On the other hand, NaCl forms a complete range of solid solution with the chemically less closely related AgCl on account of the close correspondence between the radii of the two cations.

The influence of geometrical considerations is similarly reflected in the common occurrence of solid solution among compounds with the perovskite, ilmenite and spinel structures, to which we have already referred. In these structures solid solution takes place particularly readily because in each case the structure is basically a rigid framework of large ions with small cations accommodated in the interstices. So long as neutrality is preserved, and so long as the radii of these cations lie within acceptable limits, it is a matter of little moment which particular cations they are. In intermetallic systems even the question of electrical neutrality does not arise, and between metals of comparable character and comparable atomic radius a wide range of solid solution is common. This is one of the factors which accounts for the peculiar complexity of such systems.

9.34. In all the examples so far quoted solid solution has been between two compounds with the same crystal structure. This, however, is by no means always the case, and many instances are known of partial solid solution between compounds with different crystal structures and also between compounds of quite different chemical composition. Even so, when the geometrical factors involved are taken into account the formation of such solid solutions can usually be readily understood, as we may demonstrate by some examples.

The system AgBr–AgI is an illustration of solid solution between compounds of similar composition but different structure. Silver bromide has the sodium chloride structure whereas silver iodide has the zincblende arrangement (at room temperature) owing to the predominantly covalent character of the Ag–I binding. Nevertheless, this structure can be little more stable than that of sodium chloride, as is indicated by the fact that AgBr can take up in solid solution as much as 70 per cent of AgI. On the other hand, AgBr shows little tendency to assume the zincblende arrangement, and only very limited solution of AgBr in AgI is possible. The closely analogous system AgBr–CuBr shows a different and even more interesting behaviour. Copper–halogen bonds are more covalent than the corresponding silver–halogen bonds, and in consequence the copper atoms in CuBr retain their characteristic tetrahedral co-ordination even in solid solution. As CuBr is taken up in AgBr the copper atoms do not occupy the sites of the silver atoms which they replace but instead enter previously vacant tetrahedral interstices between the bromine atoms, so that for every tetrahedral site thus occupied one empty octahedral site is created. The extent of the solid solution is limited, but if the process could be carried to completion we should have a continuous transition from the sodium chloride structure of AgBr to the zincblende structure of CuBr, the arrangement of the bromine atoms in both being identical.

As an example of solid solution between compounds which differ both in crystal structure and empirical formula we may consider the system $MgCl_2$–LiCl. The former compound has the cadmium chloride layer structure and the latter the sodium chloride arrangement, and a limited degree of solid solution of $MgCl_2$ in LiCl and of LiCl in $MgCl_2$ is possible. Although the cadmium chloride and sodium chloride structures are quite distinct there is, nevertheless, a close geometrical relationship between them. In both structures the Cl^- ions are arranged as in cubic close packing and in both structures the cations occupy octahedrally co-ordinated sites; in sodium chloride, however, all these sites are inhabited whereas in cadmium chloride it is only one-half which are filled. Starting, therefore, with $MgCl_2$ we can imagine the solution of LiCl as taking place by the successive replacement of Mg^{2+} ions by pairs of Li^+ ions, one of which occupies the site of the displaced Mg^{2+} ion and the other of which enters a vacant octahedrally co-ordinated interstice. At the other end of the series the solution of $MgCl_2$ in LiCl involves the removal of two Li^+ ions from the sodium chloride structure of LiCl and their replacement by one Mg^{2+} ion and one vacant site. If complete solid solution was possible we should have here a continuous transition between the cadmium chloride and sodium chloride structures.

A closely analogous state of affairs is seen in the systems $NiTe_2$–NiTe and $TiTe_2$–TiTe, in which the AB_2 compound has the cadmium iodide structure and the AB compound that of nickel arsenide. In both of these structures the B atoms are arranged as in hexagonal close-packing, but in other respects the relationship between them is precisely the same as that between the cadmium chloride and sodium chloride structures, so that solid solution can take place by the same mechanism. Examples such as this, and many others which could be quoted, emphasize that solid solution is in no sense a satisfactory criterion for isomorphism, and that substances may form solid solution even if their structures are formally quite different.

The chemical significance of defect structures

9.35. Apart from their crystallographic interest, defect structures are of importance in raising in acute form the question of what exactly we mean by a chemical compound and what significance is to be attached to the chemical laws of constant composition and simple proportions. The idea of a chemical compound has its foundation in the concept of the molecule, and it was originally believed that molecules existed in all compounds. We now know that this is not so, and that in the solid state molecules are found, broadly speaking, only in organic crystals and in a very limited number of other compounds. Many substances which exist only as solids never occur in molecular form at all.

In simple ionic structures, such as, say, that of sodium chloride, the conception of a definite compound is easy to understand, for in such structures the classical laws of chemistry arise directly from geometrical and electrical demands. We might thus be tempted to regard as a chemical compound any structure in which crystallographically equivalent positions are occupied by chemically identical atoms. When applied to defect structures, however, such a definition presents immediate difficulties. Thus it would demand that among the spinels only those with the 'normal' structure could be regarded as compounds, even though chemically no difference can be detected between these and the 'inversed' structures and the distinction can be made only by X-ray means. Again, is such a body as $LiFeO_2$ to be regarded as a compound, and, if so, why not also the solid solution of composition $NaCl+AgCl$ in which the distribution of the cations is exactly the same as in $LiFeO_2$? If, however, such structures are regarded as compounds we must clearly so regard all solid solution, for in a structure in which the cations are distributed statistically there can clearly be no especial significance in any particular composition. In $LiFeO_2$ the relative number of Li^+ and Fe^{3+} ions is determined by the demands of electrical neutrality, but in a solid solution of this compound with Li_2TiO_3 any relative proportion of Fe^{3+} and Ti^{4+} ions can be achieved. Nevertheless, there does not seem to be any valid reason why such a solid solution should not be regarded as being just as definite a chemical compound as its two components. Even in quite simple compounds similar difficulties arise, for 'ferrous oxide' (§8.59) has a variable composition which never conforms to the simple formula FeO. Other oxides and many sulphides, particularly those of the transition metals, show the same effect, and in intermetallic systems, where considerations of electrical neutrality do not arise, the flexibility is even greater. In the light of such arguments it seems clear that the laws of simple proportions and constant composition are valid only as applied to a limited class of chemical compounds, and that in many compounds they arise only as trivial and insignificant consequences of geometrical requirements rather than as profound and fundamental expressions of the laws of Nature. The very concept of chemical combination also needs careful interpretation in many cases. Sodium chloride differs from a mixture of sodium and chlorine not because sodium and chlorine atoms are 'combined', but because the components of the crystal are sodium and chlorine ions instead of sodium atoms and chlorine molecules. In ultimate analysis, however, it is of little moment whether or not a given body is described as a chemical compound provided that its structure is known and that the wider significance of the structure is appreciated.

STRUCTURAL 'FAULTS'

9.36. It is well known that many physical properties of crystals are 'structure sensitive' in the sense that they are profoundly dependent upon the previous history of the specimen, the presence of minute quantities of impurities, and other intangible factors. This is particularly true of the mechanical strength of crystals, which may vary by orders of magnitude from one specimen to another, of their luminescent properties and of the behaviour of semi-conductors. These properties are associated with 'faults' in the crystal structure which are 'defects' in the sense in which we have defined that term, but which differ from the defects so far discussed in that they occur relatively far apart in the structure; large numbers of unit cells have the regular structure of an ideal crystal and it is only here and there that this perfection is interrupted by a fault. There is a further and more practical reason for distinguishing between faults and defects in that the significance of faults is confined almost exclusively to the physical properties of crystals whereas defects, as we have seen, are of the utmost importance from a chemical point of view. The study of crystal faults and of the associated physical properties has been one of the most vigorously pursued and one of the most fruitful fields of investigation in recent years, but here, where our interest is primarily chemical, a very brief account must suffice.

Mosaic structure

9.37. The regular periodic structure of a crystal acts as a diffraction grating for X-rays, and, since the linear dimensions of any crystal of appreciable size are very large compared with the wavelength of X-rays, it would be expected that the resolving power of the grating would be very high. In practice it is found that the X-ray 'reflexions' from a crystal are far less sharp than such a simple picture would predict, and this leads to the view that an actual crystal possesses a 'mosaic structure' which may be likened to the arrangement of bricks in a crudely constructed wall. The crystal consists of blocks of linear dimensions of the order of 10^{-5} cm. within which the structure is perfect, but these blocks are not in perfect alinement with one another and may be misorientated by a few seconds or minutes of arc. The degree of misalinement determines the spread of the X-ray reflexions.

The mosaic structure of a crystal is intimately connected with its mechanical strength. If we consider the lattice theory of a simple ionic crystal, such as sodium chloride, it is easy to calculate the stress necessary to rupture the crystal by separating it into two halves against the forces of interionic attraction. Such calculations lead to estimates of the tensile strength which are hundreds or thousands of times greater than those actually observed. If, however, the crystal possesses a mosaic structure the mechanism of fracture will be different. The two halves of the crystal will not now be separated simultaneously at every point; instead there will be local stress concentrations at which the crystal will fail, the stress concentrations will then be transferred to other points and ultimately the crystal will break in two. The process may be likened to the tearing of a sheet of paper; it is not easy to sever a piece of paper by means of a uniformly applied stress, but if a tear is started the stress is concentrated at the end of the tear, failure at that point takes place and the tear is rapidly propagated across the sheet.

The mosaic structure of a crystal is one of its most profoundly structure-sensitive properties. As normally prepared, a crystal has a pronounced mosaic structure, but under conditions of more and more carefully controlled growth it is often possible to obtain crystals in which the degree of mosaic character is progressively reduced. As this process proceeds the X-ray reflexions become sharper and the mechanical strength increases. Conversely, by mechanical or thermal shock, it is often possible to reduce the degree of perfection of a carefully grown crystal.

Dislocations

9.38. The idea that a crystal possesses a mosaic structure was introduced soon after the discovery of X-ray diffraction. It is only in recent years, however, that the nature of the imperfections giving rise to this structure has become known, chiefly as a result of investigations into the mechanical properties of metal crystals.

We have already remarked that single crystals of the true metals are characterized by the ease with which they can be deformed by slip. Slip normally takes place only as between close-packed planes of atoms, and only in the directions of close packing in these planes, but, subject to these conditions, coherence of the structure is preserved across the slip plane in the deformed crystal. The critical shear stress required to initiate such a slip process is extremely small and cannot readily be understood in terms of any mechanism which involved the simultaneous mutual displacement of all the atoms in adjacent sheets. Let us suppose,

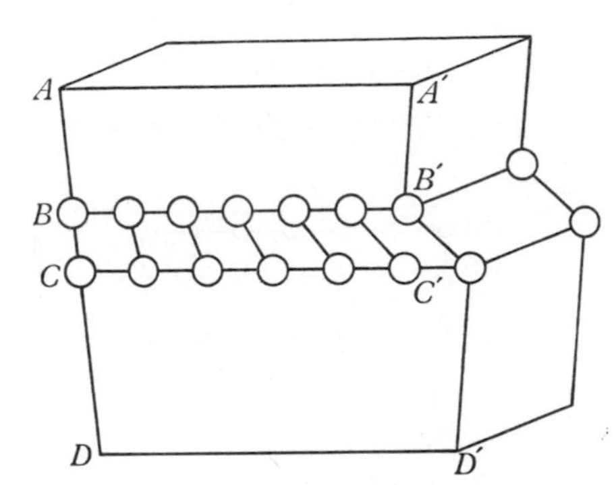

Fig. 9.05. Intermediate stage in the deformation of a crystal by slip.

however, that the atomic displacements occur not simultaneously but consecutively, so that at some intermediate stage in the process the crystal is as represented in fig. 9.05. The trace of the plane on which slip is taking place is BB', and at B' the atoms have been displaced by one atomic spacing. At B, however, no displacement has as yet taken place. If now we consider the pattern of atoms in the face $AA'D'D$ of the crystal it is clear that between B and B' there must be one more row of atoms than between C and C', so that the arrangement of atoms must be as represented schematically in fig. 9.06*b*. It will be seen that the resulting structure is 'faulted' and that the fault can be visualized as due to the insertion of an extra sheet of atoms in that portion of the crystal lying above the slip plane. Such a fault is termed an *edge dislocation*, and it is to be noted that in the neighbourhood of the dislocation edge the regularity of the structure is disturbed, but that at some distance from the dislocation the coherence of the structure across the slip plane is preserved. An essential feature of such a dislocation is that it is freely mobile, for if the strain in the crystal is increased the dislocation is displaced, as shown in fig. 9.07*c*, and ultimately passes out of the crystal (fig. 9.07*d*). On this picture the slip process takes place gradually, a row of atoms at a time, and we can readily understand in a qualitative way that such a mechanism of slip will require a much smaller shear stress than one in which all the atoms adjacent to the slip plane are displaced simultaneously. The process has been likened to that by which a heavy

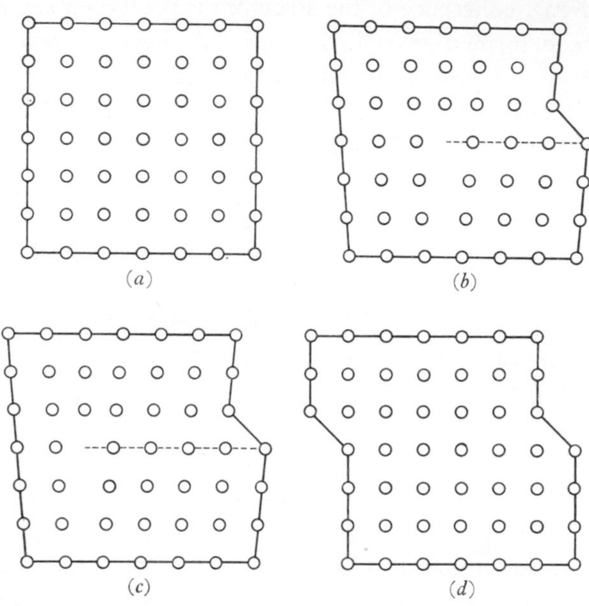

Fig. 9.06. Motion of an edge dislocation in a crystal undergoing slip deformation. (*a*) The undeformed crystal. (*b*, *c*) Successive stages in the motion of the dislocation from right to left. (*d*) The deformed crystal.

carpet may be moved by forming a 'ruck' and urging the ruck forward without at any time displacing the carpet as a whole.

The dislocation mechanism of slip propagation explains one important and hitherto surprising observation on the deformation of metal single crystals, namely, that slip does not occur between every pair of close-packed planes but only between occasional pairs of planes separated by some 10^4 atomic diameters. These privileged planes are those in which edge dislocations exist, and their separation in fact enables us to form an estimate of the density of these dislocations. Values of the order of 10^8 to 10^{12} dislocation lines per square centimetre, depending on the previous history of the crystal, are derived in this way, and correspond to a mean linear separation between dislocation lines of the order of 10^{-3} to 10^{-5} cm. This is in satisfactory agreement with the linear dimensions of the mosaic blocks estimated by X-ray diffraction. On an atomic scale the dislocations are separated by some 10^3 to 10^5 atomic diameters so that on this scale they are widely spaced, and large volumes of the crystal between the dislocations have a perfectly ordered structure.

In addition to the edge dislocation there is also another type, known as a *screw dislocation*, which plays an important part in the process of crystal growth.

Frenkel and Schottky defects

9.39. Many ionic crystals have an appreciable electrical conductivity in the solid state, due to the motion of anions or cations or both. We have already given an extreme example of this effect in the case of the high-temperature form of silver iodide (§8.05), where the silver atoms are

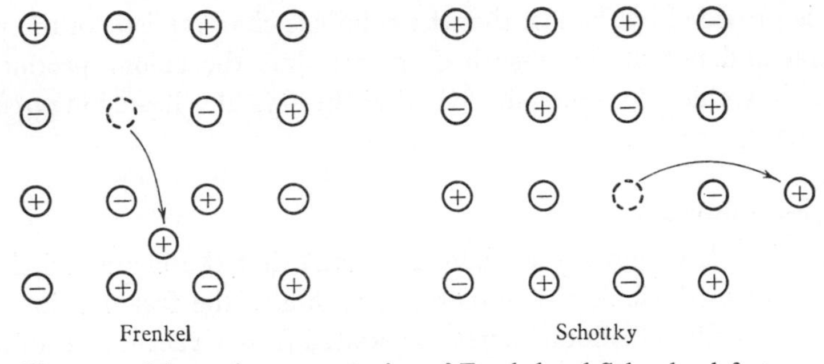

Fig. 9.07. Schematic representation of Frenkel and Schottky defects in an ionic crystal.

freely mobile. The effect, however, is not confined to such extreme examples, and is observed to a lesser degree also in other crystals such as those of the alkali halides. In these cases the conductivity is ascribed to Frenkel or Schottky defects, the nature of which is illustrated in fig. 9.07. The Frenkel defect involves the removal of an ion from its normal site in the structure and its accommodation elsewhere in an interstitial position. In the Schottky defect an ion is again removed from its normal site but is now transferred either to the surface of the crystal or to a dislocation. It is a characteristic of both, therefore, that vacant sites are created in the structure. The density of these vacant sites varies from crystal to crystal and in general increases with rising temperature. In silver chloride near the melting point the proportion of vacant sites is of the order of 1 per cent.

Colour centres

9.40. Another type of defect associated with vacant lattice sites is that responsible for the characteristic coloration developed by alkali halide crystals when heated in an atmosphere of the vapour of the alkali metal. Under these conditions the crystals develop a non-stoichiometric composition corresponding to an excess metal content of about 1 part in 10^4. It is believed that the metal atoms absorbed into the structure occupy (as ions) the normal cation sites and that the electrons derived from them are associated with the sites which would ordinarily be occupied by anions. Thus the departure from a stoichiometric composition should be described more properly as due to a deficiency of the halogen rather than to an excess of the metal. The electrons trapped in the vacant anion sites are described as F-centres, or colour centres, and it is the absorption due to them which is responsible for the distinctive coloration. Evidence for this is provided by the fact that the colour is characteristic of the parent crystal and not of the absorbed metal; thus the colour produced in potassium chloride is the same whether the crystal is heated in potassium or sodium vapour.

Semi-conductors

9.41. We have already explained (§5.22) that the characteristic properties of intrinsic semi-conductors are due to the fact that in them a completely filled Brillouin zone is separated from a vacant zone of higher energy by only a narrow energy gap (see fig. 5.17*f*); in such a crystal some electrons may possess sufficient energy of thermal agitation to surmount the gap and to enter the conduction zone, whereas in an insulator the gap is too wide to be bridged and the conduction zone is inaccessible. There is, however, another important class of *impurity semi-conductors* in which the semi-conducting properties are essentially connected with the presence of foreign atoms. Such crystals in the pure state are insulators, but the presence of foreign atoms, often in minute proportions, profoundly affects their conductivity; for example, the contamination of pure silicon by boron to the extent of only 1 part in 10^5 increases the conductivity of the silicon by a factor of 10^3.

The conductivity of impurity semi-conductors is to be attributed to the creation of new energy bands (due to the impurity atoms) lying between the filled Brillouin zone and the conductivity level of the parent insulator. Two alternative possibilities which can give rise to semi-conducting properties have to be considered (see fig. 9.08): either the impurity level lies close to the filled Brillouin zone, or it lies close to the conduction level. Both of these possibilities are realized in practice and give rise to two distinct types of impurity semi-conductor which can be best discussed further by specific reference to silicon and germanium, the semi-conductors whose properties have been most extensively studied.

Silicon and germanium

9.42. Both silicon and germanium have the diamond crystal structure, in which each atom is bound to four neighbours by covalent bonds. Since both elements are quadrivalent all the available valency electrons are utilized in the formation of these bonds, the first Brillouin zone is full and, as in the case of diamond, the crystal in the pure state is an insulator. Suppose, now, that a small quantity of arsenic is added to

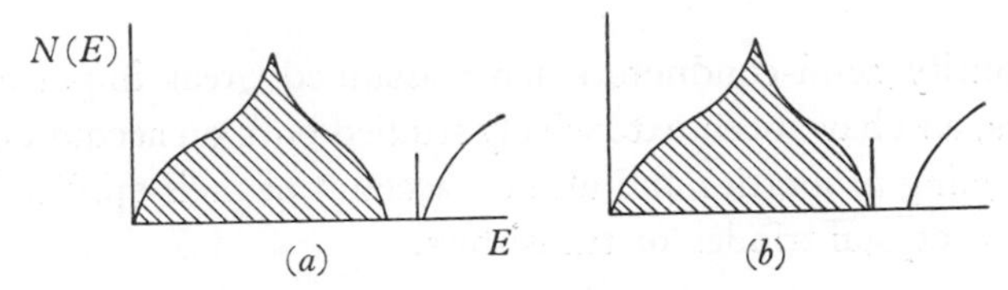

Fig. 9.08. $N(E)$ as a function of E, (*a*) for an *n*-type and (*b*) for a *p*-type semi-conductor.

a crystal of pure silicon. The arsenic atoms will replace those of silicon at random but, being quinquivalent, they will contribute one more electron per atom than is required for bond formation; for this reason they are termed *donor atoms*. The excess electrons thus provided will occupy the impurity energy level, and if this donor level is close to the conduction level (as shown in fig. 9.08*a*) promotion of these electrons into the conduction level will be possible. In this way semi-conducting properties will arise, and the conductivity will increase with temperature as more and more electrons are promoted. There may, however, be a limit to this process, for if the concentration of impurity is very small a stage may be reached at which all the available electrons have been promoted and after which the conductivity will fall again owing to the usual effect of thermal vibration. Semi-conductors containing donor impurities are termed *n-type* semi-conductors.

9.43. Let us now consider as an alternative the case in which a silicon crystal is contaminated, not by a quinquivalent atom, but by one which is trivalent, such as boron. Again the foreign atoms will replace those of silicon in the structure, but now there are insufficient electrons available to form the four B–Si bonds about each boron atom and these bonds can be formed only if the boron atoms play the role of *acceptor atoms* by acquiring electrons from elsewhere in the structure. This will be possible only if the acceptor impurity level lies close to the filled Brillouin zone (as shown in fig. 9.08*b*), but if this condition is satisfied the boron atoms will be able to acquire extra electrons by robbing the Brillouin zone. Once vacancies in this zone have been thus created conductivity will be possible. Again the conductivity will increase with temperature as more and more electrons are promoted to the impurity level, but again there may be a limit to this process when the valencies of all the impurity atoms have been satisfied. Semi-conductors containing acceptor impurity atoms are termed *p-type* semi-conductors.

9.44. Impurity semi-conductors have assumed great importance in recent years, and have been extensively studied both on account of their semi-conducting properties and also on account of their applications as rectifiers and crystal triodes or transistors.

9.45. The above discussion of defects and faults in crystal structures serves to emphasize that the character of a crystal is by no means completely specified when we know its chemical composition, the nature of the interatomic binding forces and the geometrical disposition of the atoms in the idealized crystal structure; a knowledge of the nature of the imperfections in the crystal is also essential. The imperfections which we have designated as 'faults' are of profound influence on the physical properties but are of limited chemical importance and will not be discussed further. Imperfections on a grosser scale, however, which we have termed 'defects', are of the greatest chemical significance and of them we shall encounter numerous further examples in the more complex structures to be considered later.

9.46. This concludes our discussion of some of the general principles underlying the internal architecture of crystals. As we have seen, it has been possible to illustrate these principles in terms of the relatively simple structures so far described. We now proceed to consider the crystal structures of compounds of greater chemical complexity. Such structures do, of course, display a number of features not hitherto encountered, but we shall find that these features can for the most part be interpreted in terms of the general principles already discussed, and that few further such principles will have to be invoked.

10 STRUCTURES CONTAINING COMPLEX IONS I

INTRODUCTION

10.01. In chapter 8 we discussed the structures of some compounds of composition $A_mB_nX_z$, and particularly of a number of oxides of this composition. Although these oxides included compounds which are commonly regarded as 'salts' we saw that in fact these compounds should more properly be considered as complex oxides in that the constituent atoms are all present in the ionized state and the structural arrangement is determined by the purely geometrical factors associated with the packing of ions of different sizes. In calcium 'titanate', $CaTiO_3$, with the perovskite structure, and in cadmium 'titanate', $CdTiO_3$, with that of ilmenite, there is no titanate ion TiO_3, and in fact in both arrangements the titanium ions are octahedrally co-ordinated by oxygen. It is a characteristic of these complex oxide structures that the atom B is not too strongly electronegative so that the B–O bond may be ionic in character.

We now have to consider a number of $A_mB_nX_z$ structures (and again we shall be mainly concerned with oxy-compounds) in which the B atom is less metallic and more strongly electronegative, so that the B–X binding can no longer be treated as primarily ionic. Such structures are of great importance, for they include almost all the salts of the common inorganic acids. In these structures, in marked contrast to the complex oxides, we find well defined $[B_nX_z]$ anions existing in the crystal as discrete complex groups and bound within themselves by forces very much stronger than those which bind them to the rest of the structure. Even when the other bonds are disrupted in solution, the complex group often preserves its identity and continues to exist as a single unit.

This picture of the complex group as a coherent entity often enables the structures of inorganic oxy-salts to be described very simply in terms of the disposition of these groups and of the remaining cations. Sometimes, as in the case of calcite (§10.10), this arrangement is found to simulate closely that observed in some very much simpler structure, and especially is this the case when the complex ion is in free rotation and so effectively acquires spherical symmetry. The structure can then be considered as a simple assemblage of spherical cations and anions. It is, however, important to realize that generally there will be a considerable disparity in the sizes of these two components and that simple structures can be expected only when this disparity is not too large. This point is emphasized in fig. 10.01, where the structures of a few complex anions (discussed in detail below) are represented on the same scale as some common cations and monatomic anions. It will be seen that it is only the relatively few large cations, such as Na^+, K^+, Ca^{2+} and Ba^{2+}, which are comparable in size with the anions, and that most of the common cations are far smaller. It is therefore not surprising that it is only the

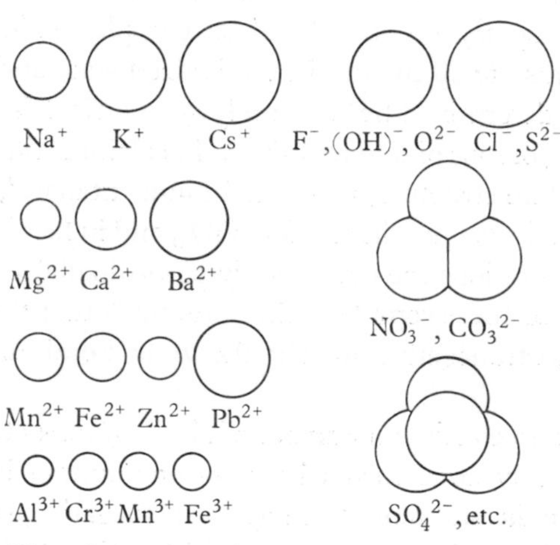

Fig. 10.01. The relative sizes of some common ions. Scale: 1 cm = 3Å.

salts of these large cations which have simple structures and that those of the other cations are more complex. This point has an important bearing on the formation of hydrates and we shall revert to it later when we come to discuss the structures of hydrated compounds in detail (§§ 12.20–12.38).

In structures containing complex ions the exact details of the structural arrangement are of less interest than the configuration of the complex group, for, while the former is necessarily characteristic of only a single substance, the latter is found to be practically the same in all structures in which the given group occurs. Here, therefore, we shall discuss primarily the structures of the complex groups of common occurrence in ionic crystals, and confine our description of the detailed features of such crystal structures to a very limited number of examples of particular interest. We exclude from our discussion at this stage the ionic salts of organic acids, since in them the configuration of the anions is peculiar to the particular acid concerned and few of the generalizations now to be considered are applicable.

THE STRUCTURE OF FINITE COMPLEX IONS

10.02. The majority of the complex groups which we shall have to discuss are the anions of simple inorganic salts, and consist of atoms such as C, N, P, S, Cl, Cr, Mn, As, Br, etc., co-ordinated by oxygen. A few ions of other types, however, and certain complex cations, also call for mention. In most of these groups the binding forces responsible for the coherence of the group are primarily covalent, but ionic binding often plays a significant part and in fact we shall find that the nature of the bonding in these groups is one of the problems on which the results of X-ray crystal structure analyses have thrown much light. For convenience we summarize in table 10.01 the configuration of a number of complex ions which we now proceed to discuss systematically.

Table 10.01. *The structures of some finite complex anions*

Type	Arrangement	Examples
XY	Linear	C_2^{2-}, O_2^{2-}, O_2^-, CN^-
XYZ	Linear	N_3^-, CNO^-, CNS^-, I_3^-, ICl_2^-
	Bent	ClO_2^-, NO_2^-
BX_3	Plane equilateral triangle	BO_3^{3-}, CO_3^{2-}, NO_3^-
	Regular trigonal pyramid	PO_3^{3-}, AsO_3^{3-}, SbO_3^{3-}, SO_3^{2-}, SeO_3^{2-}, ClO_3^-, BrO_3^-, IO_3^-
BX_4	Regular tetrahedron	SiO_4^{4-}, PO_4^{3-}, AsO_4^{3-}, VO_4^{3-}, SO_4^{2-}, SeO_4^{2-}, CrO_4^{2-}, ClO_4^-, MnO_4^-; $CoCl_4^{2-}$, BF_4^-; $Cu(CN)_4^{3-}$
	Distorted tetrahedron	MoO_4^{2-}, WO_4^{2-}, ReO_4^-
	Plane square	$PdCl_4^{2-}$, $PtCl_4^{2-}$; $Ni(CN)_4^{2-}$, $Pd(CN)_4^{2-}$, $Pt(CN)_4^{2-}$
BX_6	Regular octahedron	AlF_6^{3-}, TiF_6^{2-}, $ZrCl_6^{2-}$, ReF_6^{2-}, $OsBr_6^{2-}$, $PtCl_6^{2-}$, SiF_6^{2-}, GeF_6^{2-}, $SnCl_6^{2-}$, SnI_6^{2-}, SbF_6^-

Diatomic anions

10.03. Diatomic ions are necessarily linear, and some of those given in table 10.01 have in fact already been discussed in chapter 8. The cyanide ion $[C\equiv N]^-$ is often found in free rotation so that many cyanides of strongly electropositive metals have simple structures. Thus

NaCN, KCN and RbCN have the sodium chloride structure and CsCN and TlCN have that of caesium chloride. All of these cyanides, however, also have low-temperature forms in which the cyanide ion is no longer rotating, but these forms are closely related to the corresponding high-temperature forms from which they can be regarded as being derived by a slight distortion. Thus the rhombohedral structure of the low-temperature form CsCN (fig. 10.02) can be considered as derived from

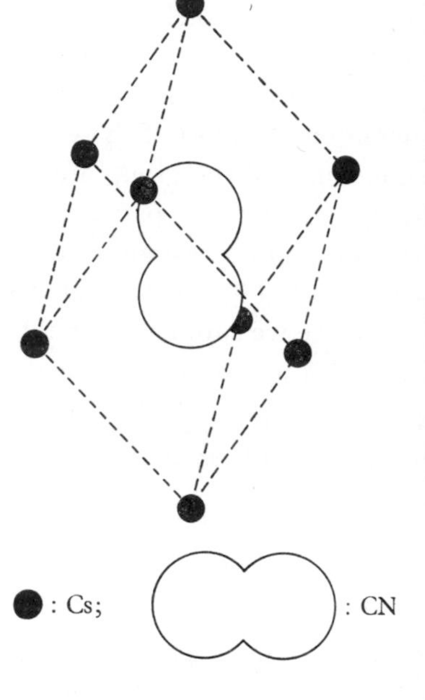

Fig. 10.02. Clinographic projection of the unit cell of the rhombohedral structure of the low-temperature form of caesium cyanide, CsCN.

the cubic caesium chloride arrangement by elongation along one of the cube diagonals in a direction parallel to the axis of the CN group. The carbide ion $[C{\equiv}C]^{2-}$ and the peroxide ion $[O{-}O]^{2-}$ do not rotate, but again simple structures closely related to that of sodium chloride are often found in compounds containing these ions (see §§8.34 and 8.40, and figs. 8.12 and 8.17). The distinction between the peroxides of di-valent metals, containing the ions O_2^{2-}, and the dioxides of quadrivalent metals, containing discrete O^{2-} ions, has already been emphasized.

10.04. The oxides KO_2, RbO_2 and CsO_2 crystallize with the same structure as CaC_2 and contain diatomic O_2^- ions. The structure of the superoxide ion, however, presents certain problems for it cannot be formulated in terms of simple electron-pair bonds since the number of electrons available is 13. We can, however, write the structure of the ion in the two equivalent forms

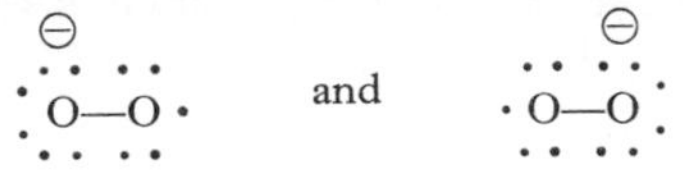

$$\overset{\ominus}{:\ddot{O}-\ddot{O}\cdot} \quad \text{and} \quad \overset{\ominus}{\cdot\ddot{O}-\ddot{O}:}$$

resonance between which gives a configuration

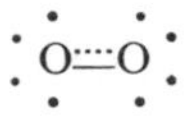

$$:\dot{O}\text{:::}\dot{O}:$$

where the bonding between the oxygen atoms is represented as due to one single bond and to one three-electron bond. The three-electron bond is not common but it is found also in nitric oxide, which may be formulated $:N\text{:::}O:$, and possibly in the molecule of oxygen itself, the paramagnetism of which cannot be explained in terms of the simple formulation $:\dot{O}=\dot{O}:$ and suggests the alternative structure $:O\text{:::}O:$. It is weaker than a normal two-electron bond, and accordingly we find that the interatomic distance of 1·28 Å in the superoxide ion is intermediate between those in the peroxide ion (1·49 Å) and in the oxygen molecule (1·21 Å).

Triatomic anions

10.05. Triatomic anions may be divided into two classes according as to whether their configuration is linear or bent.

Linear triatomic anions

10.06. The univalent ions N_3^-, CNO^-, CNS^-, Br_3^-, I_3^-, $ClICl^-$, $BrICl^-$, etc., are all linear, and many of their compounds with univalent metals have structures similar to those of the carbides, with the groups either all parallel to one another or all parallel to a plane but mutually inclined in two directions at right angles. The $[Ag(CN)_2]^-$ group in $K[Ag(CN)_2]$ is also linear.

Bent triatomic anions

10.07. A bent triatomic ion is found in the chlorite group, ClO_2^-, and in the nitrite group, NO_2^-, where the N–O bonds are inclined at an angle of 115°. This configuration is clearly not consistent with ionic binding within the groups (for if the bonds were purely ionic the oxygen atoms would be expected to be as far apart as possible) but reflects instead the characteristic spatial direction of covalent bond orbitals. The NO_2^- ion can be regarded as a resonance structure to which the configurations

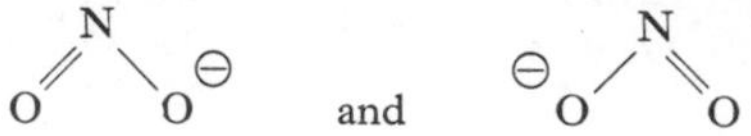

[structure: $O{=}N{-}O^{\ominus}$] and [structure: ${}^{\ominus}O{-}N{=}O$]

contribute equally, while the ClO_2^- ion can be written

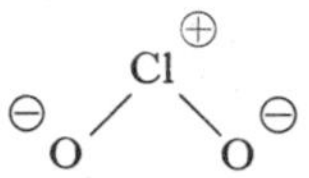

[structure: ${}^{\ominus}O{-}Cl^{\oplus}{-}O^{\ominus}$]

an arrangement which places a unit of formal positive charge on the chlorine atom and so confers some ionic character on the Cl–O bond.

A bent PO_2 group is found in ammonium hypophosphite, $NH_4H_2PO_2$, but this is due to the fact that the compound is not an acid salt but a neutral salt in which the radical must be more properly regarded as the tetrahedral $PH_2O_2^-$ group. This point is considered in more detail later (§12.18).

BX$_3$ anions

10.08. BX_3 ions fall into two classes according as to whether they have a plane or a pyramidal structure.

Planar BX_3 *anions*

10.09. The CO_3^{2-} and NO_3^- groups are the commonest ions with the plane structure, and in all the many carbonates and nitrates whose structures have been determined these groups have the same form, with the oxygen atoms symmetrically disposed at the corners of an equilateral triangle and with C–O and N–O distances of 1·31 and 1·21 Å, respectively. These distances are considerably shorter than the corresponding single C–O and N–O covalent bonds and are also much shorter than would be expected if the groups were ionic structures containing the ions C^{4+}, N^{5+} and O^{2-}. This fact, and the symmetry of the ions with respect to the three oxygen atoms, are therefore best interpreted in terms of a resonance structure to which, in the case of the carbonate ion, the three configurations

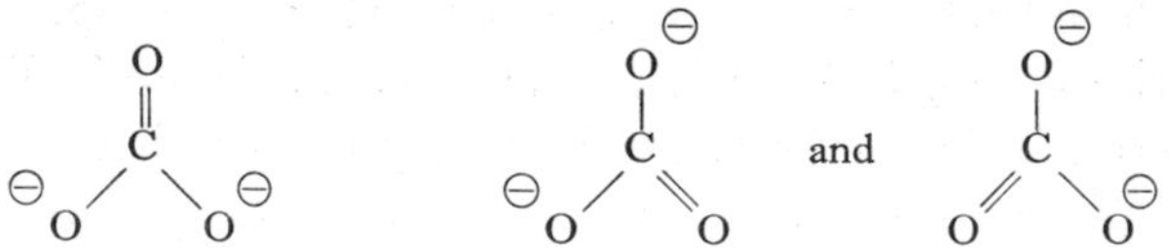

[structures: $O{=}C(O^{\ominus})_2$; ${}^{\ominus}O{-}C(O^{\ominus}){=}O$; and $O{=}C(O^{\ominus}){-}O^{\ominus}$]

contribute equally. In terms of this picture the C–O bonds are all equivalent and each possesses one-third double-bond character. For the nitrate ion the corresponding resonance structures are

and

and it will be seen that in this case a formal positive charge is placed on the nitrogen atom so that the bonding is partially ionic.*

Borates are far more complex in their structural chemistry than carbonates and nitrates, and for that reason are discussed separately later (§§ 11.02, 11.03). Here, however, we may remark that many orthoborates of composition ABO_3 contain discrete BO_3^{3-} ions with the same regular configuration as the CO_3^{2-} and NO_3^- ions and with a B–O distance of 1·37 Å.

We do not propose to describe in detail the structures of many crystals containing complex ions, for, as we have explained, the interest of such structures lies primarily in the configuration of the complex ion itself rather than in the detailed geometry of the atomic arrangement. It is, nevertheless, desirable to discuss a limited number of such structures in order that their general features may be appreciated, and at this point we may describe two of common occurrence among nitrates, carbonates and orthoborates.

* The reader may find it easier to understand these formulations if all the available valency electrons are explicity represented. In the CO_3^{2-} ion there are $4+3\times6+2 = 24$ electrons, and the first of the three configurations given above corresponds to the distribution of these electrons shown at (*a*). In this arrangement each atom has com-

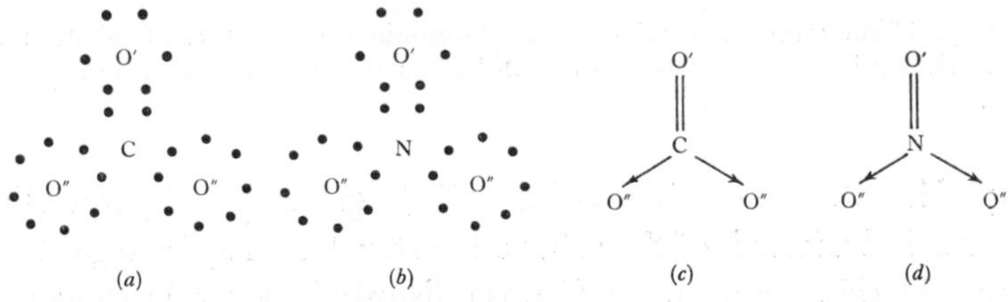

(*a*) (*b*) (*c*) (*d*)

pleted its octet, but the numbers of electrons effectively associated with the different atoms are as follows: C, 4; O′, 6; O″, 7. Thus C and O′ are neutral and the two O″ atoms each carry one unit of negative charge. Similarly, in the NO_3^- ion we have $5+3\times6+1 = 24$ electrons distributed as shown at (*b*). The numbers of electrons associated with the O′ and O″ atoms are as before, but now the quinquivalent nitrogen atom, with only 4 electrons, carries one unit of formal positive charge. The configuration (*a*) and (*b*) can alternatively be represented as in (*c*) and (*d*), respectively, where the charge distribution is implicit in the arrows and in the seemingly abnormal valencies of the atoms O″ and N.

The calcite and aragonite structures

10.10. The structure of calcium carbonate, $CaCO_3$, in the form of the mineral calcite has already been briefly described. The structure (fig. 10·03) is rhombohedral, and in terms of the cell indicated (which is not the smallest possible cell) it may be described as an arrangement in which Ca^{2+} ions occupy the corners and the centres of the faces of the cell while the CO_3^{2-} ions lie at the centre of the cell and at the mid-

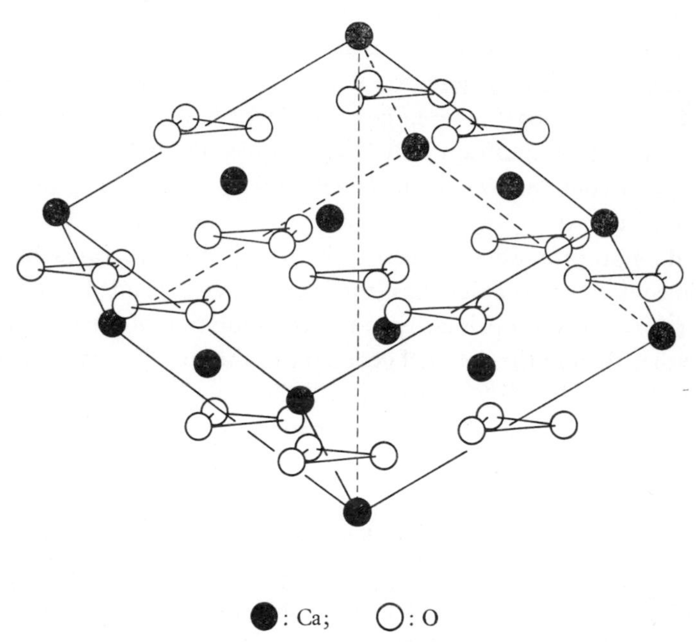

●: Ca; ○: O

Fig. 10.03. Clinographic projection of the rhombohedral structure of calcite, $CaCO_3$. The cell shown is not the smallest unit cell but corresponds to the cleavage rhombohedron.

points of its edges. The planes of the CO_3^{2-} group are all perpendicular to the vertical trial axis of the cell, so the structure may be regarded as a sodium chloride arrangement (§ 3.04) distorted by the introduction of the disk-shaped anions. Each Ca^{2+} ion is co-ordinated by six oxygen atoms of six different CO_3^{2-} groups, and each such oxygen atom is bound to two Ca^{2+} ions. If we wish to apply the electrostatic valency principle to the structure we must consider each oxygen atom of the CO_3^{2-} groups to be the seat of $\frac{2}{3}$ of a unit of negative charge. Since each of these atoms is bound to two Ca^{2+} ions the charge will be satisfied if each Ca–O bond is of strength $\frac{1}{3}$, but this is indeed the strength of these bonds, each calcium ion of charge 2 being co-ordinated by six oxygen neighbours.

10.11. The orthorhombic structure of calcium carbonate in the form of aragonite (fig. 10.04) is related to that of nickel arsenide in the same way as the calcite structure is related to that of sodium chloride; it may be regarded as a distorted NiAs structure (§ 8.12) with arsenic replaced by calcium and nickel by carbonate groups. As in calcite, all the CO_3

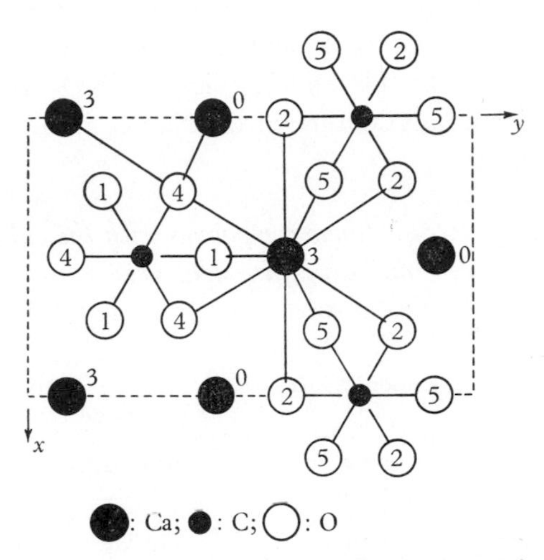

Fig. 10.04. Plan of the unit cell of the orthorhombic structure of aragonite, $CaCO_3$, projected on a plane perpendicular to the z axis. The heights of the atoms are expressed in units of $\frac{1}{6}c$.

groups are parallel, but the details of the co-ordination are different. Each calcium ion is now surrounded by nine oxygen neighbours, each of which in its turn is bound to three calcium ions. The electrostatic bond strength of each Ca–O bond is $\frac{2}{9}$ so that the total bond strength terminating on each oxygen atom is $3 \times \frac{2}{9} = \frac{2}{3}$ and the electrostatic valency rule is again satisfied in detail.

10.12. The calcite and aragonite structures are found in a large number of ABO_3 compounds, including the nitrates, carbonates and borates shown in table 10.02. Since the co-ordination of the cation is 6 and 9, respectively, in the two structures, it would be expected that the calcite arrangement would be preferred when the cation is small and that of aragonite when it is large, the transition taking place at a cation radius of about $0{\cdot}7r_O = 0{\cdot}98$ Å. It will be seen from the cation radii given in table 10.02 that this condition is, indeed, satisfied and that calcium carbonate is unique in having both structures because the radius of the Ca^{2+} ion is so close to this critical value. When the radius of the cation exceeds about 1·45 Å the aragonite arrangement in its turn becomes unstable, and $RbNO_3$ and $CsNO_3$ have other structures. There are no divalent or trivalent ions, however, sufficiently large for the corresponding transition to take place in any carbonates or borates. With cations which are not very strongly electropositive the rubidium nitrate structure is apparently not stable, and in spite of the close correspondence between the radii of the Tl^+ and Rb^+ ions, $TlNO_3$ has yet a different structure. Similarly $AgNO_3$ differs from $NaNO_3$ in that it does not have the calcite structure.

Table 10.02. *Some compounds with the calcite and aragonite structures*

Calcite structure		Aragonite structure	
Compound	Cation radius (Å)	Compound	Cation radius (Å)
$LiNO_3$	0·60	$CaCO_3$	0·99
$MgCO_3$	0·65	$SrCO_3$	1·13
$CoCO_3$	0·72	$LaBO_3$	1·15
$ZnCO_3$	0·74	$PbCO_3$	1·21
$MnCO_3$	0·80	KNO_3	1·33
$FeCO_3$	0·80	$BaCO_3$	1·35
$ScBO_3$	0·81		
$InBO_3$	0·81		
YBO_3	0·93		
$NaNO_3$	0·95		
$CdCO_3$	0·97		
$CaCO_3$	0·99		

10.13. Although the general principles of the structural chemistry of simple ionic compounds are by now sufficiently familiar to us it is fruitful at this point to consider the calcite and aragonite structures in rather more detail as examples of the application of these principles to structures more complex than those so far discussed. In the first place we see, as illustrated in the preceding paragraph, that geometrical rather than chemical considerations are the primary factor in dictating the structure adopted, so that $CaCO_3$ (as calcite) and $NaNO_3$ are isomorphous whereas the chemically more closely related $NaNO_3$ and KNO_3 have entirely different structures. Thus similarity in chemical formulae does not necessarily involve similarity in crystal structure, and the same point is illustrated even more emphatically by a comparison of the structures of $CaCO_3$ (in either form) with the quite different perovskite arrangement found in $CaTiO_3$.

A second feature of ionic structures containing complex ions is that, just as in simple ionic structures, all trace of any chemical molecule is lost. Thus in calcite each calcium ion is co-ordinated by six oxygen neighbours belonging to six different CO_3^{2-} ions, and each of these oxygen neighbours is bound to one carbon atom and to two calcium ions; and analogous relationships exist in the structure of $NaNO_3$. It is clearly impossible to represent this scheme of co-ordination within the framework of the material of a single molecule $CaCO_3$ or $NaNO_3$, and, in particular, any 'structural formulae' such as

Ca(—O—)(—O—)C=O or Na—O—N(=O)(=O)

are meaningless and misleading. Not only do they convey the impression that a single molecule has a real existence, but they also suggest that the two structures are quite different and that some physical distinction exists between one oxygen atom and the other two, implications all of which are false. Nor does the formula of calcium carbonate reveal that this substance is dimorphous and crystallizes with two quite different structures.

Structural formulae such as those which we have just quoted arise from an attempt to interpret ionic structures in terms of chemical principles which are strictly applicable only to covalent and molecular compounds. If a single molecule of calcium carbonate could exist, say as a gas, it might well have the structure represented above, but it is of little significance to discuss the structure of a compound in a state in which it does not occur. If any formula is employed to represent the structure in the solid state it must be one which simply expresses the co-ordination of each of the several atoms in turn, e.g.

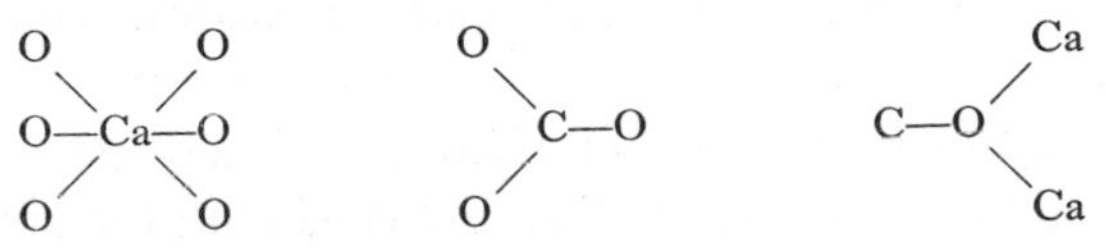

It must be remembered, however, that such a representation does not make any attempt to convey either the nature of the several bonds or their geometrical disposition, and that it is, in any case, characteristic not of the chemical compound but only of one particular structural modification of it: in calcium carbonate as aragonite the bond structure is entirely different.

Pyramidal BX_3 anions

10.14. The ions NO_3^-, CO_3^{2-} and BO_3^{3-} are the only common BX_3 ions with a planar structure. Other common ions of this composition, such as PO_3^{3-}, AsO_3^{3-}, SbO_3^{3-}, SO_3^{2-}, SeO_3^{2-}, ClO_3^- and BrO_3^-, have a low pyramidal structure in which the oxygen atoms lie at the corners of an equilateral triangle with the central atom displaced by about 0·5 Å from its plane. This configuration clearly reveals the influence of the covalent binding within the complex ion, although, as in some of the ions already discussed, the bonding may also be partially ionic. Thus the ion ClO_3^- may be formulated as (*a*), (*b*), or (*c*), with two units of formal positive

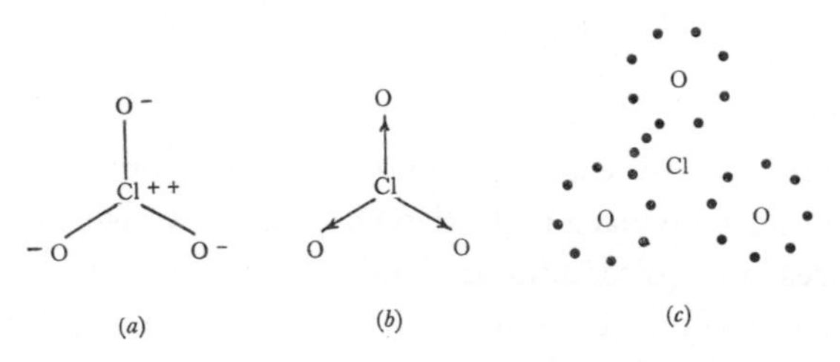

charge on chlorine atom and one unit of negative charge on each of the oxygen atoms. In (*c*) the 26 valency electrons available are explicitly shown and it will be seen that only six of the eight electrons surrounding the chlorine atom are involved in covalent bonds, the remaining two taking no part in the bonding. If these eight electrons occupy sp^3 hybrid orbitals the pyramidal form of the ion can be regarded as an arrangement in which the oxygen atoms occupy three of the four corners of a regular tetrahedron about the chlorine atom with the unshared pairs of electrons directed towards the fourth corner.

10.15. Structures containing pyramidal BX_3 ions are in general more complex than those in which planar ions are found, and will not be described in detail here. Even so, the differences are often not profound and can often be attributed to the lower symmetry of the complex group. Thus the monoclinic structure of $KClO_3$ is closely related to that of calcite and can be regarded as a distorted version of the latter structure in which the Ca^{2+} and CO_3^{2-} ions are replaced by the ions K^+ and ClO_3^-, respectively.

BX_4 anions

10.16. The BX_4 ions are the most stable and also the commonest of the complex ion groups. They fall into three classes according as the co-ordination round B is regular tetrahedral, distorted tetrahedral, or plane.

Regular tetrahedral BX_4 anions

10.17. The regularly co-ordinated tetrahedral ions are by far the most numerous and important, and include SiO_4^{4-}, PO_4^{3-}, SO_4^{2-}, ClO_4^{-}, CrO_4^{2-}, MnO_4^{-}, SeO_4^{2-} and BF_4^{-}. All these ions, with the exception of the coloured CrO_4^{2-} and MnO_4^{-} ions, are of substantially the same form and size with a B–X distance of about 1·5 Å, and, within limits, they may all be regarded as isomorphously replaceable. This is also true even of mixed complex ions such as PFO_3^{2-}, the salts of which are isomorphous with the corresponding sulphates. In such cases the isomorphism arises from the similarity in size of the two different kinds of X atom, which must be regarded as statistically distributed to conform to the tetrahedral symmetry. The chemically much more closely related sulphate and thio-sulphate ions, SO_4^{2-} and $S.SO_3^{2-}$, on the other hand, do not form isomorphous structures owing to the difference in size of the oxygen and sulphur atoms.

The bond structure of these BX_4 ions, in spite of the simplicity of their geometrical form, is by no means simple, and cannot be regarded as established beyond doubt. Let us take the SO_4^{2-} ion as an example. This ion can be formulated in a considerable number of different ways, some of which are shown at (*a*)–(*g*). Structures (*a*) and (*e*) are symmetrical and involve single and double S–O bonds, respectively, and

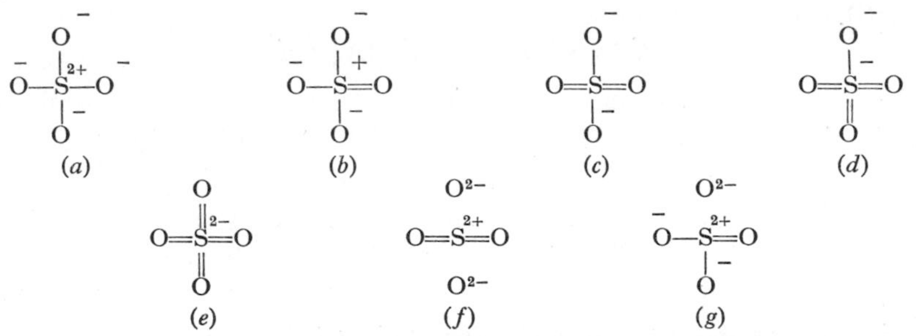

(*a*) (*b*) (*c*) (*d*)

(*e*) (*f*) (*g*)

formal charges of +2 and −2 on the sulphur atom. Structures (*b*), (*c*) and (*d*) are unsymmetrical, but if in each case resonance is assumed to take place between all the corresponding equivalent structures they give symmetrical configurations in which the degree of double-bond character in the S–O bond varies from 25 per cent in (*b*) to 75 per cent in (*d*). Structures (*f*) and (*g*) involve purely ionic bonding between the sulphur and one or more of the oxygen atoms. Structure (*a*) by itself is not satisfactory, for the observed S–O distance is considerably less than that corresponding to the known length of the S–O single bond. Beyond this, however, it is not possible to argue with certainty except to say that some or all of the other configurations must make some contribution to the structure of the ion.

The symmetrical form of the BX_4 ions considered above makes the structures in which they occur nearly isotropic in their optical and mechanical properties. A great number of different structures are found among crystals containing these ions but we shall not describe them in detail here. It is sufficient for our purpose to emphasize that, as always in ionic structures, it is primarily geometrical considerations which determine the structural arrangement adopted, and in this connexion it must be remembered that the BX_4 groups are generally very large entities compared with many of the cations of common occurrence. For this reason the geometrically simplest structures are usually found in salts containing large cations, such as NH_4^+, Rb^+, Cs^+, Sr^{2+} and Ba^{2+}, while the corresponding salts of small cations are often found to exist only as hydrates. We shall revert to this point in more detail later (§ 12.20).

Silicates are far more complex in their structural chemistry. Although there are some silicates in which isolated SiO_4^{4-} ions are found, and in which the role of these groups is analogous to that of the other BX_4 ions just described, these compounds are more conveniently discussed separately when the structures of other silicates are reviewed (§§ 11.04–11.38).

Distorted tetrahedral BX_4 anions

10.18. The most common BX_4 ions with distorted tetrahedral coordination are MoO_4^{2-}, WO_4^{2-} and ReO_4^-. The distortion takes the form of a compression along one of the diad axes which, if continued, would finally produce a flat square of X ions. Geometrically these ions are therefore intermediate in character between the tetrahedral ions which we have just described and those with a planar configuration which we are about to consider. Insufficient is known, however, about the stereochemistry of the central atom in these ions for a theoretical discussion of their configuration to be fruitful.

Planar BX_4 anions

10.19. A square BX_4 group is found in the ions $Ni(CN)_4^{2-}$, $PdCl_4^{2-}$ and $PtCl_4^{2-}$, and in its form reflects the characteristic planar configuration of dsp^2 hybrid bonds which we have already discussed in connexion with other compounds of palladium and platinum (§§8.10 and 8.28). The tetragonal structure of potassium chloroplatinite, K_2PtCl_4, is shown in fig. 10.05. Although it is not analogous to any known A_2X structure it is, nevertheless, a very simple arrangement in which each $PtCl_4$ group is surrounded by eight potassium ions nearly at the corners of a cube, and each potassium ion is co-ordinated by four $PtCl_4$ groups at the corners of a rectangle. K_2PdCl_4 and $(NH_4)_2PdCl_4$ have the same structure.

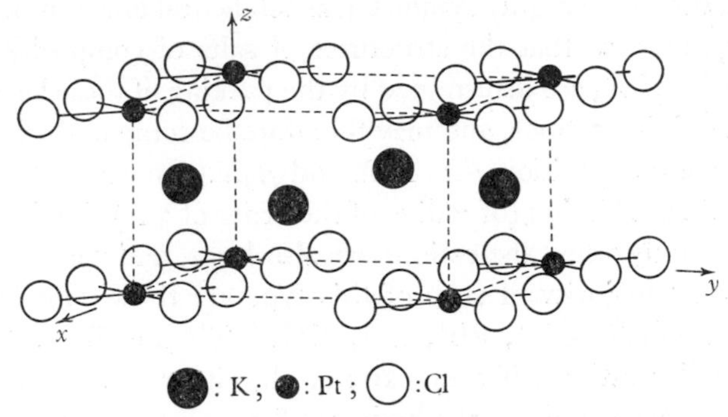

Fig. 10.05. Clinographic projection of the unit cell of the tetragonal structure of potassium chloroplatinite, K_2PtCl_4.

BX_6 anions

10.20. All the known complex BX_6 ions have a regular octahedral co-ordination, and the majority of them are halogen-containing groups such as AlF_6^{3-}, SiF_6^{2-}, $SnCl_6^{2-}$, $PtCl_6^{2-}$, $PbCl_6^{2-}$, etc.* In some of these ions the form of the group is a reflexion of the octahedral distribution of bonds characteristic of d^2sp^3 hybrid orbitals, but in others it is probable that bonds of a different kind are operative. There are also a number of BX_6 ions in which the X atoms are not all of one kind, but which, nevertheless, possess the same regular octahedral structure with the X atoms distributed statistically. Examples are the ions $NbOF_5^{2-}$ and $WO_2F_4^{2-}$, the salts of which are isostructural with the corresponding

* There are, of course, very many structures (such as the perovskite, ilmenite and spinel arrangements) in which octahedral co-ordination by oxygen is observed, but these are not structures of compounds which can be described as salts or in which the BO_6 group can be regarded as a discrete ion.

salts of the ions TiF_6^{2-} and $MoO_3F_3^{2-}$. A yet more striking example is provided by the salt $Rb_2CrF_5.H_2O$ and a limited number of other compounds of composition $A_2BX_5.H_2O$ which crystallize with the same structure as K_2PtCl_6 (described below, § 10.21). In these salts the water molecule and the five X atoms together octahedrally co-ordinate the B atom so that the water molecule is an essential constituent of the anion $[BX_5.H_2O]^{2-}$. This fact is concealed in the formula as given but can be expressed if the formulation $A_2B(X_5.H_2O)$ is employed.

10.21. The structures of a large number of substances containing BX_6 ions have been investigated, and furnish elegant illustrations of the principles of the crystal chemistry of ionic compounds. To a first approximation these highly symmetrical octahedral ions can be regarded as large spheres, so that the structures of salts of composition ABX_6, A_2BX_6 and A_3BX_6 are determined by the packing of these large spheres and the remaining cations, and may therefore be compared with those of substances of composition AX, A_2X and A_3X. On account of the very large size of the BX_6 ion (of radius of the order of 3 Å) simple structures are possible only when the cations are also large, and we therefore find that the great majority of salts of this type are those containing large positive ions such as K^+, Rb^+, Cs^+, NH_4^+, Tl^+ and Ba^{2+}. Salts containing smaller cations, if they exist at all, either have more complex structures or occur only in the hydrated form.

Among ABX_6 compounds we find counterparts of both the caesium chloride and sodium chloride structures of AX compounds and, as we should expect, the former obtains when the cation is very large and the latter when it is smaller. Thus $RbSbF_6$ has a deformed version of the CsCl structure, with Rb^+ and SbF_6^- ions each 8-co-ordinated by ions of the other kind, whereas $NaSbF_6$ has a deformed version of the NaCl structure in which the ions are 6-co-ordinated.

Many A_2BX_6 compounds have the cubic K_2PtCl_6 structure (fig. 10.06) when the A ion is sufficiently large. This is an anti-fluorite structure in that the K^+ and $PtCl_6^{2-}$ ions occupy the F^- and Ca^{2+} sites of the fluorite structure (fig. 8.04) so that these ions are 4- and 8-co-ordinated, respectively. Each K^+ ion is surrounded by twelve chlorine neighbours belonging to four different $PtCl_6$ groups and each chlorine atom is linked to four different K^+ ions. Thus the electrostatic valency rule is satisfied in detail. When the A ion is too small to form this structure other arrangements of greater complexity are found.

Compounds A_3BX_6 similarly show a resemblance to A_3X compounds in their structures when the A ion is sufficiently large.

Polynuclear finite anions

10.22. All the anions we have so far discussed have been of the form BX_z, containing a single atom B. There are, however, also polynuclear anions in which two or more B atoms occur, and these ions may be either finite or infinite in extent. For the moment, however, we shall

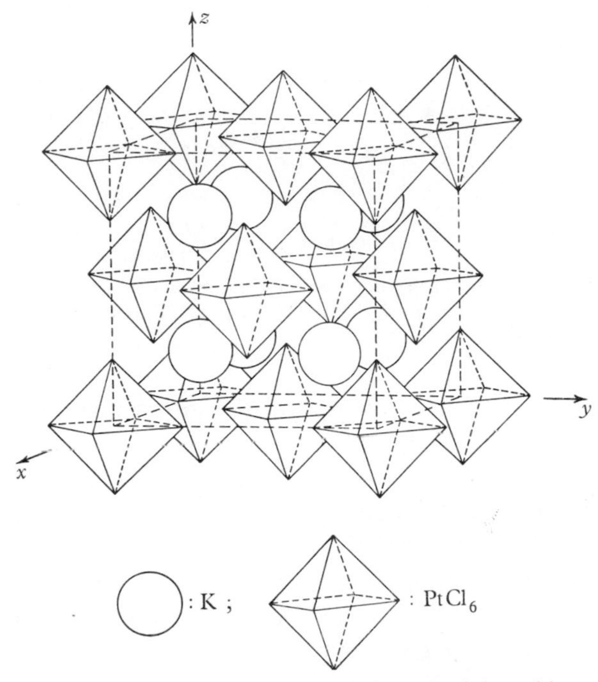

Fig. 10.06. Clinographic projection of the unit cell of the cubic structure of potassium chloroplatinate, K_2PtCl_6.

consider only those of finite size, but even with this limitation the number of possible configurations is large. The complex may contain several B atoms, and these B atoms may be bonded to one another or they may be united only through the medium of X atoms.

Sulphur oxy-ions

10.23. Among the most common of the polynuclear complex ions (apart from borate and silicate ions, which will be discussed later) are the sulphur oxy-ions. These include the metabisulphite ion, $S_2O_5^{2-}$, the thionate ions, $S_2O_6^{2-}$–$S_6O_6^{2-}$, and the perdisulphate ion, $S_2O_8^{2-}$, the structures of some of which have been established by crystallographic methods. The dithionate ion, $S_2O_6^{2-}$ (fig. 10.07*a*), consists essentially of two sulphur atoms each tetrahedrally co-ordinated by the other sulphur atom and by three oxygen neighbours. In the trithionate group, $S_3O_6^{2-}$ (fig. 10.07*b*), the co-ordination round the end sulphur atoms is again tetrahedral, but these atoms are now united through the central sulphur atom, the bonds to which are inclined at an angle of about 103°. In the perdisulphate group, $S_2O_8^{2-}$ (fig. 10.07*c*), on the other hand,

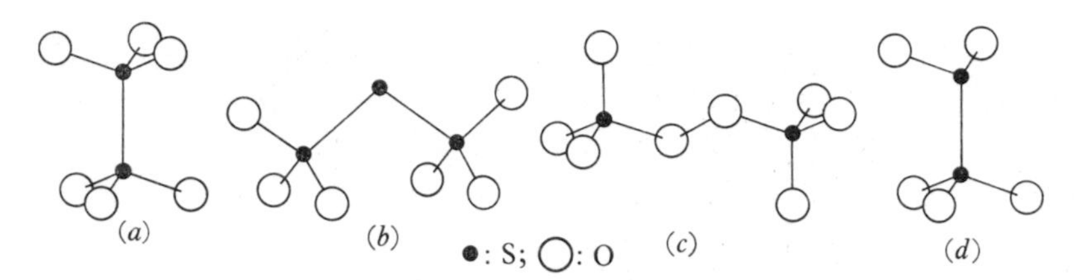

Fig. 10.07. The structures of some polynuclear sulphur oxy-ions. (*a*) The dithionate ion, $S_2O_6^{2-}$. (*b*) The trithionate ion, $S_3O_6^{2-}$. (*c*) The perdisulphate ion, $S_2O_8^{2-}$. (*d*) The metabisulphite ion, $S_2O_5^{2-}$.

linkage is through oxygen atoms, and the group may be regarded as consisting of two sulphate groups linked in this way. The metabisulphite ion, $S_2O_5^{2-}$ (fig. 10.07*d*), may be described geometrically as a dithionate ion from which one oxygen atom has been removed.

The 'kinked' structures of the trithionate and perdisulphate ions is clear evidence that the bonds within these groups are not ionic, and in fact both ions can be satisfactorily formulated in terms of a picture in which the binding throughout is by single covalent bonds:

```
    –     –                    –           –
    O     O                    O           O
 –  |2+   |2+  –            –  |2+         |2+  –
 O—S—S—S—O                  O—S—O—O—S—O
    |–    |–                   |–          |–
    O     O                    O           O
```

The S–S distances of 2·15 Å in the trithionate ion (and also of 2·01 Å in the dithionate ion) are in reasonably satisfactory agreement with the S–S distance of 2·04 Å in the S_8 molecule of elementary sulphur, in which, also, the sulphur atoms are linked by single covalent bonds.

The ions of the polyacids

10.24. Finite polynuclear anions of far greater complexity are found in the extensive series of salts of the iso- and hetero-polyacids. The salts of the isopolyacids contain ions $[N_yO_z]$, where N is W, Mo or V, and y and z are large; an example of such an ion is $Mo_7O_{24}^{6-}$. The salts of the heteropolyacids contain ions $[M_xN_yO_z]$, where M can be one of many elements (B, Si, P, As, S, Te, I and others), N is again W, Mo

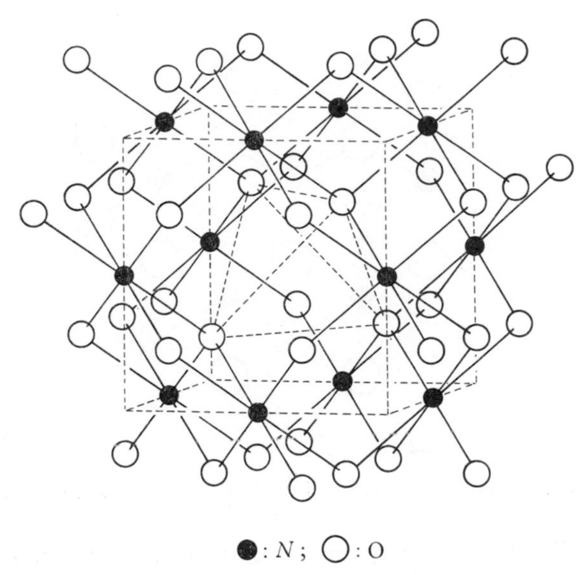

Fig. 10.08. Clinographic projection of the structure of the 12-isopolyacid ion $[N_{12}O_{40}]$. The cube is drawn to emphasize the disposition of the NO_6 octahedra; it does not, of course, represent a unit cell. The tetrahedrally co-ordinated interstice at the centre of the ion is indicated.

or V, and y/x varies between 6 and 12. A few examples of the ions are $TeMo_6O_{24}^{6-}$, $As_2Mo_{18}O_{62}^{6-}$ and $PW_{12}O_{40}^{3-}$. These compounds have been the subject of very extensive chemical investigations but only a limited number have been studied by X-ray methods. It now seems clear, however, that the essential feature of the structure of the complex ion is an association of NO_6 octahedra, which, by sharing edges or vertices, form a finite cage or basket. In the heteropolyions an interstice in the cage accommodates the M atom.

These principles are illustrated by the structures of the two ions shown in figs. 10.08 and 10.09. Fig. 10.08 represents the ion $[N_{12}O_{40}]$, which consists of twelve NO_6 octahedra centred about the mid-points of the edges of a cube and packed together in such a way that each octahedron shares two edges and also two vertices with neighbouring octahedra. At the centre of the ion is a tetrahedrally co-ordinated interstice which is vacant in the ions of the 12-isopolyacids (i.e. those with $y = 12$) but is occupied by the M atom in the ions of the corresponding heteropoly-acids (i.e. those with $y/x = 12$).

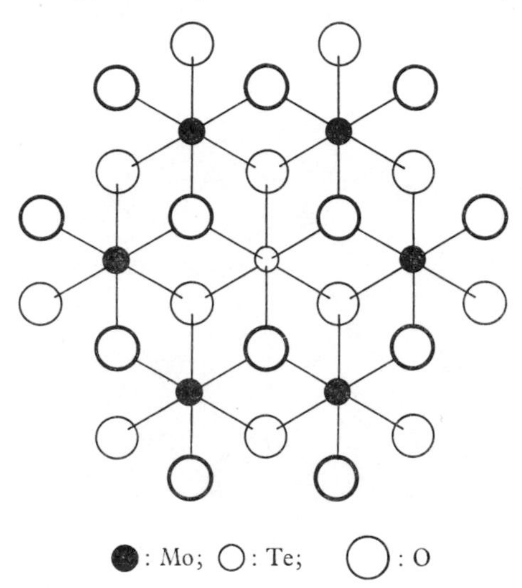

Fig. 10.09. Plan of the structure of the 6-heteropolyacid ion $TeMo_6O_{24}^{6-}$.

The ion $TeMo_6O_{24}^{6-}$ has the structure shown in fig. 10.09. Here the MoO_6 octahedra are linked by sharing two edges (but no other vertices) into a hexagonal ring. The interstice in this ring, in which the Te atom is located, is now octahedrally co-ordinated, and this distinction between the co-ordination of the interstices in the 6- and 12-acids is reflected in the fact that the former are known only with M = Te or I whereas in the latter M can be any of ten or more different elements. The ion $Mo_7O_{24}^{6-}$ has a structure similar to that of $TeMo_6O_{24}^{6-}$; it should therefore be more properly formulated as $Mo.Mo_6O_{24}^{6-}$ and regarded as the ion of a 6-heteropolyacid in which, by chance, the M and N atoms are the same.

Complex cations

10.25. By comparison with the complex anions, the number of complex cations is very small and is practically limited, among simple compounds, to the ammonium ion and its derivatives, and to the complex cations, such as $Ni(NH_3)_6^{2+}$, etc., which occur in co-ordination compounds. These latter compounds are closely related to the hydrates and are most conveniently considered under that head (§12.39).

The structures of a number of ammonium salts have already been described. In many of them the NH_4^+ ion is in free rotation and can be treated as a spherical ion of radius 1·48 Å so that the structures observed are determined by the usual geometrical considerations. Thus most of the ammonium halides have either the sodium chloride or the caesium chloride structure. In ammonium fluoride, however, the cation is not in free rotation and the wurtzite structure results.

FREE ROTATION OF COMPLEX GROUPS

10.26. Rotation of complex ions is of frequent occurrence, and structures are known in which, at sufficiently high temperatures, the ions NH_4^+, CN^-, CO_3^{2-}, NO_3^-, SO_4^{2-}, etc., are freely rotating. As we have already seen, rotation is likely to take place most readily in groups of pseudo-cylindrical or pseudo-spherical form, and in the former case to be restricted to rotation about the axis of pseudo-symmetry. Thus in the case of ions such as CO_3^{2-}, NO_3^-, etc., rotation is usually possible only about the triad axis, and in consequence such rotation often sets in without any increase in the symmetry of the structure as a whole; in sodium nitrate, for example, with the calcite structure, the transition to the rotating form occurs gradually over an extended temperature range without alteration in the structure but with a small change in the angle of the rhombohedral unit cell. On the other hand, groups of more nearly spherical form, such as the NH_4^+ group, can rotate about any axis, and the same is occasionally true even of less highly symmetrical groups if the temperature is sufficiently high and the bonding not too strong. The particular case of ammonium nitrate, in which both ions are capable of free rotation and in which six polymorphs correspond to the successive excitation of the various rotational degrees of freedom of the two ions, has already been described (§9.17).

11 STRUCTURES CONTAINING COMPLEX IONS II

INTRODUCTION

11.01. In chapters 8 and 10 we have discussed the structures of some compounds of composition $A_mB_nO_z$, and we have seen that these may be broadly divided into two classes. On the one hand we have complex oxides, in which B is an electropositive element and in which the binding forces are predominantly ionic throughout the structure. On the other hand we have the salts of the inorganic acids, which are characterized by the existence of discrete complex anions bound to the cations A by ionic bonds but bound within themselves by covalent bonds. These complex anions are generally relatively simple in form, and although occasionally polynuclear (as in the thionates) they are always finite in extent.

We now turn to consider compounds of the same general composition $A_mB_nO_z$ but in which the B atom is neither strongly electropositive nor strongly electronegative; in particular we shall discuss those compounds in which this atom is boron, silicon, phosphorus or germanium. The oxy-salts containing these atoms have long been noted for the peculiar complexity of their chemistry, and structural studies have revealed that this complexity arises from the fact that in them anions of many different forms can occur. This is not to imply that simple anions are not found, for there are borates and silicates containing the ions BO_3^{3-} and SiO_4^{4-} just as there are nitrates and sulphates containing the ions NO_3^- and SO_4^{2-}. What is relevant, however, is that in borates and silicates not only BO_3^{3-} and SiO_4^{4-} ions but also many others of far greater complexity are of common occurrence. We can best illustrate this point by considering the various salts separately, and we shall begin with a discussion of the borates. These salts are less important than the silicates, and have been less extensively studied, but they serve well as a relatively simple introduction to the principles which we wish to exemplify.

BORATES

11.02. The boron atom is normally trivalent, the bonds involved being sp^2 hybrids directed in a plane towards the corners of an equilateral triangle. The simplest oxy-anion is therefore BO_3^{3-}, which may be formulated*

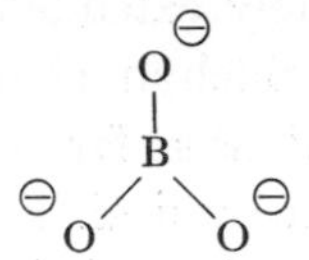

⊖
O
|
B
⊖ O / \ O ⊖

Since, however, the oxygen atom can form two bonds it is possible for two such groups to be linked together, by sharing a common oxygen atom, to give the polynuclear ion $B_2O_5^{4-}$:

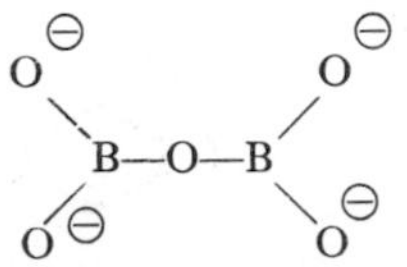

⊖ O O ⊖
\ /
B—O—B
/ \
⊖ O O ⊖

This, however, is not the only way in which BO_3 triangles can be joined, for more than one of the oxygen atoms may be shared and more than two BO_3 groups may be united to give an indefinite number of possible arrangements, some of which are shown in fig. 11.01.

Fig. 11.01 *a* represents the isolated BO_3^{3-} ion. The salts in which this ion occurs are the *orthoborates* with an oxygen:boron ratio of 3:1, and structurally they are analogous to nitrates and carbonates. Thus $ScBO_3$, $InBO_3$ and YBO_3 have the calcite structure and $LaBO_3$ has that of aragonite. In (*b*) two BO_3 groups are linked by sharing a common oxygen atom to form the ion $B_2O_5^{4-}$, but this has not as yet been found in the crystal structure of any borate. If two of the three oxygen atoms of two BO_3 groups are shared the arrangement (*c*) results, but this, too, is unknown in any crystal structure; nor indeed, is it a likely configuration for it involves an abnormally small angle between the two bonds of each of the shared oxygen atoms and a very close approach of the two boron atoms. It is not, however, necessary that the number of BO_3 groups in association should be limited to two, and once this point is accepted it becomes possible for an indefinite number of these groups to be bonded together in the form of closed rings. In (*d*) four groups are so bonded to give the ion $B_4O_8^{4-}$, but, whatever the number of the groups, the arrangement corresponds to an oxygen:boron ratio of 2:1, as found in the *metaborates*. In potassium metaborate three BO_3 groups are linked in this way to give the ion $B_3O_6^{3-}$, and the formula should therefore be written $K_3B_3O_6$, rather than KBO_2, to emphasize this point.

* It is, however, probable that the B–O bond also possesses some degree of ionic character.

If the number of BO_3 groups in the ring is increased indefinitely we reach in the limit the infinitely extended open chain of composition $[BO_2]_n^{n-}$ shown in fig. 11.01 *e*. Such an arrangement differs fundamentally from all the other complex anions so far described in that it is no longer possible to point to any discrete radical. The repeat period in the crystal may extend over several units of the chain, at different points of which different atoms may be attached, so that in chemical terminology such a

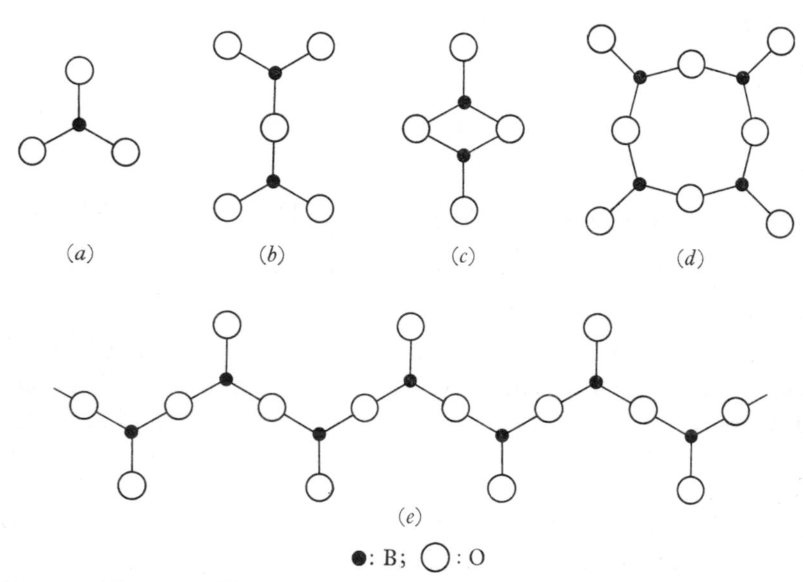

Fig. 11.01. Some possible boron–oxygen groupings in borates. (*a*) Isolated triangles, BO_3^{3-}, e.g. orthoborates. (*b*) Triangles sharing one oxygen atom, $B_2O_5^{4-}$. (*c–e*) Triangles sharing two oxygen atoms to form closed rings or an open chain, $[BO_2]_n^{n-}$ e.g. metaborates.

chain ion is indefinitely polybasic. An infinite ion of this type (but somewhat distorted from the idealized form shown at (*e*)) is found in the crystal structure of calcium metaborate, CaB_2O_4. The ions extend parallel to one another through the structure and are linked together by ionic bonds between the calcium and the unshared oxygen atoms.

11.03. Although boron is normally 3-co-ordinated by oxygen there are also structures in which 4-co-ordination is found, as, for example, in the heteropolyacids (§ 10.24). The same co-ordination is found in a limited number of borate ions. Thus in borax, normally formulated as $Na_2B_4O_7 . 10H_2O$, the anion has the configuration shown in fig. 11.02, in which two of the boron atoms are 3-co-ordinated and two are 4-co-ordinated by oxygen. The boron atoms are linked by shared oxygen atoms and the unshared oxygen atoms belong to hydroxyl groups. These hydroxyl groups are essential constituents of the complex anion, the composition of which is $B_4O_5(OH)_4^{2-}$, so that the formula of the salt should more properly be written $Na_2[B_4O_5(OH)_4].8H_2O$. Examples such as this, and many others which could be quoted, serve to emphasize that the peculiar complexity of the chemistry of borates stems from the multiplicity of possible configurations for the complex ion, and that many of the difficulties which have arisen in discussing these compounds

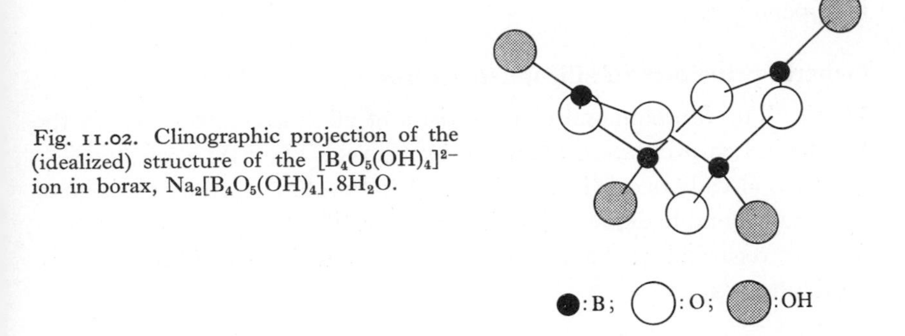

Fig. 11.02. Clinographic projection of the (idealized) structure of the $[B_4O_5(OH)_4]^{2-}$ ion in borax, $Na_2[B_4O_5(OH)_4].8H_2O$.

have been due to attempts to interpret them as the salts of acids which often have no existence. We shall not, however, pursue the question further at this stage because the same point is illustrated even more strikingly by the structures of the silicates, to which we now turn.

SILICATES

11.04. From both a structural and a chemical point of view the silicates are of far greater importance than the borates, and, on account of their great petrological interest as the most abundant component of the Earth's crust, they have been more exhaustively studied by X-ray methods than any other class of ionic crystals. The peculiar complexity of the chemistry of silicates has long been recognized, the enormous variety of composition, the difficulty of assigning significant formulae, and the close morphological association of seemingly quite unrelated compounds all being characteristics difficult to interpret in terms of conventional chemical principles. Structural studies have thrown much light on these topics, and, in fact, in the structures of the silicates we find not only some of the most elegant examples of the general principles of structural chemistry but also what is perhaps, as yet, the most impressive illustration of the power of such studies in rationalizing a previously intractable field of investigation. A discussion at some length is therefore justified, but, even so, it will be possible for us to describe only a limited number of the many known silicate structures, and our choice of examples will be confined to those which advance our arguments by the general principles which they exemplify. Many structures of which the interest is primarily mineralogical will have to be ignored, but the reader who is particularly concerned with this aspect of silicate chemistry will find an exhaustive and readily accessible account in the work by Bragg quoted in appendix 1.

General principles of silicate structures

11.05. The fundamental characteristic of all silicate structures is the tetrahedral co-ordination of silicon by oxygen. The nature of the Si–O bond has already been discussed (§8.33) and we have seen that this bond has a considerable degree of ionic character. The SiO_4 group can therefore be regarded as having a structure to which both the extreme configurations

$$\begin{array}{ccc} & \overset{\ominus}{O} & \\ \overset{\ominus}{O}— & Si & —\overset{\ominus}{O} \\ & \underset{}{\overset{\ominus}{O}} & \end{array} \qquad \text{and} \qquad \begin{array}{ccc} & O^{2-} & \\ O^{2-} & Si^{4+} & O^{2-} \\ & O^{2-} & \end{array}$$

make a contribution, and it is for this reason that silicates occupy a position in respect of their structures intermediate between the salts and the complex oxides. If we picture a simple silicate such as Mg_2SiO_4 (to be described shortly) as consisting of the ions Mg^{2+} and SiO_4^{4-} it would be regarded as a salt, but if we think of it in terms of the separate ions Mg^{2+}, Si^{4+} and O^{2-} it would more properly be classified as a complex oxide. Either view is equally acceptable provided that we understand the implication that neither alone is wholly adequate.

Although in a limited number of silicates isolated SiO_4^{4-} ions are found, just as isolated SO_4^{2-} ions occur in sulphates, the great majority of these compounds contain SiO_4 groups linked together through common oxygen atoms to form complex polynuclear anions of finite or infinite extent. The silicates may therefore be classified in a manner formally analogous to that used for the borates in terms of the way in which the co-ordinating tetrahedra are united. The higher co-ordination of silicon, however, leads to a greater number of structural types, while at the same time additional complexities due to other causes arise.

In the first place the isomorphous replacement of cations by others of the same charge and similar size, or even the replacement of several different cations by others of different valencies but with the same aggregate charge, although possible in many ionic structures, is particularly common among the silicates. We give in table 11.01 a list of the ions of most frequent occurrence in silicate minerals, together with their radii and the oxygen co-ordination numbers commonly observed. Within wide limits we may say that those ions which have the same oxygen co-ordination (e.g. Mg^{2+}, Fe^{2+}; Na^{+}, Ca^{2+}; etc.) are readily interchangeable, and it is this isomorphous replacement which confers on silicate structures their variable and indefinite composition, the chemical interpretation of which presented many difficulties before its structural origin was understood. It is, moreover, not only the cations in silicate structures which can experience isomorphous replacement; the anions O^{2-}, OH^{-} and F^{-} are closely comparable in size and they, too, can replace each other statistically.

Table 11.01. *Some common ions in silicate structures*

Ion	Radius (Å)	Oxygen co-ordination	Ion	Radius (Å)	Oxygen co-ordination
Be^{2+}	0·31	4	Fe^{2+}	0·80	6, 8
Si^{4+}	0·41	4	Mn^{2+}	0·80	6, 8
Al^{3+}	0·50	4, 6	Na^{2+}	0·95	6, 8
Cr^{3+}	0·63	6	Ca^{2+}	0·99	8
Fe^{3+}	0·64	6	K^{+}	1·33	6–12
Mg^{2+}	0·65	6, 8	F^{-}	1·36	—
Ti^{4+}	0·68	6	O^{2-}	1·40	—
Zn^{2+}	0·74	4	OH^{-}	1·53	—

A second feature of silicate crystal chemistry is peculiar to silicates alone, and arises accidentally from the particular value of the radius of the aluminium ion. The Al:O radius ratio of 0·36 is so close to the critical value of 0·3 for transition from 6- to 4-co-ordination that this ion can occur in both conditions, sometimes in the same structure. When 4-co-ordinated the aluminium ion replaces silicon, and such replacement is purely random and may be of indefinite extent. For every aluminium ion so introduced a corresponding substitution of Ca^{2+} for Na^{+}, Al^{3+} for Mg^{2+} or Fe^{3+} for Fe^{2+} must simultaneously occur elsewhere in the structure to preserve neutrality. The appearance of aluminium in one structure in two entirely different roles is a feature of the silicates which chemical analysis alone cannot reveal, and which has given rise to many difficulties in the interpretation of such systems.

The classification of silicates

11.06. The classification of silicate structures is immediately analogous to that of the borates, and is most conveniently made in terms of the way in which the silicon tetrahedra are linked together. Such a classification, originally proposed by Machatschki, must be regarded as one of the most valuable contributions to silicate chemistry, for it has introduced order in a field where previous classifications, based on hypothetical acids of which the silicates were supposed to be salts, led to considerable confusion; and it is interesting to find that this new classification supports in almost every detail that developed by the mineralogist from purely morphological and physical properties. We shall now proceed to discuss briefly a number of silicate structures in order to make clear the basis of the classification and to illustrate the characteristic features which these structures show.

Isolated SiO_4 groups: orthosilicates

11.07. The simplest possible structural arrangement is that in which isolated SiO_4^{4-} groups of the form shown in fig. 11.03*a* are linked together only through the medium of other cations. Such groups are found in the *orthosilicates*, and correspond to an oxygen:silicon ratio of 4:1 or greater.

Olivine

11.08. A typical structure of this kind is that of the olivine series of minerals, of which forsterite, Mg_2SiO_4, may be taken as an example. A somewhat idealized form of this structure is shown in fig. 11.04, in which, for clarity, the SiO_4 tetrahedra have been outlined although the lines shown do not of course represent bonds between oxygen atoms. The silicon atoms at the centres of these tetrahedra are not shown in the figure.

The main points of importance to note in this structure are:

(1) The SiO_4 tetrahedra are isolated and occur pointing alternately up and down.

(2) The tetrahedra are linked together only by O–Mg–O bonds.

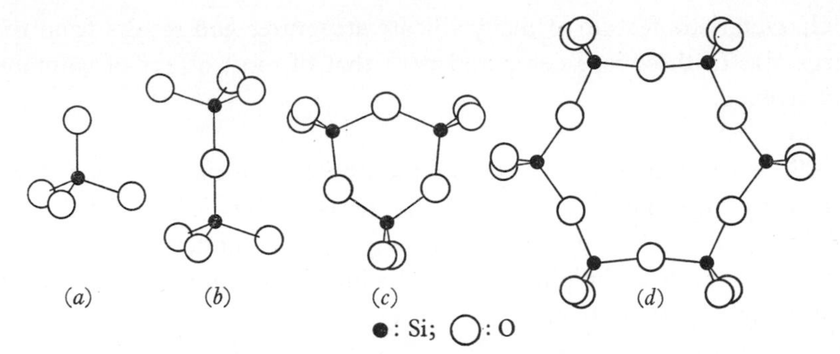

Fig. 11.03. Some possible closed silicon–oxygen groupings in silicates. (*a*) Isolated tetrahedra, SiO_4^{4-}, e.g. orthosilicates. (*b*) Tetrahedra sharing one oxygen atom, $Si_2O_7^{6-}$, e.g. thortveitite. (*c*) Tetrahedra sharing two oxygen atoms to form a three-membered ring, $Si_3O_9^{6-}$, e.g. benitoite. (*d*) Tetrahedra sharing two oxygen atoms to form a six-membered ring, $Si_6O_{18}^{12-}$, e.g. beryl.

In some of the diagrams superimposed oxygen atoms have been displaced slightly for clarity.

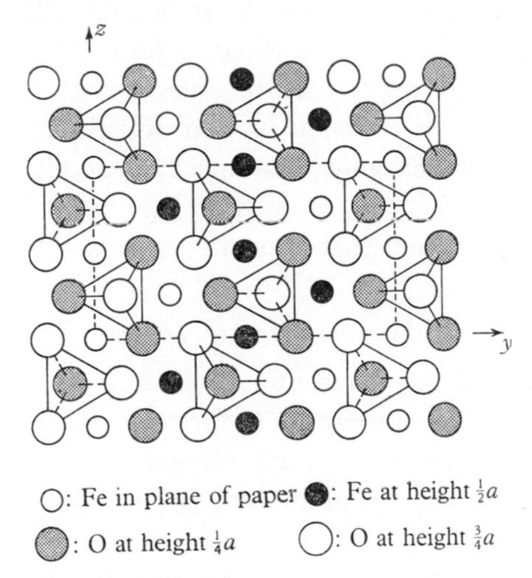

Fig. 11.04. Plan of the idealized orthorhombic structure of forsterite, Mg_2SiO_4, projected on a plane perpendicular to the *x* axis. The silicon atoms lie at the centres of the tetrahedra of oxygen atoms, and are not shown.

(3) The magnesium ions are co-ordinated by six oxygen ions lying at the corners of a very nearly regular octahedron, so that the whole structure can be described as a packing together of tetrahedra and octahedra.

(4) The disposition of the oxygen ions alone is approximately that of hexagonal close packing. A close-packed arrangement of oxygen ions is

a characteristic feature of many silicate structures and results from the large size of these ions compared with that of most others of common occurrence.

11.09. Other members of the olivine group have the same structure, but with Mg^{2+} ions partly or wholly replaced by Fe^{2+} or Mn^{2+}. In olivine itself about 10 per cent of the Mg^{2+} ions of forsterite are replaced by Fe^{2+}; fayalite and tephroite have the idealized compositions Fe_2SiO_4 and Mn_2SiO_4, respectively.

Chondrodite

11.10. The minerals of the chondrodite series are closely related to olivine, and illustrate a type of mixed structure of frequent occurrence among silicates. These minerals may be represented by the general formula $Mg(F, OH)_2 . nMg_2SiO_4$, and members are known for all values of n from 1 to 4. The exact details of the structures will not be considered here, but briefly they may be described as formed by the superposition of sheets of the forsterite structure and of that of $Mg(OH)_2$. The structure of $Mg(OH)_2$ is the same as that of cadmium iodide, with the OH^- ions arranged in hexagonal close packing, and it is the similarity of the arrangement of the oxygen ions in olivine and of the hydroxyl groups in $Mg(OH)_2$ which enables a 'fit' to be achieved.

Phenacite and willemite

11.11. The structure of phenacite, Be_2SiO_4, is of interest in that in spite of its similarity in composition to forsterite the atomic arrangement is entirely different, and a relatively complex rhombohedral structure obtains in which the oxygen atoms are far from close packed. In this structure both the silicon and beryllium atoms are tetrahedrally co-ordinated, and each oxygen atom is common to two BeO_4 tetrahedra and to one SiO_4 group; the complexity of the resulting structure is merely an expression of the difficulty of linking these tetrahedra together in a manner satisfying this condition.

The difference between the structures of forsterite and phenacite can readily be attributed to the difference in size of the Mg^{2+} and Be^{2+} ions, the radii of which are appropriate to oxygen co-ordination of 6 and 4, respectively. It is not, however, possible in this way to explain why willemite, Zn_2SiO_4, has the phenacite structure rather than that of olivine, for the radii of the Zn^{2+} and Mg^{2+} ions are very nearly the same, and it seems likely that in both phenacite and willemite the tetrahedral co-ordination of the metal atoms is due to a considerable degree of covalent character in the metal–oxygen bonds. The difference between the structures of Mg_2SiO_4 and Zn_2SiO_4 may therefore be compared with that between the structures of MgO and ZnO; the former has the sodium chloride structure with 6-co-ordination of magnesium, whereas in the latter the more covalent character of the bonding gives rise to the 4-co-ordinated wurtzite arrangement.

Garnets

11.12. The garnets comprise a wide range of minerals of very variable composition which may be represented by the general formula $A_3^{2+} B_2^{3+} Si_3O_{12}$, where A^{2+} is Ca^{2+}, Mg^{2+}, Fe^{2+} or Mn^{2+}, and B^{3+} is Al^{3+}, Fe^{3+} or Cr^{3+}. They are typical orthosilicates, with independent SiO_4 groups, and the co-ordinations of the *A* and *B* ions are in accordance with their relative sizes; the larger *A* ions are 8- and the smaller *B* ions 6-co-ordinated by oxygen. The readiness with which the garnets form solid solution with each other, and the variable composition which thus results, are clearly to be attributed to the closely comparable sizes of most of the alternative substituent *A* and *B* ions. In this connexion however, it is noteworthy that the calcium garnets differ from the others in showing only a limited range of solid solution, owing to the significantly larger radius of the Ca^{2+} ion in comparison with the ions Mg^{2+}, Fe^{2+} and Mn^{2+}. We have already quoted the garnet $Ca_3Al_2Si_3O_{12}$ as an illustration of the application of Pauling's rules to silicate structures (§9.10).

Other orthosilicates

11.13. Many other orthosilicates with an oxygen:silicon ratio of 4:1 and isolated SiO_4 groups are known, but they are mostly among substances of primarily mineralogical interest. When the oxygen:silicon ratio exceeds 4:1 some oxygen atoms are necessarily not linked to silicon at all, and such atoms may occur either as isolated O^{2-} ions co-ordinated only by metallic cations, as in cyanite, Al_2SiO_5, or as hydroxyl ions OH^-, as in topaz, $(F, OH)_2Al_2SiO_4$. In either case these oxygen atoms must be ignored in deriving the oxygen:silicon ratio, for otherwise the value of this ratio characteristic of the isolated SiO_4 tetrahedra will be concealed. Thus the formula of euclase is often given as $HBeAlSiO_5$, but X-ray analysis has shown that it is actually a normal orthosilicate and therefore to be regarded as $(OH)BeAlSiO_4$.

Structures with Si_2O_7 groups

Thortveitite and melilite

11.14. When the SiO_4 groups in a silicate are not separate, but occur linked together through one or more shared oxygen ions, the number of possible arrangements is very large. The simplest of these is that shown in fig. 11.03*b*, where two such groups share a single oxygen ion to form the composite group $Si_2O_7^{6-}$. Such an arrangement is found in thortveitite, $Sc_2Si_2O_7$, melilite and a number of other minerals. The idealized composition of melilite may be represented as $Ca_2MgSi_2O_7$, but the mineral frequently contains considerable quantities of aluminium and is our first example of a structure in which aluminium replaces silicon in 4-co-ordination. For every Si^{4+} ion thus replaced by Al^{3+} a compensatory change must be made elsewhere in the structure to maintain electrical neutrality, and this is achieved by the replacement of one Mg^{2+} ion by an Al^{3+} ion in 6-co-ordination. The composition of the aluminium-containing mineral may thus be represented as $Ca_2(Mg_{1-x}Al_x)$-$(Si_{2-x}Al_xO_7)$, and such a formula emphasizes the importance of distinguishing between the aluminium atoms in their two different roles and the difficulty in assigning a significant formula on the basis of chemical analysis alone.

Hemimorphite and vesuvianite

11.15. A further interesting structure containing Si_2O_7 groups is that of hemimorphite. On the basis of chemical analysis this mineral is usually represented by the formula $H_2Zn_2SiO_5$. The structure analysis reveals, however, that the mineral is not an orthosilicate and that the oxygen ions are of three kinds, those co-ordinating the silicon ions, those contained in hydroxyl groups and those in water molecules. The true arrangement is made clear by writing the formula $(OH)_2Zn_4Si_2O_7.H_2O$ so that the characteristic oxygen:silicon ratio is revealed. In vesuvianite a combination of independent SiO_4 tetrahedra and Si_2O_7 groups is found: $(OH)_4Ca_{10}Al_4(Mg, Fe)_2\{(Si_2O_7)_2.(SiO_4)_5\}$. On the basis of Pauling's fifth rule, however, such a combination of two different types of co-ordination is not likely to be of common occurrence.

Ring structures: metasilicates

11.16. Closed ring-groups of composition $(SiO_3)_n^{2n-}$ containing an indefinite number of members are formed when two oxygen ions are shared between neighbouring SiO_4 tetrahedra. Such groups occur in the *metasilicates*, and correspond to an oxygen:silicon ratio of 3:1. In view of the partially ionic character of the Si–O bond the two-membered ring Si_2O_6 is unlikely to be found in any structure because it would involve an edge common to the two tetrahedra;* of the other possible ring anions only those with $n = 3$ and $n = 6$ have so far been observed. The three-membered ring (fig. 11.03*c*) occurs in benitoite, $BaTiSi_3O_9$, and the ring of six SiO_4 groups (fig. 11.03*d*) is found in beryl,

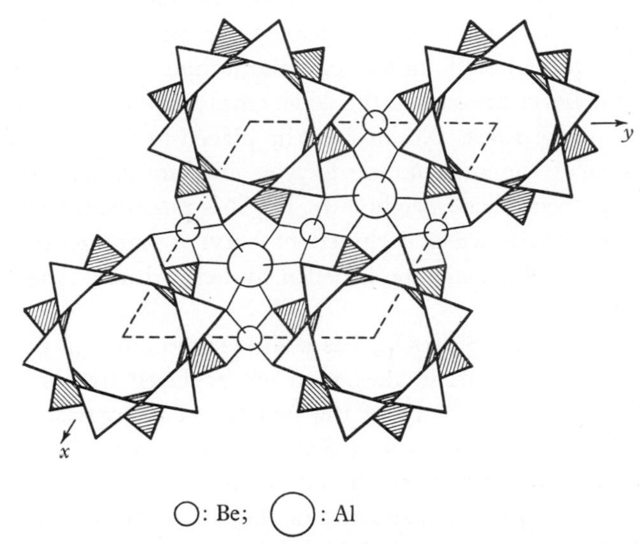

Fig. 11.05. Plan of the unit cell of the hexagonal structure of beryl, $Be_3Al_2Si_6O_{18}$, projected on the basal plane. The Si_6O_{18} rings (see fig. 11.03*d*) at two different heights are represented as rings of shaded and unshaded tetrahedra. The beryllium and aluminium ions lie in a plane midway between the two sets of rings.

$Be_3Al_2Si_6O_{18}$, a structure which we may describe in somewhat more detail as a particularly beautiful illustration of many of the principles of silicate crystal chemistry.

Beryl and cordierite

11.17. A plan of the beryl structure is shown in fig. 11.05. For clarity the SiO_4 groups are here shown as solid tetrahedra, and only the beryllium and aluminium ions are separately indicated. The Si_6O_{18} rings are clearly revealed and it will be seen that these rings are bound together

* On the other hand, it will be remembered that in the structure of SiS_2 (§8.39) the SiS_4 tetradehra do, indeed, share edges, and that this fact was used as an argument pointing to the covalent character of the Si–S bond.

by the metallic cations. The oxygen co-ordination of these cations is that demanded by the radius ratio, and each beryllium ion is tetrahedrally surrounded by four oxygen ions belonging to four different rings. Similarly, the aluminium ions occur between six different rings and are octahedrally co-ordinated by oxygen. The separate rings are thus bound together both laterally and vertically by the oxygen–cation bonds. The strength of each of these bonds, both from the aluminium and beryllium, is $\frac{1}{2}$, and the electrostatic valency rule is therefore satisfied in detail, since every oxygen ion not shared between two SiO_4 tetrahedra is linked to one silicon, one beryllium and one aluminium ion. A characteristic feature of the beryl structure is the wide, empty tunnels passing through the centres of the rings. No ions can be placed in these tunnels, since the valencies of the surrounding oxygen ions are already completely expended on their silicon neighbours, but it is possible that the helium so often found physically occluded in beryl is here accommodated. Such occluded helium may be expelled by heat without damage to the structure.

Cordierite, $Al_3Mg_2(Si_5Al)O_{18}$, has a similar structure to beryl. One-sixth of the silicon atoms in the anion are replaced by aluminium to give the ion $[(Si_5Al)O_{18}]^{13-}$, and the balance of charge is maintained by replacing $(3Be^{2+}+2Al^{3+})$ by $(3Al^{3+}+2Mg^{2+})$. In this case all the aluminium ions are 4-co-ordinated, but the distinction between their two roles, as cations and as constituents of the anion, must still be recognized and should be represented in the formula, as indicated above.

Chain structures

The pyroxenes

11.18. When the closed rings of SiO_4 groups, each sharing two oxygen ions, contain an infinite number of members, they degenerate into indefinitely extended straight chains of the type shown in fig. 11.06*a*. Chains of this kind are found in the important group of pyroxene minerals, the structure of the simplest of which, diopside, $CaMg(SiO_3)_2$, is illustrated in an idealized form in fig. 11.07. Here the chains are seen

---

Fig. 11.06. Some possible open silicon–oxygen groupings in silicates. (*a*) Tetrahedra sharing two oxygen atoms to form open chains, $[SiO_3]_n^{2n-}$, e.g. pyroxenes. (*b*) Tetrahedra sharing alternately two and three oxygen atoms to form open chains, $[Si_4O_{11}]_n^{6n-}$, e.g. amphiboles. (*c*) Tetrahedra sharing three oxygen atoms to form open sheets, $[Si_4O_{10}]_n^{4n-}$, e.g. talc.

Some oxygen atoms have been displaced slightly to reveal the silicon atoms.

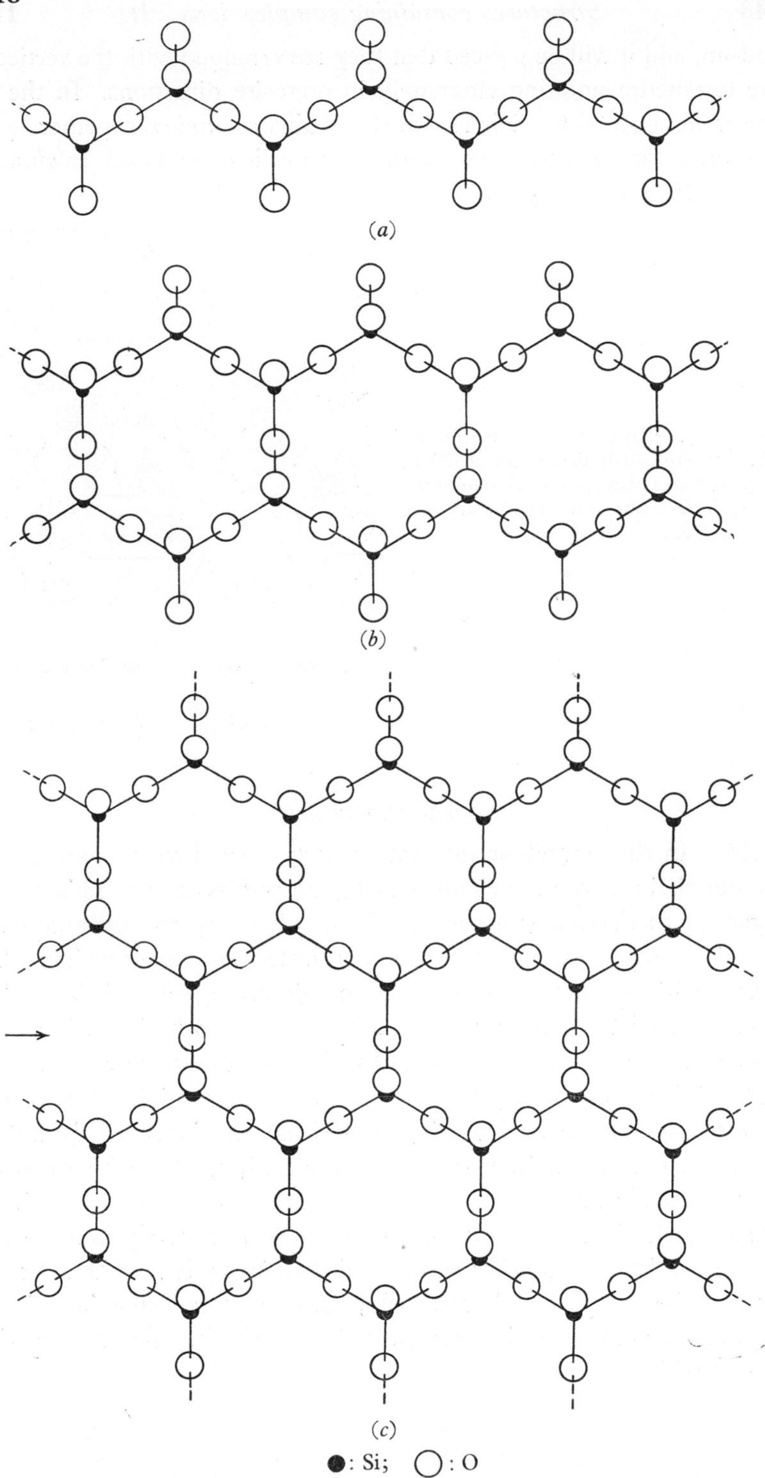

For legend see facing p. 246

end-on, and it will be noticed that they are arranged with the vertices of the tetrahedra pointing alternately in opposite directions. In the two directions normal to their length the chains are linked together by the oxygen–cation bonds. Each magnesium ion is 6- and each calcium ion 8-co-ordinated by oxygen.

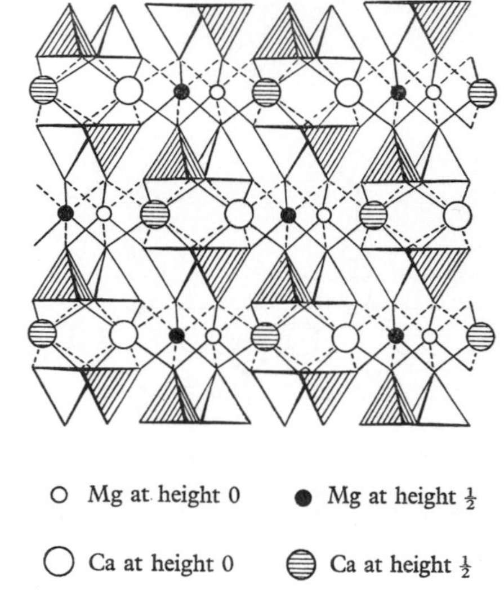

Fig. 11.07. Plan of the idealized monoclinic structure of diopside, $CaMg(SiO_3)_3$, projected on a plane perpendicular to the *z* axis. The chains of fig. 11.06*a* are here seen end-on.

The amphiboles

11.19. In the amphibole minerals, closely related to the pyroxenes, a double chain of the type shown in fig. 11.06*b* is found. This may be regarded as derived from the pyroxene chain by the operation of a reflexion plane passing through the outermost oxygen ions. The oxygen:silicon ratio corresponding to this arrangement is 11:4. The structures of the amphiboles resemble very closely in their general features those of the pyroxenes, and will not be discussed in detail here. They differ from the pyroxenes, however, in containing hydroxyl or fluorine as an essential constituent, so that the idealized formula of tremolite may be written $(OH, F)_2Ca_2Mg_5Si_8O_{22}$. The hydroxyl and fluorine ions in such structures can never occur directly linked to silicon, since the single Si–OH or Si–F bond would completely saturate their valency and leave no other bonds available for their attachment to the rest of the structure. Accordingly, oxygen atoms contained in the hydroxyl group must be disregarded in deducing the characteristic oxygen:silicon ratio.

The crystal chemistry of pyroxenes and amphiboles

11.20. It is instructive to compare the now accepted formula of tremolite, $(OH, F)_2Ca_2Mg_5Si_8O_{22}$, with the composition $CaMg_3(SiO_3)_4 = Ca_2Mg_6Si_8O_{24}$ previously assigned on the basis of chemical analysis. The latter formula not only ignores the small quantity of water found as an essential constituent of the mineral, but also indicates a considerable excess of magnesium compared with the amount usually observed. Nevertheless, six magnesium ions are necessary to achieve electrical neutrality. In the light of the structure analysis it at once becomes clear that the characteristic oxygen:silicon ratio is 22:8 and not 24:8, and that there are only five magnesium ions present. We cannot, however, express the composition in the form $O_2Ca_2Mg_5Si_8O_{22}$, since such a structure, containing two oxygen ions other than those in the silicon chains, is not neutral. Only if these oxygen ions occur as hydroxyl groups is the electrostatic valency rule satisfied and neutrality preserved.

The pyroxenes and amphiboles are characterized by the very extensive isomorphous substitution which they show. In the amphiboles we find another example of the replacement of silicon by aluminium in 4-co-ordination, and in hornblende this occurs up to the extent of about $Si_6Al_2O_{22}$ compared with Si_8O_{22} in tremolite. The corresponding balance of charge is achieved either by the replacement of some Mg^{2+} ions by Al^{3+} in 6-co-ordination, or by the introduction of additional alkali or alkaline earth metal ions into the crystal. There is sufficient space between the chains in the structure to accommodate these additional ions. Even when no replacement of silicon occurs, extensive substitution of cations by others of the same charge and comparable size may take place. Thus Fe^{2+} or Mn^{2+} may replace Mg^{2+}, the ions Fe^{2+}, Mg^{2+} or $2Na^+$ may replace Ca^{2+}, and F^- may replace OH^-. In view of the complexities of these structures it is not surprising that their interpretation in terms of chemical analysis alone presented great difficulties. Even when we know what characteristic oxygen:silicon ratios to expect, care must be taken in deducing such ratios to exclude from the oxygen ions any which occur in hydroxyl groups or as water, and to include with the silicon ions those aluminium ions which are in 4- but not those in 6-co-ordination.

Sheet structures

11.21. If SiO_4 tetrahedra are linked together in such a way that three of the four oxygen atoms are shared with neighbouring tetrahedra a two-dimensional network of the type shown in fig. 11.06*c* is one of a number of possible arrangements which may result. Such a network corresponds to an oxygen:silicon ratio of 10:4, and is found in talc and the micas.

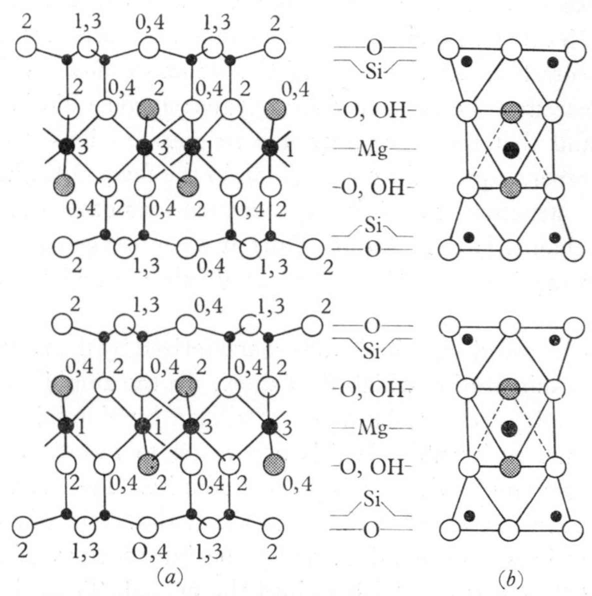

Fig. 11.08. (*a*) Plan of the idealized monoclinic structure of talc, $Mg_3(OH)_2Si_4O_{10}$, projected on a plane perpendicular to the *x* axis. The sheets of fig. 11.06*c* are here seen edge-on, as viewed in the direction of the arrow. For clarity only one ring of each sheet is drawn and only some of the magnesium atoms are shown, but the sheets do, of course, extend indefinitely. The numbers indicate the heights of the various atoms above the plane of projection, expressed in units of $\frac{1}{4}a$. The heights of the silicon atoms are not marked; they are the same as those of the unshared oxygen to which they are attached. (*b*) Schematic representation of the same structure, showing the co-ordinating tetrahedra about the silicon atoms and the co-ordinating octahedra about the magnesium atoms.

Talc

11.22. The structure of talc, $Mg_3(OH)_2Si_4O_{10}$, is illustrated in idealized form in fig. 11.08*a*. Here the sheets of SiO_4 tetrahedra are seen edge-on, and it will be observed that they are arranged in pairs with the vertices and bases of the tetrahedra alternately adjacent. The bases of the sheets are held together only by weak van der Waals forces, but the unshared oxygen atoms at the vertices are strongly cross-linked by magnesium ions, each of which is octahedrally co-ordinated by four of these oxygen atoms (two from each sheet) and by two hydroxyl ions. These hydroxyl ions lie at the centres of the hexagonal rings of unshared oxygen atoms and, with them, form a close-packed (O+OH) layer. If the distinction between O and OH is ignored, the magnesium ions are co-ordinated in the same manner as in $Mg(OH)_2$. The structure can thus alternatively be regarded as consisting of sheets of $MgO_4(OH)_2$ octahedra sandwiched between sheets of SiO_4 tetrahedra and bound to them by the common oxygen atoms which they share, an arrangement conveniently represented, purely conventionally, in the way shown in fig. 11.08*b*.

The micas

11.23. The structure of muscovite mica, $KAl_2(OH)_2(Si_3Al)O_{10}$, is related to that of talc in the manner shown in fig. 11.09. One-quarter of the silicon atoms in the sheets are replaced by aluminium in tetrahedral co-ordination and the three magnesium ions are replaced by two aluminium ions octahedrally co-ordinated. To maintain electrical neutrality one potassium ion is introduced for every silicon atom replaced by aluminium, and these potassium ions are located between the faces of the sheets, where they are co-ordinated by twelve oxygen neighbours.

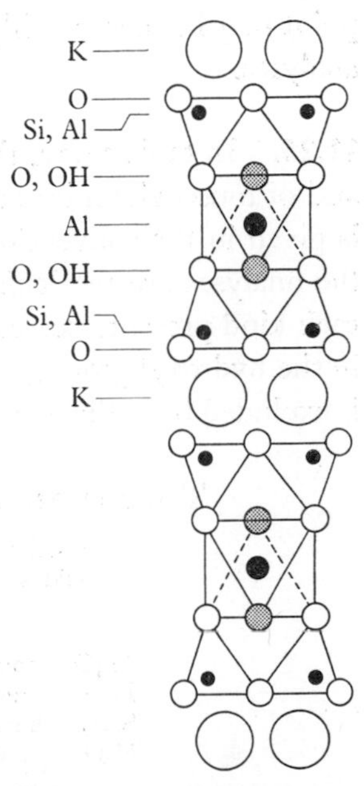

Fig. 11.09. Schematic representation of the structure of muscovite mica, $KAl_2(OH)_2$-$(Si_3Al)O_{10}$.

11.24. The structure of other common micas may be readily derived from that described for muscovite. In phlogopite, $KMg_3(OH)_2(Si_3Al)O_{10}$, the 4-co-ordinated aluminium of muscovite is preserved but the 6-co-ordinated sites are now occupied by magnesium, as in talc. In margarite, $CaAl_2(OH)_2(Si_2Al_2)O_{10}$, two instead of only one of the silicon atoms are replaced by aluminium and the divalent Ca^{2+} ion is therefore required in place of K^+ to restore neutrality. As with the pyroxenes and amphiboles, however, isomorphous substitution in the micas is very common, and idealized compositions of the type just considered are rarely realized. Thus the extent to which silicon is replaced by aluminium is very variable, while the cations which may occur in 6-co-ordination include

Al^{3+}, Fe^{3+}, Fe^{2+}, Mg^{2+}, Mn^{2+} and Li^{+}. The alkali metal, although most commonly K^{+}, may be Na^{+} or Ca^{2+}, while finally OH^{-} may be replaced by considerable quantities of F^{-}. In spite of this complexity of composition, however, the structures can all be interpreted in terms of the simple idealized formula provided only that the primary significance of co-ordination number rather than valency is clearly appreciated. This point may be most readily illustrated by considering a number of typical analyses.

11.25. In table 11.02, the chemical analyses of some specimens of talc and of muscovite mica are shown. In the first column the composition is given in the conventional form as oxides, and in the second column the analyses are recalculated to show the relative number of atoms of each kind present, expressed on a basis of 10 atoms of oxygen. Oxygen in the hydroxyl group, which appears as water in the chemical analysis, is excluded from these 10 oxygen atoms and expressed separately as OH.

Table 11.02. *Analyses of talc and muscovite mica**

Composition by oxides		Composition by atoms			
		Talc			
MgO	30·22	Mg	2·92	—	3·06 (Mg + Fe)
FeO	2·66	Fe	0·14	—	
SiO_2	62·24	Si	4·05		
H_2O	4·94	OH	2·14		
	100·06	O	10·00		
		Muscovite I			
Na_2O	0·90	Na	0·12	—	1·03 (Na + K)
K_2O	10·70	K	0·91	—	
MgO	0·38	Mg	0·05		1·98 (Mg + Al 1·93)
Al_2O_3	37·15	Al	2·90	{1·93, 0·97}	
SiO_2	45·54	Si	3·03		4·00 (Al 0·97 + Si)
H_2O	4·80	OH	2·13		
	99·47	O	10·00		
		Muscovite II			
Na_2O	1·01	Na	0·12	—	0·61 (Na + K)
K_2O	6·05	K	0·49	—	
MgO	0·50	Mg	0·05		2·08 (Mg + Ca + Al 2·01)
CaO	0·27	Ca	0·02		
Al_2O_3	34·70	Al	2·61	{2·01, 0·60}	
SiO_2	53·01	Si	3·40		4·00 (Al 0·60 + Si)
H_2O	4·67	OH	1·99		
	100·21	O	10·00		

* Idealized formulae: talc $(OH)_2Mg_3Si_4O_{10}$; mica $(OH)_2KAl_2(Si_3Al)O_{10}$.

The figures in the second column are therefore immediately comparable with the idealized formulae of the minerals.

The analysis of the talc shows that this particular specimen corresponds closely to the ideal composition $(OH)_2Mg_3Si_4O_{10}$. The O:Si ratio of 10:4 is closely satisfied, but there is some replacement of magnesium by iron. The total of (Mg + Fe) slightly exceeds 3 and the resulting excess of positive charge is compensated by a corresponding excess of OH.

The analysis of muscovite I may be compared with the formula $(OH)_2KAl_2(Si_3Al)O_{10}$. The number of silicon ions is 3·03, so that 0·97 of the 2·90 aluminium ions must occur in tetrahedral co-ordination. The remaining 1·93 aluminium ions, together with 0·05 magnesium, are 6-co-ordinated. The loss of 0·97 unit of positive charge, brought about by the substitution of Al^{3+} for Si^{4+}, is compensated by the introduction of (Na + K), and there is therefore a close correspondence between the amount of aluminium in 4-co-ordination and the total alkali content. In muscovite II an unusually low replacement of silicon occurs and only 0·60 aluminium ion is found in tetrahedral co-ordination. Corresponding to this the total alkali content is only 0·61 ion of (Na + K).

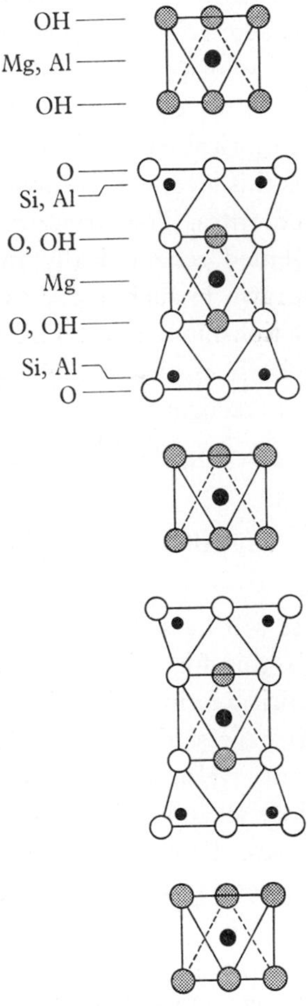

Fig. 11.10. Schematic representation of the structure of chlorite, $Mg_2Al(OH)_6.Mg_3(OH)_2(Si_3Al)O_{10}$.

The chlorite minerals

11.26. The minerals of the chlorite group closely resemble the micas in many of their physical properties, but differ from them chemically in that they contain a very much larger proportion of hydroxyl and no alkali or alkaline earth metal; a typical member has the approximate empirical composition $Al_2Mg_5Si_3O_{10}(OH)_8$. The structure of these minerals can most easily be described by starting from the structure of muscovite shown in fig. 11.09, which, however, for our present purpose may now be considered to represent phlogopite, $KMg_3(OH)_2(Si_3Al)O_{10}$.

As we have seen, this structure may be regarded as consisting of two-dimensional 'sandwich' anions $[Mg_3(OH)_2.(Si_3Al)O_{10}]^-$ held together by potassium cations. In the chlorites the same anions are found, but now the cations themselves are also complex two-dimensional units the structure of which is closely related to that of $Mg(OH)_2$. In the $Mg(OH)_2$ structure (§8.30) $Mg(OH)_6$ octahedra are linked together in the form of layers, and the layers are, of course, electrically neutral. If, however, we replace one-third of the Mg^{2+} ions by Al^{3+} we obtain a positively charged layer of empirical composition $[Mg_2Al(OH)_6]^+$, and it is this layer which constitutes the cation in the chlorites. The structure is therefore as represented schematically in fig. 11.10, which is drawn in such a way as to emphasize the relationship to mica. This same relationship may also be made clear by expressing the formulae in analogous styles, e.g.

$K.Mg_3(OH)_2(Si_3Al)O_{10}$ for phlogopite

and

$Mg_2Al(OH)_6.Mg_3(OH)_2(Si_3Al)O_{10}$ for chlorite.

It will be seen how much more illuminating is this formulation of chlorite than the empirical formulation given earlier; not only does it express the correct O:(Si, Al) ratio of 10:4 but it also shows that Mg^{2+}, Al^{3+} and OH^- ions all appear in the structure in two distinct roles. Few more striking examples could be given of the value of X-ray structural studies in resolving the complex problems of silicate chemistry.

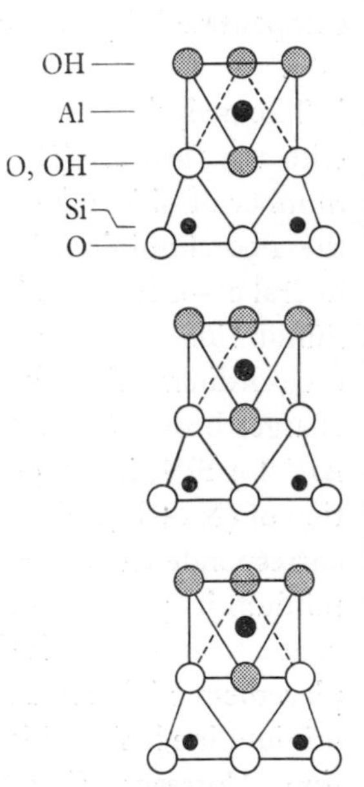

Fig. 11.11. Schematic representation of the structure of kaolinite, $Al_2(OH)_4Si_2O_5$.

The clay minerals

11.27. The clay minerals comprise a series of complex silicates which again resemble the micas in some of their physical properties but which differ from them in forming sizeable crystals only with extreme difficulty. As an example, the structure of kaolinite, $Al_2(OH)_4Si_2O_5$, may be described. In this structure, illustrated in fig. 11.11, there is no replacement of silicon by aluminium and we find single sheets of SiO_4 tetrahedra linked by three corners, just as in talc. These sheets, however, are no longer grouped in pairs, but instead all point the same way. Hydroxyl groups lie inside the hexagonal ring of unshared oxygen atoms at the vertices of the SiO_4 tetrahedra to form a close-packed (O + OH) layer, again as in talc, and above this layer is a complete close-packed sheet of hydroxyl groups. Aluminium ions occupy positions of octahedral co-ordination between these two sheets so that each Al^{3+} ion is co-ordinated by four hydroxyl groups and two oxygen atoms. The structure may therefore be regarded as made up by the superposition of composite layers each consisting of a sheet of SiO_4 tetrahedra and a sheet of $AlO_2(OH)_4$ octahedra united by the common oxygen atoms. These layers are, of course, electrically neutral and are therefore held together only by van der Waals forces. It will be noted, however, that they are unsymmetrical in character, and therefore probably polar, and it may be this fact which is responsible for the reluctance of the clay minerals to form large crystals.

The physical properties of the sheet structures

11.28. Many of the very distinctive physical properties of the micas and related minerals find a ready explanation in their structures. The pseudo-hexagonal symmetry is an immediate consequence of the hexagonal form of the separate sheets, and the eminent cleavage results from the weakness of the bonds by which these sheets are held together. These bonds, however, are not uniformly weak, and corresponding to gradations in strength we find gradations in hardness and in quality of cleavage. Thus in talc and the clay minerals the binding is due only to van der Waals forces, and these minerals are extremely soft and cleave readily; talc is in fact the softest of all known minerals and its cleavage finds application in the use of French chalk as a lubricant. In muscovite mica, on the other hand, the bonds are ionic links through the K^+ ion, and these forces, although still weak compared with the other bonds in the structure, are strong enough to occasion a marked increase in hardness. Finally, in margarite and the other so-called brittle micas, the bonds are yet stronger, owing to the presence of Ca^{2+} or other doubly charged ions between the layers, and a further increase in hardness, accompanied by less perfect cleavage, is observed.

Framework structures

11.29. The final class of silicate structures is that which arises when the SiO_4 tetrahedra are so linked that every oxygen ion is shared between two tetrahedra. Such an arrangement gives rise to an indefinitely extended three-dimensional framework with an oxygen:silicon ratio of 2:1, and is, of course, found in its simplest form in the various modifications of silica already discussed. It is clear that no metallic silicate can be based on such a structure unless some of the silicon is replaced by aluminium, for only then does the framework acquire the negative charge required to balance that of the cations introduced.

In contrast to the other types of tetrahedron linkage already described, the three-dimensional frameworks commonly exist in a number of entirely different forms which give rise to several quite distinct families of petrologically important minerals. A few silicates have structures based on the frameworks found in the various forms of silica. Thus the structures of kalsilite, $K(AlSi)O_4$, and nepheline, $Na_3K(Al_4Si_4)O_{16}$, can be regarded as derived from that of tridymite by the replacement of one-half of the silicon ions by aluminium and the introduction of the appropriate number of alkali ions to restore neutrality. The structures of carnegeite, $Na(AlSi)O_4$, and eukryptite, $Li(AlSi)O_4$, are similarly related to those of cristobalite and quartz, respectively. It will be noted that in all these minerals aluminium is, of course, always present and that the O:(Si, Al) ratio is always 2:1.

In the majority of the framework silicates, however, the framework is far more open than that found in silica, and this open character, coupled with the intrinsic rigidity of the three-dimensional network, confers on these structures a quite exceptional degree of flexibility of composition. The framework silicates are also notably less dense, as a class, than those (such as olivine and the micas) in which the oxygen atoms are more nearly close packed. We shall confine our discussion to three important groups of such structures.

The felspars

11.30. Mineralogically by far the most important framework silicates, and indeed the most important of all rock-forming minerals, are the felspars, the structures of which are based on the framework shown in fig. 11.12. It will be seen that the $(Si, Al)O_4$ tetrahedra are grouped in rings of four members and that these rings in their turn form links of a kinked chain running vertically in the diagram. This appearance is, however, somewhat deceptive, for it must be remembered that all the oxygen atoms of the tetrahedra are shared so that the bonding in the structure extend not only vertically, as shown, but also transversely to build up a three-dimensional network.

In orthoclase, $K(AlSi_3)O_8$, one-quarter of the silicon is substituted by aluminium, and for each silicon ion thus replaced one potassium ion is introduced. The potassium is accommodated in large interstices in the structure and is surrounded by ten oxygen atoms; the co-ordination is, however, irregular and the structure cannot conveniently be regarded as based on a packing of polyhedra. In albite, $Na(AlSi_3)O_8$, sodium replaces the potassium of orthoclase, and in anorthite, $Ca(Al_2Si_2)O_8$, two of the silicon ions are replaced by aluminium and a divalent cation is therefore required to restore neutrality. Complete solid solution occurs between albite and anorthite, on account of the close correspondence between the radii of the Na^+ and Ca^{2+} ions (0·95 and 0·99 Å, respectively), but not between albite and orthoclase, despite their closer chemical similarity, owing to the difference in size between the Na^+ and K^+ ions (0·95 and 1·33 Å, respectively). The close morphological and physical association between albite and anorthite has long been recognized by mineralogists as more significant than their chemical dissimilarity, and the structural justification for this association provides us with another elegant example of the geometrical principles determining the structures of ionic crystals.

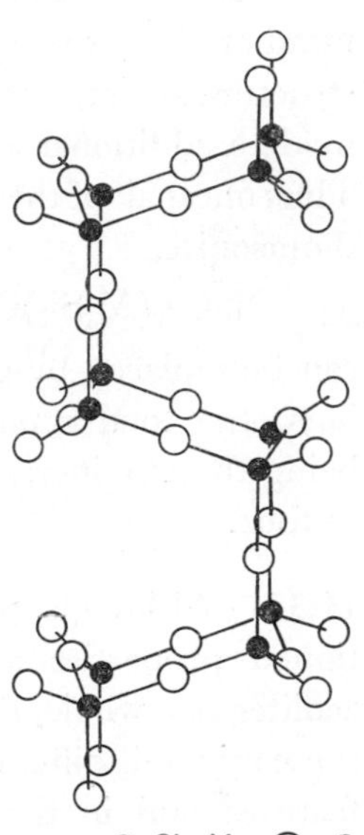

Fig. 11.12. Idealized representation of the silicon–oxygen framework in felspars.

The zeolites

11.31. A second group of minerals based on a framework structure is the zeolite family of hydrated silicates, but here the framework is very much more open than that found in the felspars and is intersected by wide channels in which the cations and water molecules are located. These water molecules, however, are very loosely bound in the crystal and it is a distinctive feature of the zeolites that they can be readily expelled by heat without destroying the structure; they can, moreover, be reversibly replaced or even substituted by other neutral molecules such as those of ammonia, mercury, alcohol or iodine provided that these molecules are not too large to penetrate the interstices of the framework. Zeolites therefore possess the remarkable property of acting as 'molecular sieves' permeable only to small molecules.

Another distinctive characteristic of the zeolites is that the cations, too, can be readily and reversibly exchanged without damage to the structure as a whole, often by simply immersing the crystal in a solution of the appropriate salt, and it is, of course, this base-exchange property which is the foundation of the permutite system of water softening. Naturally, if ions are removed they must be replaced by others of equal aggregate charge, but it is not necessary that they should be equal in number since the interstices in the structure are large enough to accommodate additional ions if required. Thus one-half of the calcium ions in thomsonite,

$$NaCa_2(Al_5Si_5)O_{20}.6H_2O,$$

can be replaced by pairs of sodium ions, the total number of cations being thereby increased from three to four.

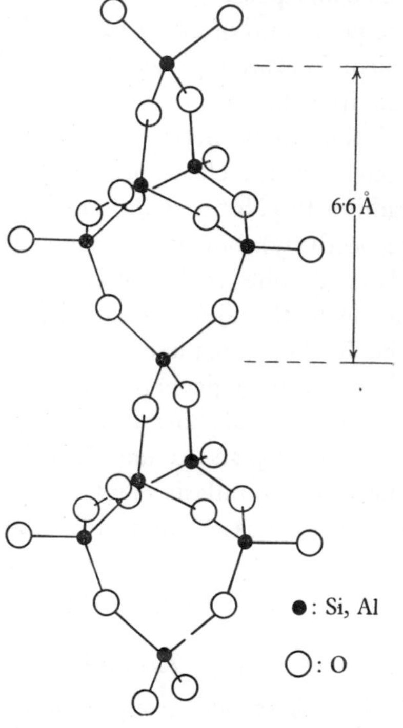

Fig. 11.13. The silicon–oxygen framework in the fibrous zeolites.

11.32. Although the above-mentioned properties are common to zeolites as a whole, the structures of these minerals differ in their detailed features and in the nature of the framework which they contain. Those which have been most extensively studied are the so-called *fibrous zeolites*, which tend to grow as needles or as fibrous aggregates. In these minerals the (Si, Al)O_4 tetrahedra are linked in the manner shown in fig. 11.13 to form chains extending indefinitely through the structure with a repeat unit of length 6·6 Å made up of five such tetrahedra. It is therefore a characteristic of all fibrous zeolites (of which thomsonite is an example) that the anion contains the unit $(Si, Al)_{5x}O_{10x}$ and that one dimension of the unit cell is 6·6 Å or a multiple of this distance. As with the felspars, the description in terms of chains of tetrahedra is in part deceptive, for these chains are, of course, cross-linked through the unshared oxygen atoms in fig. 11.13 to form coherent three-dimensional frameworks. Nevertheless, these cross-links are relatively few in number, so that the structures are bonded more firmly along the lengths of the chains than transversely. In this respect they resemble the amphiboles and pyroxenes, and their fibrous character may be ascribed to the ease with which cleavage parallel to the chains can occur.

The ultramarines

11.33. The ultramarines comprise a group of naturally occurring and artificial framework silicates which differ from all silicate structures so far considered in that they contain not only the (Si, Al)–O complex anion but also the negative ions Cl^-, SO_4^{2-} or S_2^{2-}. Ultramarine itself has the idealized formula $Na_8(Al_6Si_6)O_{24}.S_2$, and sodalite and noselite have analogous compositions with the S_2^{2-} groups replaced respectively by $2Cl^-$ and by SO_4^{2-}. The ultramarines differ from the zeolites in being anhydrous, but they resemble these minerals in the readiness with which ion exchange can be achieved. Nor is this ion exchange confined to the cations, for the ions Cl^-, SO_4^{2-} and S_2^{2-} can also be interchanged, and sodalite, for example, can be artificially transformed into noselite. Moreover, other ions, both positive and negative, not found in the naturally occurring silicates may be readily introduced, giving an extensive series of synthetic ultramarines containing the elements Li, Ca, Ag, Tl, Se, Te and others. Wide variations in colour, ranging throughout the spectrum from red to blue, accompany these substitutions, and the artificial ultramarines accordingly find extensive use as pigments.

In spite of the complexity of the ultramarines their structures are basically quite simple. The minerals are all cubic, and contain basket-like frameworks of $(Si, Al)O_4$ tetrahedra of the form shown in fig. 11.14*a*. This representation may appear complex but in fact it can be very simply described as consisting of rings of four tetrahedra on the faces of the cubic unit cell and rings of six tetrahedra encircling its corners, a description which is perhaps more easily seen in the alternative presentation of fig. 11.14*b*. Here the structure is shown in plan projected on the base of the unit cell and the tetrahedra in the lower half of the cell have been omitted. The four tetrahedra in the top face of the cell and the four groups of six about the corners of this face, are clearly visible. The 'baskets' in the structure are not, of course, isolated but are linked with their neighbours into a very open three-dimensional framework in the cavities of which the remaining ions are located. As with the zeolites, it is the large size of these cavities, coupled with the rigidity of the framework, which makes ion exchange possible without destruction of the crystal.

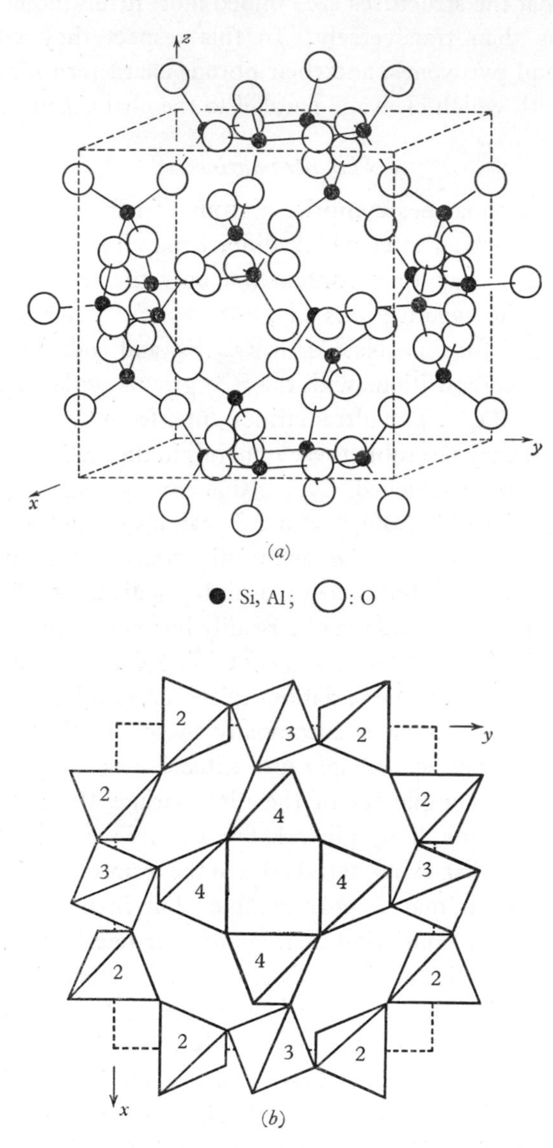

Fig. 11.14. (*a*) The silicon–oxygen framework in the ultramarines. (*b*) The same, represented as a plan projected on the base of the cubic unit cell. Only the SiO_4 tetrahedra in the upper part of the cell are shown, and these have been displaced slightly to reveal more clearly the linkage between them. The numbers indicate the heights of the centres of the tetrahedra above the plane of projection, expressed in units of $\frac{1}{4}c$.

11.34. This concludes our systematic description of silicate structures, and for reference we give in table 11.03 a summary of the several classes which we have considered, together with the corresponding O:(Si, Al) ratios. Before we pass on to other topics there remain a number of general points common to all silicates which call for brief consideration.

Table 11.03. *Some silicate structure types*

Structural arrangement	Examples	O:(Si, Al) ratio
Independent tetrahedra	Orthosilicates, e.g. olivine, garnets	4:1
Two tetrahedra sharing one oxygen atom	Thortveitite	7:2
Closed rings of tetrahedra each sharing two oxygen atoms	Metasilicates, e.g. benitoite, beryl	3:1
Infinite chains of tetrahedra each sharing two oxygen atoms	Pyroxenes	3:1
Infinite double chains of tetrahedra sharing alternately two and three oxygen atoms	Amphiboles	11:4
Infinite sheets of tetrahedra each sharing three oxygen atoms	Micas, chlorites, clays	5:2
Infinite framework of tetrahedra each sharing all four oxygen atoms	Felspars, zeolites, ultramarines	2:1

The crystal chemistry of the silicates

The replacement of silicon by other ions

11.35. Although Al^{3+} is the only ion which commonly replaces Si^{4+} in tetrahedral co-ordination in the silicates, other ions occasionally appear in the same role. The most important of these is As^{5+}, which is found in berzeliite, $Mg_2(Ca_2Na)As_3O_{12}$. This mineral is isomorphous with the orthosilicate garnet, of ideal composition $Al_2Ca_3Si_3O_{12}$, and it is seen from a comparison of the formulae that the three units of positive charge acquired by the substitution of As^{5+} for Si^{4+} are compensated by the replacement of $2Al^{3+}$ by $2Mg^{2+}$ and of Ca^{2+} by Na^{+}. Small but appreciable quantities of silicon are often found replacing arsenic in berzeliite.

In the ultramarine helvite, $(Mn, Fe, Zn)_8Be_6Si_6O_{24}.S_2$, one-half of the tetrahedrally co-ordinated sites in the framework are occupied by beryllium.

Model structures of the silicates

11.36. We have already defined model structures as isostructural substances in which ions occur closely equivalent in size but different in charge. In this sense Li_2BeF_4 is a model of the structure of willemite,

Zn_2SiO_4. The radii of the several ions in these two structures are

Li^+	0·60 Å	Be^{2+}	0·31 Å	F^-	1·36 Å
Zn^{2+}	0·74	Si^{4+}	0·41	O^{2-}	1·40

and the close correspondence between the two sets of radii results in isomorphous structures with closely similar cell dimensions and crystallographic properties. On the other hand, the increased strength of the polar binding in the silicate is reflected in the physical properties, as the following data show:

	Li_2BeF_4	Zn_2SiO_4
Hardness (Mohs scale)	3·8	5·5
Melting point (°C)	470	1509
Solubility in water	Readily soluble	Insoluble

The chemical classification of silicates

11.37. The purely structural classification of the silicates which we have here adopted naturally challenges comparison with older and more orthodox chemical classifications. Long before the application of X-ray structure analysis, the mineralogist had classed together, on purely morphological grounds, many silicates of very different constitution, while the chemical classification of substances of such uncertain and variable composition necessarily presented great difficulties. Nevertheless, we can now see that the failure of the chemical classification was due not so much to the inherent difficulty of the material as to the falsity of the principles on which it was based. On the one hand, the dual role of aluminium, which enables it to appear either as a normal metallic cation or in place of silicon, could be detected only by X-ray methods, and, on the other, the chemical classification sought always to refer all silicates to some corresponding acid. We cannot here emphasize too strongly that structurally there is no relation whatsoever between acids and their salts, and that in so far as acids exist it is a structural accident which does not represent any structural association. The properties of the strongly polarizing and vanishingly small H^+ ion are so entirely different from those of any other cation that they demand entirely separate treatment, and we deliberately refrain until chapter 12 from discussing the structures of any acids or acid salts. In most of the silicates the disorganization of the structure which would result from the substitution of hydrogen for the other cations is so drastic that it is unthinkable that such an arrangement should be stable. The acids simply have no existence.

11.38. This concludes our account of the silicates. While we have tried to discuss the more important structural and chemical features of these minerals we would again emphasize that our treatment has been in no sense exhaustive from a mineralogical point of view; (Si, Al)O_4 tetrahedra can be linked together in many different ways and numerous types of silicate exist in addition to those which we have discussed. The reader who wishes for a more detailed account of these structures is referred to the work by Bragg already mentioned.

GERMANATES AND PHOSPHATES

11.39. Other oxy-salts in which large polynuclear anions are found are the germanates and phosphates. The radius of the Ge^{4+} ion, which is very nearly the same as that of Si^{4+}, again corresponds to tetrahedral co-ordination by oxygen, and GeO_2 (in one of its forms) has the quartz structure. A number of germanates are known with structures analogous to those of some of the silicates which we have discussed, but they have not been widely studied.

11.40. Phosphorus is characterized by the complexity of its oxygen chemistry (compared, say, with that of nitrogen), and the salts of many oxy-acids are known. X-ray structural studies have been concerned principally with the salts of quinquivalent phosphorus, in all of which the phosphorus atom is tetrahedrally co-ordinated by oxygen. Isolated PO_4^{3-} ions, analogous to SO_4^{2-} and SiO_4^{4-} ions, exist in many orthophosphates, some of which have structures closely related to those of orthosilicates when due allowance is made for the difference in valency. The PO_4 groups, however, may also be linked, by sharing one or more oxygen atoms, to give polynuclear complexes. Thus the $P_2O_7^{4-}$ ion (analogous to the $Si_2O_7^{6-}$ ion in thortveitite) is found in pyrophosphates, and four-membered ring-ions $P_4O_{12}^{4-}$, of the form shown in fig. 11.15, occur in aluminium metaphosphate, $Al_4(P_4O_{12})_3$. (This ring may be compared with the ions $Si_3O_9^{6-}$ and $Si_6O_{18}^{12-}$ in benitoite and beryl.) The association of PO_4 groups into infinite chains (analogous to those in the pyroxenes) has not been established in any metaphosphate but may occur in metaphosphoric acid and be responsible for the fact that this acid normally exists only as a glass.

The possibility of more complex associations of PO_4 groups, analogous to those found in the micas and felspars, does not arise in phosphates because any arrangement in which these groups share three of their four oxygen atoms is already electrically neutral, while the sharing of all oxygen atoms would yield a positive ion of composition $[PO_2]_n^{n+}$. In the various forms of phosphorus pentoxide, however, PO_4 groups are found united by sharing three oxygen atoms. In one form the association is into discrete P_4O_{10} molecules (fig. 11.16), bound to one another only by van der Waals forces, and in this form the compound is soft, volatile and reactive. In a second form, however, the PO_4 groups are linked into a coherent three-dimensional network extending throughout

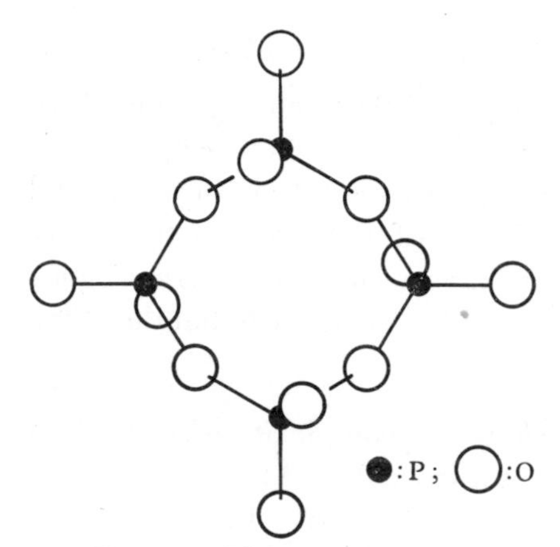

Fig. 11.15. The structure of the $P_4O_{12}^{4-}$ ion.

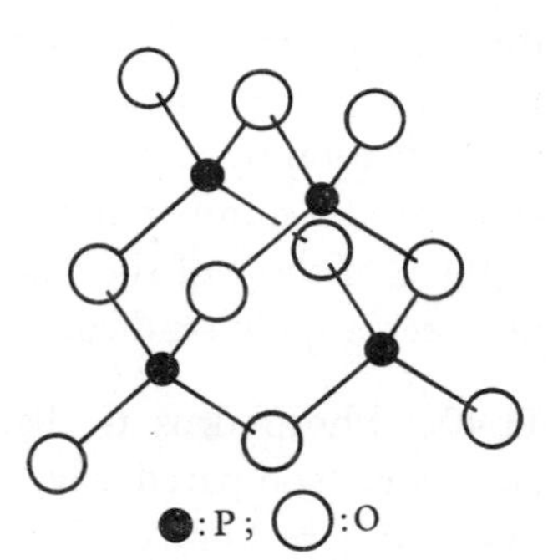

Fig. 11.16. The structure of one form of the P_4O_{10} molecule.

the structure to give a harder, less volatile and less reactive crystal. This latter arrangement may be regarded as the analogue of the various silica structures, but clearly no silica analogue of the molecular structure is possible since the union of SiO_4 groups by a sharing of all oxygen atoms can never yield a finite complex.

12 THE STRUCTURES OF SOME COMPOUNDS CONTAINING HYDROGEN

INTRODUCTION

12.01. We have already discussed the structures of a limited number of hydrogen-containing compounds, notably the hydrides of the metals, certain hydroxides and some ammonium salts. In the hydrides the hydrogen atom appears as the anion H^-, comparable in character with a halogen ion, and the structures of the hydroxides and of the ammonium salts which we have so far considered can similarly be described in terms of spherical OH^- and NH_4^+ ions. We now turn to discuss some structures in which the hydrogen atom plays a much more distinctive part and in most of which it appears as the positive ion H^+. The peculiar characteristics of these structures arise from the unique position of the hydrogen cation as an ion of vanishingly small size and one lacking any extranuclear electronic structure. In terms of the simple geometric picture of ionic structures such an ion would be expected to be 2-co-ordinated, and in the majority of the structures to which we now turn we find two atoms X and Y linked by a hydrogen atom to form the group X–H–Y. Such a link is termed a hydrogen bond.

THE HYDROGEN BOND

The hydrogen bond in chemistry

12.02. The conception of a binding between two atoms as being possible through the medium of a hydrogen atom was first advanced some fifty years ago and has since been widely applied, especially in organic chemistry. Thus the fact that certain types of organic compound, such as the sugars, for example, are harder and more brittle, and melt at higher temperatures, than the majority of organic substances is attributed to hydrogen bonds between the molecules in the crystal stronger than the normal van der Waals forces. Similarly, the tendency to polymerization shown by compounds containing the groups —NH_2, —OH, —COOH, =NOH, =NH, etc., may also be ascribed to hydrogen-bond formation between two or more molecules; the fact that polymerization does not occur in these compounds when the hydrogen is replaced by other atoms or groups confirms that hydrogen plays an essential part in the polymerization process. In chelate compounds hydrogen bonds occur between atoms in the same molecule.

The hydrogen bond was originally interpreted in terms of two covalent bonds to a divalent hydrogen atom. The hydrogen atom, however, with only one stable orbital, cannot form more than one covalent link, and for this reason it is now generally accepted that the bond must be essentially ionic in character. Additional evidence for this view is provided by the fact that its strength is sensitive to the electronegativity of the atoms involved, and that in practice it is found only between atoms of the strongly electronegative elements fluorine, oxygen, nitrogen and (occasionally) chlorine. In spite of this limitation, however, the common occurrence of these elements, and particularly of oxygen and nitrogen, in both inorganic and organic compounds is sufficient to emphasize the importance of this type of binding.

The hydrogen bond is weak by comparison with the covalent link, its energy being of the order of 5 kcal/mole, or some ten times smaller than that of a covalent bond; on the other hand it is distinctly stronger than the van der Waals bond. For this reason, atoms united by hydrogen bonds are usually separated by distances (lying with in the range 2·5–3·0 Å) considerably greater than if covalent bonds were involved but, nevertheless, appreciably smaller than if residual forces alone were operative.

The H^+ ion is unique not only in respect of its size but also in respect of the fact that it has no extranuclear electronic structure. When attracted by another atom it can therefore penetrate within the electron cloud of that atom until finally arrested in an equilibrium position by the repulsion of the nucleus. This is believed to occur when a hydrogen bond is formed, so that if such a bond exists between two atoms X and Y the hydrogen ion is unsymmetrically disposed and is specifically associated, say, with the atom X. The bond is then due to the attraction between this ion and a lone electron pair of the atom Y, a state of affairs which we may express by the formulation $X\text{–H}\cdots Y$.

The hydrogen bond in crystal structures

12.03. The location of the hydrogen atom by the conventional methods of crystal structure analysis presents considerable difficulty, owing to its very small scattering power for X-rays. Although in a limited number of cases refined techniques have enabled hydrogen atoms to be directly located by X-ray diffraction, and although in a few other instances their positions have been found by neutron scattering, it nevertheless remains true to say that in the great majority of the structures of hydrogen-containing compounds so far studied the location of these atoms has been determined only by inference. For this reason many of the properties of the hydrogen bond (for example, its asymmetry) have been studied primarily by other physical methods, such as those of infra-red spectroscopy, rather than by X-ray techniques. This is not to imply, however, that there is any uncertainty in the detection of hydrogen bonds in crystal structures, for such bonds are unequivocally revealed by the abnormally close approach of the atoms between which they operate as well as by the distinctive physical properties to which they give rise.

We proceed now to describe some inorganic crystal structures in which hydrogen bonds are found. Examples of hydrogen-bond formation in organic crystals will be given in chapter 14.

Ammonium halides

12.04. We have already seen (§8.06) that all the ammonium halides, except the fluoride, have either the caesium chloride or the sodium chloride structure and that the NH_4^+ ion is in free rotation and behaves as a sphere of effective radius 1·48 Å. (They are, in fact, all dimorphous with a transformation from the former to the latter structurc as the temperature is raised.)

On the basis of purely geometrical considerations it would be expected that ammonium fluoride, with a radius ratio $r^-/r^+ = 0{\cdot}92$, would also have the caseium chloride structure, with an NH_4–F distance of $1{\cdot}48 + 1{\cdot}36 = 2{\cdot}84$ Å. This, however, is not the case, and instead the fluoride has the structure of wurtzite (fig. 4.03), in which each ion is tetrahedrally co-ordinated by only four of the other kind and in which the NH_4–F distance is only 2·66 Å. We have here clear evidence of the existence of hydrogen bonds between the nitrogen and fluorine atoms which effectively 'lock' the NH_4 group in position and reduce the NH_4–F distance. It is to be noted that it is only in the fluoride that these bonds exist, confirming the view that the hydrogen bond can operate only between strongly electronegative atoms.

Ice

12.05. It is well known that water is quite exceptional in many of its physical properties. In the series of compounds H_2Te, H_2Se and H_2S the melting points and boiling points fall progressively with decreasing molecular weight, but in H_2O this trend is sharply reversed, as the following figures show:

	H_2Te	H_2Se	H_2S	H_2O
Melting point (°C)	−48	−63	−83	0
Boiling point (°C)	− 4	−43	−62	100

Moreover, water has an exceptionally high dielectric constant, a high specific heat, high latent heats of fusion and evaporation, and a low density.

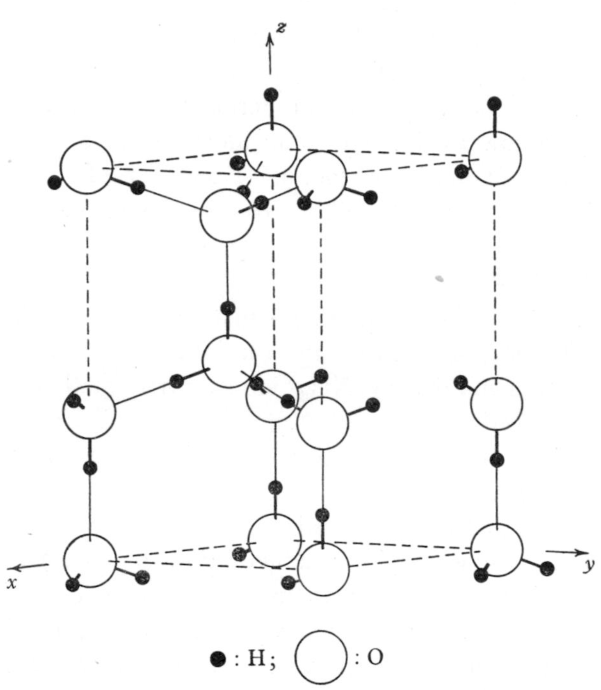

Fig. 12.01. Clinographic projection of the hexagonal structure of ice. The unit cell is indicated by broken lines. The distribution of the hydrogen atoms is arbitrary (see text).

In respect of its crystal structure water is similarly anomalous. In the solid state H_2S has a cubic close-packed structure with an S–S distance of about 4·0 Å, as would be expected for an assembly of spherical molecules bound together by van der Waals forces. Water, however, as ice, has the very different hexagonal structure shown in fig. 12.01. If the distribution of the hydrogen atoms is for the moment ignored, this arrangement may be described as one in which the water molecules occupy the same positions as the zinc and sulphur atoms in wurtzite, or alternatively, as the silicon atoms in the structure of β-tridymite.

Each molecule is tetrahedrally co-ordinated by only four others and the intermolecular distance is only 2·76 Å, so that the effective radius of the water molecule is 1·38 Å. The distinctive features of this structure—the low co-ordination and the short intermolecular distance—as well as the relatively high melting point of ice, all point to the existence of hydrogen bonds binding the molecules together.

The position of the hydrogen atoms in ice has been established unequivocally by both infra-red spectroscopy and neutron diffraction, and it is found that each lies along an O–O bond at a distance of 1·01 Å from one of the oxygen atoms. The bond is therefore unsymmetrical, and each oxygen atom has associated with it two hydrogen atoms which

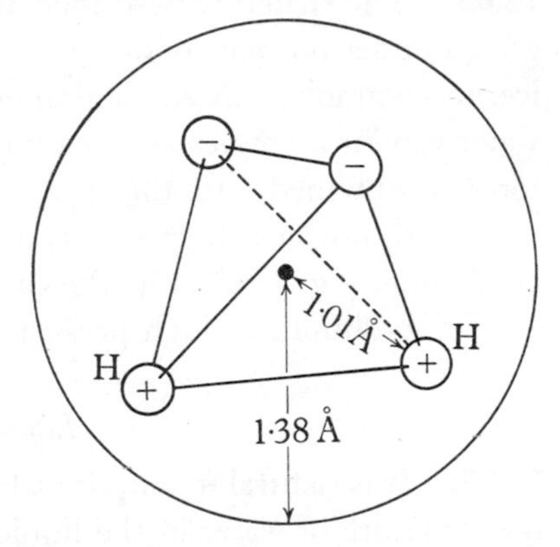

Fig. 12.02. The tetrahedral distribution of charge on the water molecule.

are effectively embedded within its sphere of influence. We can thus picture the water molecule as a sphere which, although neutral, is nevertheless polar, with the tetrahedral charge distribution shown in fig. 12.02; the two corners of the tetrahedron occupied by the hydrogen atoms are positively and the remaining corners negatively charged. It is the packing together of these polar molecules in such a way that seats of opposite charge are adjacent which determines the structural arrangement.

A distribution of hydrogen atoms consistent with this scheme is shown in fig. 12.01, and it will be observed that when these atoms are taken into account the cell shown is not a true unit cell in the strict crystallographic sense in that the environment of all the atoms at the corners of the unit cell is no longer identical. The hydrogen atom distribution shown, however, is not unique, for (subject to the condition that each oxygen atom has two close hydrogen neighbours) the hydrogen atoms in any one bond may be associated with either of the two oxygen atoms to give the alternative configurations O–H···O or O···H–O. There is, therefore, a multiplicity of different ways in which the hydrogen atoms may be arranged, and the fact that the dielectric constant of ice is comparable with that of water implies that either by rotation of the molecules or by displacement of the hydrogen atoms along the O–O bonds an effectively random distribution is achieved. While each individual molecule retains the structure already described, the crystal as a whole may be envisaged as an assembly of molecules each of which behaves statistically as an oxygen atom tetrahedrally co-ordinated by four 'half-atoms' $\frac{1}{2}$H. The cell of fig. 12.01 then once again becomes the true cell.

12.06. The structure described above is that which obtains when ice is grown under normal conditions. It is interesting to note, however, that ice, in common with zinc sulphide and silica, is polymorphous. When water vapour is condensed on a surface cooled to about −100 °C a cubic form is obtained. In this the molecules are again tetrahedrally co-ordinated, but now they are arranged as the zinc and sulphur atoms in zincblende (fig. 3.04) or as the silicon atoms in β-cristobalite (fig. 8.07). Under conditions of high pressure yet other structures are found.

Liquid water

12.07. It is natural to enquire what light X-ray studies of ice throw on the structure of water in the liquid state.

The characteristic feature of what we may term an 'ideal liquid' is a close contact between spherical molecules, leading to a relatively dense, approximately close-packed structure, but one in which any regularity of packing is confined to only a few molecular diameters. We may illustrate this structure by its analogy to a sack full of spheres: there is no regularity extending over large distances, so that it is not possible to deduce the arrangement of spheres in one region given that in another, but the assemblage as a whole has a density not very different from that of a close-packed structure, and there is a high probability that the immediate environment of any given sphere is nearly the same as in a close-packed crystal. Ideal liquid structures of this type are found in the molten metals, even including those of the *B* sub-groups which are not close packed in the solid state. Thus in bismuth the directional tendence of the binding in the crystal is lost in the liquid, which is in consequence denser than the solid so that a contraction in volume occurs on melting.

In liquids in which the molecules depart considerably from spherical symmetry, or in which the molecules are strongly polar, a less regular packing of lower co-ordination is found, and in methyl alcohol, for example, the distribution in the liquid indicates a tendency towards the formation of clusters of molecules bound by hydrogen bonds. Such clusters, however, do not have any permanent existence in the liquid but are in a continuous state of formation and dissolution, and must not be regarded as in any way indicative of chemical association. In compounds (such as acids and alcohols) which contain active groups the structure of the liquid will therefore be determined both by the directed character of the intermolecular forces and by the shape of the molecules, whereas in neutral compounds (such as the hydrocarbons) the latter factor alone can operate.

If water had a pseudo-close-packed structure of molecules of radius 1·38 Å its specific gravity would be about 1·8. It is therefore clear that the structure must be far from close packed, and that the co-ordination characteristic of ice must be in large measure retained in the liquid state. This view is confirmed by the fact that the X-ray diffraction pattern from liquid water is entirely different from that from a close-packed liquid, and also by the relatively low latent heat of fusion (compared with that of evaporation), which argues that few hydrogen bonds can be ruptured when ice melts. We are thus led to a picture of water as a pseudo-crystalline liquid consisting of small regions in which a more or less regular tetrahedral arrangement of molecules obtains. Such a tetrahedral arrangement is not, however, sufficient completely to specify the structure of the liquid, for the tetrahedra can be linked in a number of different ways just as the SiO_4 tetrahedra in silica can give rise to the different structures of cristobalite, tridymite and quartz. In ice, as we have seen, the arrangement of the water molecules is the same as that of the silicon ions in tridymite, but the same structure cannot occur in liquid water since the density would in that case be lower than that of ice. If, however, the structure of water is a pseudo-crystalline arrangement based on the more compact quartz structure the higher density can be understood, and, indeed, the observed X-ray diffraction curves for water at room temperature are consistent with this structure. At high temperatures, however, the diffraction effects approximate more and more closely to those of an ideal liquid, so that between, say, 150 °C and the critical temperature water has a pseudo-close-packed structure. Similarly at temperatures below about 4 °C, and in super-cooled water, a change in structure again occurs, and a transition towards a tridymite- or ice-like arrangement is found. We may thus recognize a total of three different forms of water, between which a continuous transition with temperature takes place:

Water I	⇌	Water II	⇌	Water III
Ice-like		Quartz-like		Close packed
Stable below about 4 °C		Stable between 4 °C and about 150 °C		Stable above about 150 °C

It is not to be considered, however, that the liquid is ever in any way heterogeneous, consisting of a mixture of regions of different structure, or that any specific polymers such as $(H_2O)_2$ or $(H_2O)_3$ exist, but only that the average mutual arrangement of the molecules resembles more closely water I, II or III according to the temperature.

In terms of this picture the anomalous contraction in volume with increasing temperature between 0 and 4 °C is readily understood, for in this temperature range the ice-like arrangement is rapidly breaking down to give place to the more compact quartz-like structure and the normal expansion of the latter is completely masked by the larger volume change accompanying the morphotropic transformation. Only when the ice-like arrangement is completely transformed does the normal expansion become apparent.

It should be added that the detailed features of the water structure outlined above are not universally accepted and that the problem of the structure of water in the liquid state certainly cannot be regarded as solved beyond doubt. What is clear is that hydrogen bonds continue to exert a dominating influence in the liquid and that as the temperature is raised these bonds are progressively ruptured; the high specific heat is a measure of the energy required for this process. Even at the boiling point, however, strong hydrogen bonding must persist, for only in this way can the very high latent heat of evaporation be understood.

12.08. In conclusion it is interesting to consider briefly the structure of some other simple compounds which might be expected to show resemblances to water in the liquid state. In H_2S the sulphur atom is insufficiently electronegative for hydrogen bonding to be possible. As a result, the structure of the solid is close packed and the liquid has a pseudo-close-packed arrangement with a much lower boiling point than water. In hydrogen fluoride, on the other hand, hydrogen bonding can occur, and the structure of the solid consists of infinite zigzag chains of the form

F F F
F F F

in which the hydrogen bonds are even stronger than those in water. Some of these bonds persist in the liquid and also in the vapour, where polymers $(HF)_n$, with $n = 3$ or more, are observed. It is to be noted, however, that these bonds, being limited in number to two for each fluorine atom, can never form a coherent array in three dimensions. In ammonia the polar molecule has a tetrahedral charge distribution as in water, but now with three positive poles, and again a structure bound by hydrogen bonds in three dimensions is impossible. The melting and boiling points of both hydrogen fluoride and ammonia are high compared with those of the analogous compounds HCl and PH_3 but they are, nevertheless, much lower than those of water. It is to the special combination of circumstances that oxygen is both divalent and sufficiently electronegative to form hydrogen bonds that water owes its unique properties.

Hydroxides

12.09. In chapter 8 (§§8.07 and 8.30) we described the structures of a number of simple hydroxides, of composition AOH and $A(OH)_2$, in which we were able to treat the hydroxyl group as a spherical anion of radius 1·53 Å, analogous in its structural properties to the ions of the halogens. Thus, for example, we saw that the hydroxides of the divalent metals Ca, Mg, Mn, Co, Ni, Fe and Cd all have the cadmium iodide layer structure, as do the bromides and iodides of these same elements. We should now note, however, that to regard the anion in a layer structure as a spherical entity is something of an oversimplification; the asymmetrical co-ordination of this ion must inevitably produce considerable polarization in it, and a more faithful picture is therefore that of a cylindrical dipole. Neighbouring layers of the structure will then be held together not only by purely residual forces but also by further forces between these dipoles. The influence of these latter forces is revealed in a progressive decrease in the effective radius of the hydroxyl groups towards adjacent hydroxyl groups with increasing polarizing power of the metallic cation, so that in the hydroxides of the series of metals listed above the characteristic OH–OH distance falls continuously from 3·36 Å in $Ca(OH)_2$ to 2·98 Å in $Cd(OH)_2$. In LiOH (which also has a layer structure but with the cations tetrahedrally co-ordinated by OH groups) the less strong polarizing power of the univalent ion results in the very large OH–OH distance of 3·61 Å.

Aluminium hydroxide

12.10. We now turn to consider some hydroxide structures in which the hydroxyl group can no longer be treated as a simple spherical or cylindrical unit and which differ from the structures so far discussed in two important respects: the hydroxyl groups approach one another far more closely, and the mutual disposition of these groups is such as to

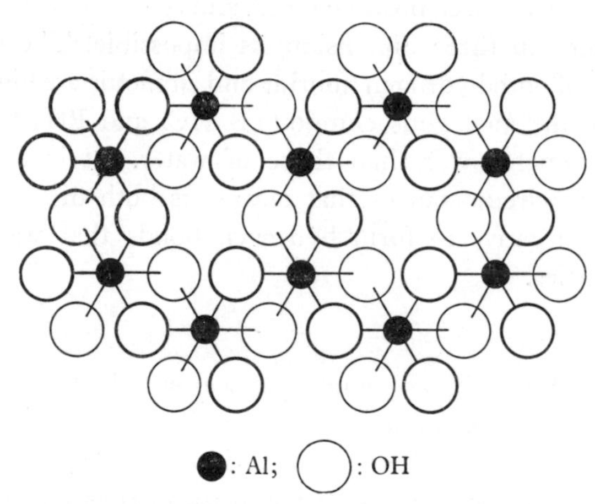

●: Al; ○: OH

Fig. 12.03. Idealized plan of a single layer of the structure of aluminium hydroxide, $Al(OH)_3$.

indicate the existence of directed bonds between them. We may illustrate the features of these structures in terms of the specific example of aluminium hydroxide, $Al(OH)_3$. This compound, too, has a layer structure, with the atomic arrangement in each layer as shown, in somewhat idealized form, in fig. 12.03. It will be seen that this arrangement is similar to that of the layers in the cadmium iodide structure but with one out of every three cation sites vacant. The aluminium atoms are still octahedrally co-ordinated by OH groups but each of these groups is now co-ordinated by only two cations. Here, however, the resemblance to cadmium iodide ends, for in the mutual disposition of the layers the structures are quite different. Instead of being arranged in such a way as to give a close-packed distribution of OH groups, the layers are now so superimposed that OH groups fall directly above one another. Moreover, the binding between the layers is exceptionally strong, as is revealed by the very small OH–OH distance of 2·78 Å and also by the relatively small coefficient of thermal expansion normal to the plane of the layers.

It is clear that such a structure is not compatible with the picture of the OH group as a simple dipole, for an arrangement of such groups end-to-end would clearly be unstable and would give rise to repulsion between the layers rather than particularly strong attraction. The structure can, however, be satisfactorily explained if we assume that under the strong polarizing influence of the aluminium ion the hydroxyl group develops a tetrahedral structure analogous to that of the water molecule, with three of the corners of the tetrahedron occupied by concentrations of negative charge, each of value $-\frac{1}{2}e$, and the fourth, representing the hydrogen atom, carrying a positive charge $+\frac{1}{2}e$. Under these conditions it will be possible for hydroxyl groups to attract one another strongly, provided that they are so disposed that regions of opposite charge are adjacent. The groups can then be regarded as united by hydrogen bonds just as bonds of this type unite the water molecules in ice.

When we come to consider the detailed features of the $Al(OH)_3$ structure in terms of this picture we see that each OH group must be so disposed that two of its negative poles are directed towards the two aluminium atoms by which it is co-ordinated, and it is interesting to note that the electrostatic valency principle is satisfied in detail by such an arrangement since the strength of each Al–OH bond is $\frac{3}{6} = \frac{1}{2}$. The remaining two poles of the OH group are directed to oppositely charged poles of adjacent OH groups, one in the same layer and one in the adjacent layer, and give rise to the hydrogen bonds between these groups. We thus see that all the links between OH groups in different layers are due to hydrogen bonds, but that in addition hydrogen bonds operate within each layer between each OH group and one only of its neighbours. These hydrogen bonds are of course shorter than the distance between OH groups between which no such bonds exist, and it is the asymmetrical distribution of the hydrogen bonding within the layers which accounts for their distortion from the idealized form shown in fig. 12.03.

Other hydroxides

12.11. Another hydroxide in which hydrogen bonding is observed is $Zn(OH)_2$. This structure differs from that of the hydroxides of the other divalent metals so far described in that each zinc atom is co-ordinated by only four OH groups, while each of these groups, in its turn, is co-ordinated by two other such groups and by two zinc atoms tetrahedrally disposed about it. The existence of hydrogen bonding is established not only by this tetrahedral bond distribution but also by the short OH–OH distance of only 2·83 Å.

12.12. It is natural to enquire what factors determine whether a given hydroxide $A(OH)_n$ will have a relatively simple structure or a less symmetrical arrangement in which hydrogen bonds operate. The answer to this question is implicit in what we have already said, for we have seen that as the polarization of the OH groups increases so also does the strength of the bond between them. It is only when the group is very strongly polarized, however, that the tetrahedral configuration is developed and hydrogen bonding becomes possible. Although the polarizing power of a cation is dependent not only on its charge but also on its radius and electronic configuration we may for our present purposes take the electrostatic valency strength of the *A*–OH bond as a measure of this quantity. It then seems that if the strength of this bond is less than $\frac{1}{2}$ a simple structure results, whereas if the strength is $\frac{1}{2}$ or more a hydrogen-bonded arrangement is found. This conclusion is illustrated by the data given in table 12.01, from which it will be seen that the structures can be divided into two distinct groups. On the one hand we have those structures in which the bond strength is less than $\frac{1}{2}$ and in which the shortest OH–OH distance is about 3 Å or more, and on the other we have the structures in which a bond strength of at least $\frac{1}{2}$ is achieved and in which hydrogen bonding results is a significantly shorter OH–OH distance of about 2·8 Å. The structure of $B(OH)_3$, included in the table for reference, is described below (§12.14).

In the light of the above remarks it is now easy to see what characteristic features may be expected in hydroxide structures containing

Table 12.01. *Some data on hydroxide structures*

Compound	Electrostatic strength of *A*–OH bond	Shortest OH–OH distance between layers (Å)	
LiOH	$\frac{1}{4}$	3·61	No H bonds
$Ca(OH)_2$	$\frac{1}{3}$	3·36	
$Cd(OH)_2$	$\frac{1}{3}$	2·98	
$Zn(OH)_2$	$\frac{1}{2}$	2·83	H bonds
$Al(OH)_3$	$\frac{1}{2}$	2·78	
$B(OH)_3$	1	2·72	

hydrogen bonds. In the first place the strength of the *A*–OH bond must be adequate to develop the tetrahedral structure of the OH group, but in addition the electrostatic valency principle must be satisfied as applied to the bonds from this group. Thus each OH group must be bound not only to two other OH groups by hydrogen bonds, but also either to two cations by bonds of strength $\frac{1}{2}$ or to one cation by a single bond of strength unity. In the former case the co-ordination round each OH group must be tetrahedral; in the latter case a planar arrangement is to be expected.

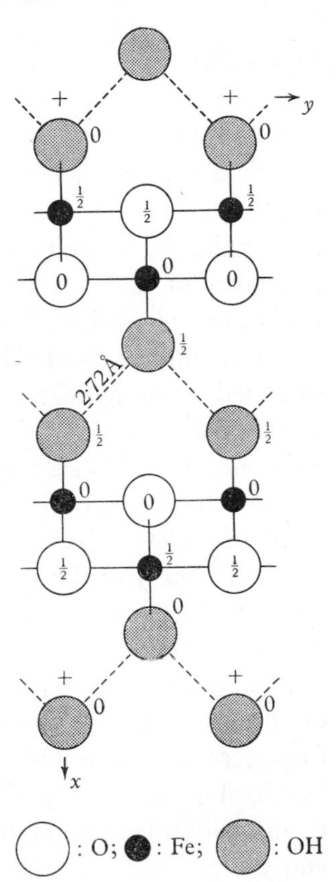

Fig. 12.04. Idealized plan of the orthorhombic structure of lepidocrocite, FeO(OH), projected on a plane perpendicular to the *z* axis. The heights of the atoms are indicated in fractions of the *c* translation. Hydrogen bonds are represented by broken lines, and crosses indicate the corners of the unit cell.

Oxy-hydroxides

12.13. Hydrogen bonds are also found in the mineral lepidocrocite, FeO(OH), the structure of which is shown in fig. 12.04. This is a layer structure in which each layer (here seen edge-on) is composed of four planes of atoms. The two inner planes contain iron and oxygen atoms, while the two outer planes contain only hydroxyl groups. Each iron atom is octahedrally co-ordinated by four oxygen atoms and two OH groups, and each hydroxyl group is tetrahedrally co-ordinated by two iron atoms in its own layer and by two OH groups, to which it is bound by hydrogen bonds of length 2·72 Å, in the adjacent layer. In this structure the hydrogen bonds are therefore exerted only between OH groups in different layers, and hydrogen bonding within the layers, as in $Al(OH)_3$, does not occur. It is interesting to compare the structure of lepidocrocite with that of the directly related FeOCl. The separate layers of this latter structure are identical with those in FeO(OH) but the packing of the layers is completely different; the arrangement of the chlorine atoms does not have to satisfy the distinctive spatial distribution characteristic of the hydrogen bonds from the OH groups, and in adjacent layers they are therefore close packed. The difference between the two structures is thus analogous to that between the structures of $Ca(OH)_2$ and $Al(OH)_3$.

Oxy-acids

Boric acid

12.14. In the oxy-acids of general formula H_nXO_y it is possible, in principle, for hydrogen bonds to be formed between the oxygen atoms, but the extent of the bonding achieved will clearly be dependent on the H:O ratio. If, for example, this ratio is 1:1 there will be a sufficiency of hydrogen for each oxygen atom to be involved in two hydrogen bonds (as in the hydroxides), if it is 1:2 there could be one hydrogen bond to each oxygen atom, and if it is less than 1:2 some oxygen atoms must be excluded from hydrogen-bond formation altogether. Unfortunately very few structures of this type have been studied and we shall describe only one, that of boric acid.

The structure of boric acid, H_3BO_3, is essentially a molecular arrangement of isolated $B(OH)_3$ molecules. These molecules have a planar configuration with the OH groups surrounding the boron atom at the corners of an equilateral triangle and with a B–OH distance of 1·36 Å. In the crystal the molecules lie in flat sheets, each of the form shown in fig. 12.05, and the structure as a whole is built up by the superposition of these sheets. The sheets are held together only by van der Waals forces, as is shown by the large separation of 3·18 Å between the hydroxyl groups of adjacent sheets and the excellent cleavage between them. Within the sheets, however, the OH–OH distance is only 2·72 Å and the molecules are held together by hydrogen bonds between the hydroxyl groups, each of which is linked to two others. It is interesting to note that if residual forces alone operated between the molecules a more closely packed arrangement would be expected with each molecule surrounded by six rather than three neighbours. Such an arrangement, however, is not compatable with any scheme of hydrogen bonding and we therefore see in the relatively open structure of boric acid (as in the open structure of ice) an elegant example of the profound structural influence of the characteristically directed distribution of the hydrogen bonds. It is of interest, too, to contrast the structure of $B(OH)_3$ with that of $Al(OH)_3$. The latter is also a layer structure but it is not a molecular arrangement, and in it each aluminium atom is co-ordinated by six OH groups. In both structures each hydroxyl group is linked by hydrogen bonds to two others, but in $Al(OH)_3$ one of these lies in the same layer and one in an adjacent layer whereas in $B(OH)_3$ both are in the same sheet. In $Al(OH)_3$ cohesion between the layers is due to hydrogen bonds but in $B(OH)_3$ van der Waals forces alone are operative.

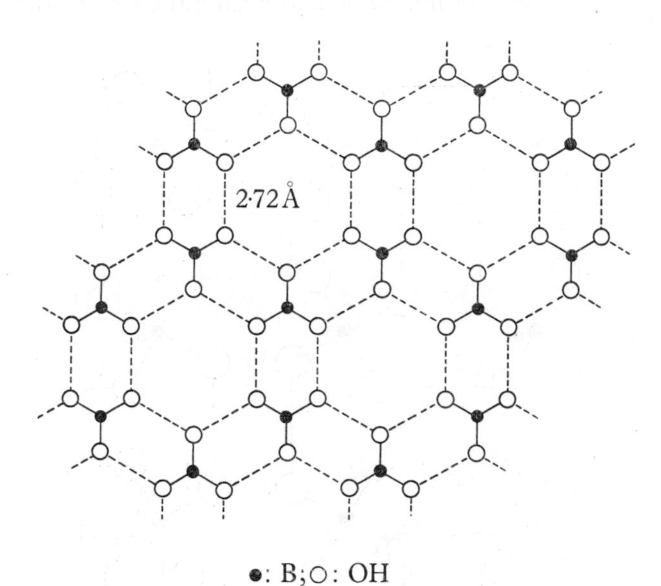

●: B; ○: OH

Fig. 12.05. Plan of a single layer of the structure of boric acid, $B(OH)_3$. Hydrogen bonds are represented by broken lines.

Acid salts of oxy-acids

12.15. In the acid salts of the oxy-acids hydrogen bonding is possible, just as in the oxy-acids themselves, and again the extent of the bonding will be determined by the H:O ratio. Among structures of this type which have been investigated are those of potassium dihydrogen phosphate, KH_2PO_4, and some bicarbonates.

Potassium dihydrogen phosphate

12.16. In KH_2PO_4 the H:O ratio is 1:2 and there are therefore enough hydrogen atoms for every oxygen atom to be involved in one hydrogen bond. The structure, shown schematically in fig. 12.06, consists of tetrahedral PO_4 groups with the potassium atoms disposed between them in such a way that each is somewhat irregularly co-ordinated by eight oxygen atoms. Every oxygen atom belongs to one PO_4 group and to two KO_8 groups. The PO_4 groups are so arranged that each oxygen atom of each group is linked to one oxygen neighbour in another group by a hydrogen bond of length 2·49 Å, the remaining O–O distances being considerably greater. The pairs of oxygen atoms linked in this way are at the same height in fig. 12.06 and a careful consideration of this

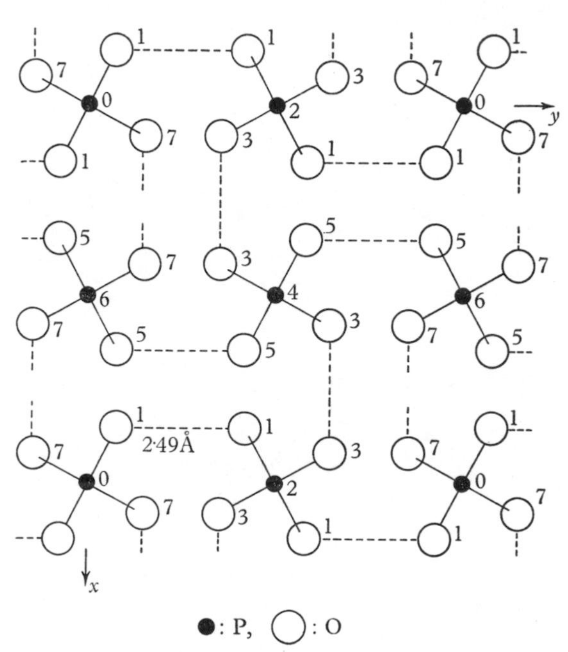

Fig. 12.06. Schematic plan of the unit cell of the tetragonal structure of potassium dihydrogen phosphate, KH_2PO_4, projected on a plane perpendicular to the z axis. The phosphorus atoms lie at the centres of the tetrahedra of oxygen atoms, and the potassium atoms (not shown) lie midway between pairs of tetrahedra in the z direction. The heights of the phosphorus and oxygen atoms are indicated in units of $\frac{1}{8}c$. Hydrogen bonds are represented by broken lines. The corners of the unit cell coincide with the phosphorus atoms at height 0.

illustration will reveal that the PO_4 tetrahedra are joined by the hydrogen bonds to form a coherent three-dimensional framework in the interstices of which the potassium atoms are located. In this respect the structure may therefore be compared with that of the framework silicates. In the silicates, however, the linkage of SiO_4 tetrahedra is achieved by a sharing of common oxygen atoms, whereas here no sharing takes place and union is only through the medium of hydrogen bonds.

Some bicarbonates

12.17. In the bicarbonates $AHCO_3$ the H:O ratio is 1:3, and the number of hydrogen atoms available is now sufficient to form hydrogen bonds between only two-thirds of the oxygen atoms, i.e. two such bonds for each CO_3 group. It is therefore no longer possible for these groups to be linked into three-dimensional frameworks of the type found in KH_2PO_4, but linkage in one or two dimensions can still take place. In the structure of $NaHCO_3$ each CO_3 group is bound to two others by hydrogen bonds of length 2·55 Å to form the infinite zigzag chain anion

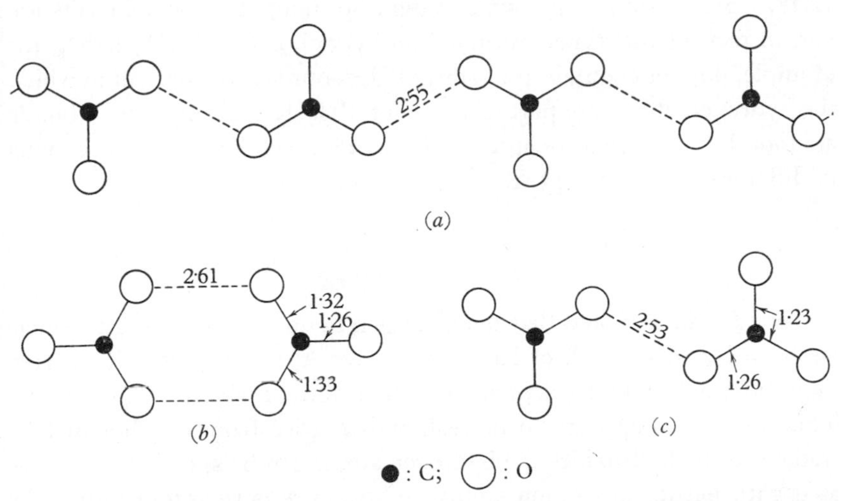

Fig. 12.07. Hydrogen bonding between CO_3 groups in bicarbonates: (*a*) sodium bicarbonate, $NaHCO_3$; (*b*) potassium bicarbonate, $KHCO_3$; (*c*) sodium sesqui-carbonate, $NaHCO_3.Na_2CO_3.2H_2O$. Distances are in Ångström units and hydrogen bonds are represented by broken lines.

of composition $[HCO_3]_n^{n-}$ shown in fig. 12.07*a*, and these chains are held together laterally by sodium ions situated between them; the structure therefore bears a formal resemblance to that of the pyroxenes (§11.18). In $KHCO_3$ quite a different arrangement is found, and now the CO_3 groups are linked only in pairs to form 'dimer' ions of composition $[CO_3.H_2.CO_3]^{2-}$ (fig. 12.07*b*). In sodium sesqui-carbonate $NaHCO_3.Na_2CO_3.2H_2O$, with H:O ratio of 1:6, only one hydrogen bond per CO_3 group is possible, and these groups are now arranged, as shown in fig. 12.07*c*, to give the ion $[CO_3.H.CO_3]^{3-}$. In all these examples the presence of the hydrogen bonding is clearly revealed in the crystal structure by the abnormally small separation of about 2·5 Å between the oxygen atoms concerned, and also by the distortion of the CO_3 group to which it gives rise. Thus in calcite this group has the form of a strictly regular equilateral triangle with a C–O distance of 1·31 Å, whereas in $KHCO_3$ and in sodium sesqui-carbonate it has the less symmetrical configuration revealed by the dimensions quoted in fig. 12.07. In each case the C–O bond to oxygen atoms involved in hydrogen-bond formation is slightly but significantly longer than the corresponding bond to atoms not so involved.

12.18. Some compounds which might be thought to be acid salts are not, in fact, of this type. Ammonium hypophosphite, $NH_4H_2PO_2$, for example, does not contain the radical PO_2^{3-} but is a normal salt in which the hydrogen atoms are part of the anion $PH_2O_2^-$. No hydrogen bonds are found in the structure and the closest approach between oxygen ions in different anions is 3·45 Å.

The hydroxyl bond

12.19. It will be noted that the hydrogen bonds in the acid salts which we have just discussed, of length about 2·5 Å, are appreciably shorter than those in ice and the hydroxides, the length of which is about 2·7 Å. This distinction appears to be real, and to arise from the fact that in water and the hydroxides the hydrogen atoms can be specifically associated with particular oxygen atoms. A structure is therefore formed in which the identity of the H_2O or OH group is preserved and in which the O–H···O bond is unsymmetrical. On the other hand, in, say, sodium sesqui-carbonate the hydrogen atom cannot be assigned preferentially to one or other of the two CO_3 groups which it unites, and resonance between the alternative configurations O–H···O and O···H–O is possible. Some authors emphasize this distinction by using the designation hydrogen bond to describe only the shorter links; the longer bonds, formed by hydroxyl groups and water molecules, are then termed *hydroxyl bonds*. We shall find that there are many structures in which bonds of both types occur.

HYDRATES

Introduction

12.20. It is not in general possible for neutral atoms or molecules to be bound in ionic crystal structures, but the molecule of water, and to a lesser extent that of ammonia, are exceptions to this general rule. The reason for this anomalous behaviour lies primarily in the electrical polarity and small size of these molecules. The particular relevance of these two factors will appear as our discussion proceeds.

In many hydrates the function of water is simply to co-ordinate the cations and thereby effectively surround them with a neutral shell which increases their radius and enables their charge to be distributed over a greater number of anions. In most ionic structures, and particularly in those containing complex anions, the disparity in radius between the anion and cation is too great for simple structures of high co-ordination to obtain, and a tendency towards layer structures or purely molecular arrangements often results. This tendency becomes the more marked the higher the ionic charges and therefore the greater the disparity in radii. The possibility of co-ordinating the cations by water molecules, however, completely alters this state of affairs, for a small and strongly polarizing cation can be thereby transformed into a feebly polarizing ion with a size comparable with that of many complex anions. Thus the 6-co-ordination of the aluminium ion by water changes this ion of radius 0·50 Å into the $[Al(H_2O)_6]^{3+}$ complex of approximate radius 3·3 Å, which is considerably greater than that of the anions I^- and SO_4^{2-}. Such large cations may be expected to give rise to stable structures of much less complexity than those in which small cations occur in isolation. For the same reason we may expect hydrates to be commonest and most stable among salts with small cations. This is in accordance with the well-known fact that lithium and sodium salts show a much greater tendency to crystallize as hydrates than do the corresponding compounds of the larger ions potassium, rubidium, caesium and ammonium. Very small ions, such as Be^{2+}, are 4-co-ordinated by water and so form tetrahydrates, while the salts of ions of intermediate size, such as Mg^{2+}, Na^+, Ca^{2+}, etc., with radii appropriate to 6-co-ordination, commonly occur as hexahydrates.

The function of water in the hydrates, however, is not quite as simple as such an elementary picture would suggest, for we cannot regard the water molecules merely as neutral conducting spheres whose function is to distribute more widely the cation charge. Indeed, if they could be so regarded, the forces of attraction between the neighbouring ions and the induced charges on the water molecules would be insufficient to give coherence to the structure, and hydrates would not exist. On the contrary, as we have already seen in discussing the structure of ice, the water molecule is a highly polar entity the polarity of which is consistent with a tetrahedral distribution of two regions of positive and two of negative electrification over its surface. It follows from this that the packing of water molecules into a crystal must satisfy certain physical requirements, for a structure can be stable only if these molecules are so disposed that their charged areas are appropriately directed towards ions of opposite sign. We therefore find in hydrate structures that the water molecules are normally co-ordinated in one or other of two characteristic ways: either

(i) each molecule is tetrahedrally co-ordinated by four neighbours, of which two present positive and two present negative charges towards it; or

(ii) each molecule is co-ordinated by only three neighbours, of which one (always a cation) presents a positive charge and two present negative charges. This arrangement may be regarded as a degeneration of the tetrahedral configuration of the molecule when the two seats of negative charge coincide, and it therefore gives rise to a planar distribution of neighbours.

These restrictions on the co-ordination about the water molecule often preclude simple structural arrangements which would otherwise be possible.

A second consequence of the tetrahedral structure of the water molecule is that these molecules may also occur in a crystal structure in a different capacity, without any cation neighbours; i.e. they may be bound only to other water molecules (as is, of course, the case in ice) or to some water molecules and to some anions, provided that such an arrangement can be achieved in a way which satisfies the characteristic charge distribution. These distinctive roles of water provide a convenient basis for the classification of crystalline hydrates into two groups, as we now explain.

Classification of hydrates

12.21. The first class of hydrates comprises those in which the primary function of the water molecules is to co-ordinate the cations, and so to increase their effective size and distribute more widely their electric charge; these we shall describe as hydrates containing co-ordinating water. In such crystals the water molecules play an essential part in determining the stability of the structure as a whole and cannot be removed without its complete breakdown. The anhydrous compound, if it exists, can bear no structural relationship to the hydrate. Our second class of hydrate comprises those structures in which the cations are not directly co-ordinated by water, and these may be said to contain structural water. In some of these compounds the water molecules merely occupy interstices in the structure where they can add to the electrostatic energy without upsetting the balance of charge, and sometimes the function of the water is so trivial that it can be removed without any breakdown of the crystal or alteration in its structure. This, however, is not always so, and in other hydrates of this type, especially those containing a high proportion of water, the arrangement of the water molecules is the dominating factor in determining the structure adopted; these molecules by themselves form a coherent three-dimensional framework and now it is the remaining atoms in the structure which occupy the interstitial sites. In some hydrate structures water is found in both a co-ordinating and a structural capacity.

A number of hydrates of which the structures have been determined, classified in the manner just described, are shown in table 12.02. The subdivision of the first class is explained below. Hydrates of organic compounds are not included in this table, but some are described in chapter 14.

Hydrates containing co-ordinating water

12.22. The hydrates of the salts of inorganic acids may be represented by the general formula $A_xX_y.zH_2O$, where A is the cation and X is the anion, Cl^-, CO_3^{2-}, SO_4^{2-}, etc. The structure adopted will depend on the relative number of water molecules and cations present, i.e. on the ratio z/x. If (*a*) this ratio is equal to the oxygen co-ordination number of the cation a structure will be possible in which each cation is co-ordinated by z/x water molecules and in which these co-ordinating polyhedra are independent. If z/x is smaller than this value there are insufficient water molecules to form independent polyhedra, and two possibilities now arise: (*b*) the co-ordination of the cations by water is achieved by a sharing of water molecules between polyhedra so that some, at least, of these molecules co-ordinate two cations; (*c*) the cations are co-ordinated partly by water molecules and partly by anions. It is on the basis of these distinctions that the first class of hydrates in table 12.02 has been subdivided into three groups.

Table 12.02. *The classification of hydrate structures*

Hydrates containing only co-ordinating water		
(*a*)	(*b*)	(*c*)
$AlCl_3.6H_2O$	$LiClO_4.3H_2O$	$CuCl_2.2H_2O$
$NiSO_4.6H_2O$	$SrCl_2.6H_2O$	$CaSO_4.2H_2O$
$MgCl_2.6H_2O$	$Na_2B_4O_5(OH)_4.8H_2O$	$KB_5O_8.4H_2O$
$BeSO_4.4H_2O$		$CsAl(SO_4)_2.12H_2O$
$NiSnCl_6.6H_2O$		
The alums*		
Hydrates containing only structural water		
	Zeolites	
	Hydrates of polyacids	
	Gas hydrates	
Hydrates containing both co-ordinating and structural water		
	$NiSO_4.7H_2O$	
	$CuSO_4.5H_2O$	

* See text.

If z/x exceeds the oxygen co-ordination of the cation more water molecules are available than are required to co-ordinate that ion, and the excess can be present only in a structural role. Hydrates of this type are considered separately later.

Cations co-ordinated by independent polyhedra

12.23. The characteristic features of hydrate structures of this type may be best illustrated by describing in detail one typical such compound and by then considering only briefly a number of others which have been studied. The structure which we select for discussion is that of beryllium sulphate tetrahydrate.

Beryllium sulphate tetrahydrate

12.24. The structure of $BeSO_4.4H_2O$, projected on the face of the tetragonal unit cell of dimensions $a = 8{\cdot}02$, $c = 10{\cdot}75$ Å, is illustrated in fig. 12.08*a*. It will be seen that the structure consists of discrete $Be(H_2O)_4$ and SO_4 tetrahedra, of very nearly the same size, and that the tetrahedra of each kind lie in columns parallel to the z axis of the cell. Successive tetrahedra, separated by a distance $\frac{1}{2}c$, have their bonds approximately in anti-parallel orientation, but if this distinction is ignored, and if each tetrahedron is regarded simply as a spherical unit, it is possible to describe the structure in terms of the alternative cell, of height $\frac{1}{2}c$, shown in fig. 12.08*b*. This cell, of dimensions $a = 5{\cdot}68$, $c = 5{\cdot}37$ Å, is pseudo-cubic, and has an SO_4 group at each corner and a $Be(H_2O)_4$ group at the centre. Looked at in this way, the structure may be described in very simple terms as a slightly distorted caesium chloride arrangement in which $[Be(H_2O)_4]^{2+}$ replaces Cs^+, and SO_4^{2-} replaces Cl^-. This co-ordination emphasizes that the hydrate is better formulated as $[Be(H_2O)_4]SO_4$, rather than in the conventional manner as $BeSO_4.4H_2O$.

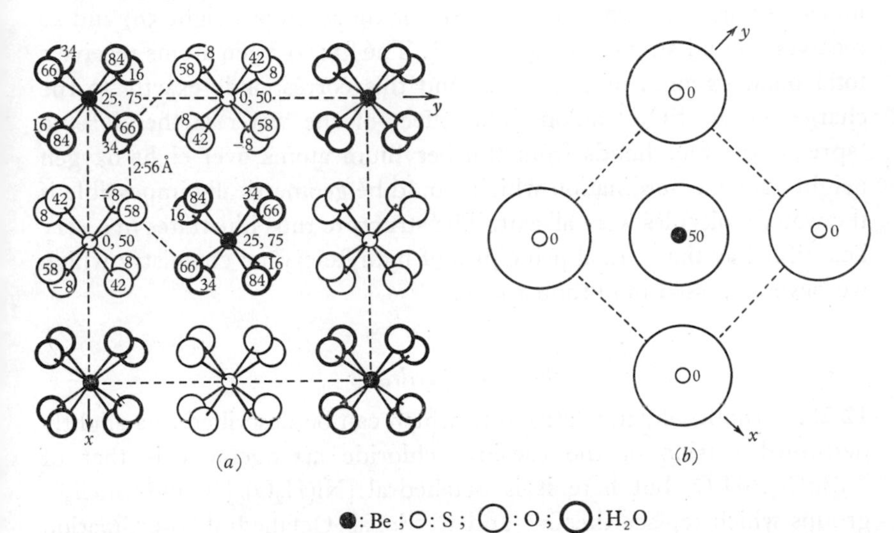

Fig. 12.08. (*a*) Plan of the unit cell of the tetragonal structure of $BeSO_4.4H_2O$ projected on a plane perpendicular to the *z* axis. The heights of the atoms are expressed in units of *c*/100. Hydrogen bonds from one of the water molecules and from one of the oxygen atoms of an SO_4 group are represented by broken lines. (*b*) The corresponding pseudo-cubic pseudo-cell showing the relationship of the structure to that of caesium chloride. The co-ordinating tetrahedra $Be(H_2O)_4$ and SO_4 are symbolized by large circles.

Before we dismiss the structure, however, it is important also to consider whether the co-ordination of the water molecules is consistent with their characteristic structure. Each water molecule, in addition to its beryllium neighbour, has two oxygen neighbours of two different SO_4 groups to which it is attached by hydrogen (or hydroxyl) bonds of length 2·56 Å. These bonds, for the molecule at height 66, are shown in fig. 12.08*a*, and it will be seen that they give rise to an approximately planar configuration of neighbours.

The water molecules must therefore be considered to be disposed in the crystal in such a way that the two seats of negative charge are directed towards a beryllium atom while the positive charges are presented towards two different oxygen atoms, corresponding to the second alternative configuration described in § 12.20. It is interesting to note that on this basis the electrostatic valency principle is satisfied in detail. Each Be–H_2O bond is of strength $\frac{2}{4} = \frac{1}{2}$, and therefore each H_2O–O bond must be of strength $\frac{1}{4}$ to accord with the neutrality of the water molecule. Each oxygen atom of the SO_4 groups, however, is linked to two water molecules (as is shown in fig. 12.08*a* for the atom at height 58) and so receives a bond strength of $2 \times \frac{1}{4} = \frac{1}{2}$. The four oxygen atoms receive a total bond strength of $4 \times \frac{1}{2} = 2$, and this corresponds exactly to the charge on the SO_4^{2-} anion. The effect of the water is therefore to 'spread' the four bonds from the beryllium atoms over eight oxygen neighbours, a co-ordination which would be geometrically impossible if the water molecules were absent. The structure thus illustrates in a very beautiful way the general principles of hydrate crystal chemistry which we have discussed in outline above.

Some hexahydrates

12.25. Another hydrate structure which can be described as a slightly deformed version of the caesium chloride arrangement is that of $NiSnCl_6.6H_2O$, but here it is octahedral $[Ni(H_2O)_6]^{2+}$ and $SnCl_6^{2-}$ groups which replace the Cs^+ and Cl^- ions. Octahedral co-ordination is also found in $MgCl_2.6H_2O$ and $MgBr_2.6H_2O$, and, on the basis of the size of the complex $Mg(H_2O)_6$ group compared with that of the halogen ion, a fluorite structure might be anticipated. Such an arrangement, however, would necessitate each water molecule being in contact with four chlorine ions and is clearly not compatible with the tetrahedral water structure, which would allow only two such contacts. The actual structure observed is closely related to the fluorite arrangement and may be described as a distortion of that structure whereby the number of chlorine contacts of each water molecule is reduced from four to two. The structure of $Mg(H_2O)_6Cl_2$ is of especial interest by reason of its relationship to that of the ammine $Mg(NH_3)_6Cl_2$ discussed below (§ 12.39).

The alums

12.26. In the structures of the alums, $A^+B^{3+}(SO_4)_2 . 12H_2O$, both the *A* and *B* cations are surrounded by six water molecules. The arrangement of the resulting $[A(H_2O)_6]^+$, $[B(H_2O)_6]^{3+}$ and SO_4^{2-} groups, however, is not always the same, but gives rise to three different structures, the transition between which is determined by the size of the univalent ion. The two commonest of these, typified by $NaAl(SO_4)_2 . 12H_2O$ and $KAl(SO_4)_2 . 12H_2O$, correspond to ions of small and intermediate size, respectively. In $CsAl(SO_4)_2 . 12H_2O$, however, the Cs^+ ion is too large to have only six neighbours, and a structure accordingly results in which this ion is co-ordinated not only by six water molecules but also by six oxygen atoms of the SO_4^{2-} groups. Strictly speaking, therefore, this alum does not belong to the class now under consideration but instead to that discussed in §§12.29–12.32 below.

It is interesting to note that although the alums have three structures they are all cubic and morphologically indistinguishable. We have, therefore, in these salts an example of a case where the lack of isomorphism can be revealed only by X-ray methods.

Cations co-ordinated by shared polyhedra

Borax

12.27. In borax, $Na_2B_4O_5(OH)_4 . 8H_2O$, the H_2O:Na ratio of 4:1 is insufficient to permit octahedral co-ordination of the sodium atoms by independent water molecules. Octahedral co-ordination, however, can still be achieved if four of the six molecules in each polyhedra are shared with other polyhedra, and this is realized in the structure of the salt in the way illustrated in fig. 12.09*a*. It will be seen that the $Na(H_2O)_6$ octahedra are linked into endless zigzag chains by sharing two edges, and that half the water molecules are therefore co-ordinated by two sodium ions and half by only one. The anions $[B_4O_5(OH)_4]^{2-}$, of the form already discussed (§11.03 and fig. 11.02), lie between these chains and are themselves also linked into chains by hydrogen bonds of length 2·74 Å between hydroxyl groups and oxygen atoms (fig. 12.09*b*). Finally, hydrogen bonds link these anions to the water molecules of the $Na(H_2O)_6$ chains.

It is difficult to represent this system of bonding in a structure diagram, but the environment of the water molecules, hydroxyl groups and oxygen atoms is shown schematically in fig. 12.10, where the atoms of fig. 12.09

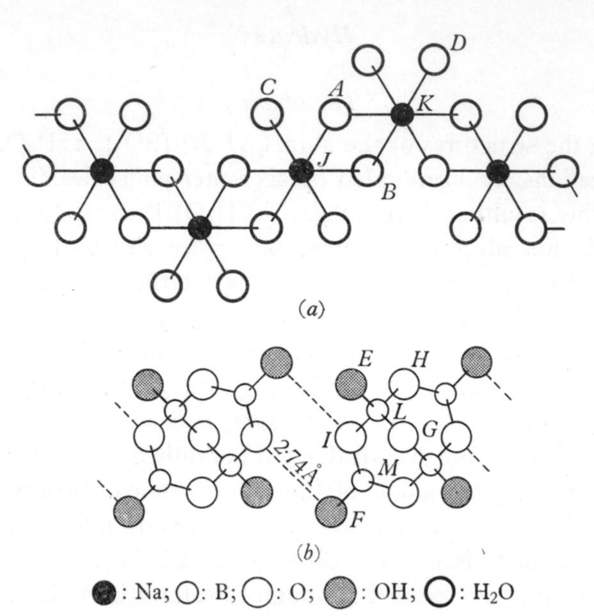

Fig. 12.09. The structure of borax, $Na_2B_4O_5(OH)_4.8H_2O$: (*a*) linkage of $Na(H_2O)_6$ octahedra into chains of composition $Na(H_2O)_4$ by sharing water molecules; (*b*) linkage of $B_4O_5(OH)_4$ anions into chains by means of hydrogen bonds.

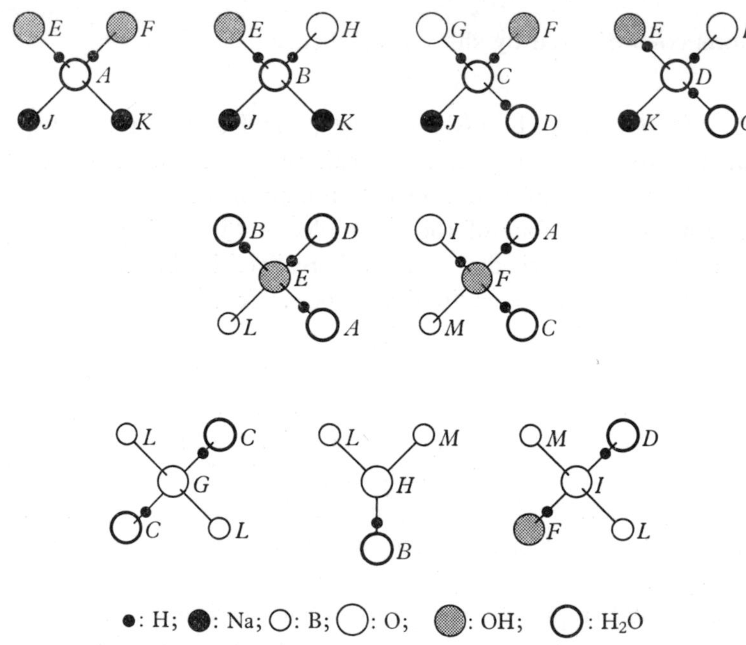

Fig. 12.10. Co-ordination of the water molecules, hydroxyl groups and oxygen atoms in the structure of borax, $Na_2B_4O_5(OH)_4.8H_2O$.

are identified by letters *A*, *B*, ... and where the polarity of the hydroxyl groups and water molecules is symbolically indicated by dots representing the hydrogen atoms.

If we consider first the water molecules it will be seen that they are of four types, with co-ordinations as follows:

(*A*) 2Na+2OH, (*B*) 2Na+1OH+1O,
(*C*) 1Na+1OH+1O+1H_2O, (*D*) 1Na+1OH+1O+1H_2O.

In each case, however, the water molecule is tetrahedrally co-ordinated and in each case the polarity of its neighbours is consistent with a tetrahedral charge distribution of two regions of positive and two regions of negative charge over its surface.

Turning next to the hydroxyl groups, we find that they are of two types:

(*E*) 1B+3H_2O, (*F*) 1B+1O+2H_2O.

Again, both are tetrahedrally co-ordinated, but now the polarity of the neighbours corresponds to a tetrahedral distribution of one positive and three negative charges on the hydroxyl group, just as we have found in those hydroxides, such as $Al(OH)_3$, in which hydrogen bonds are operative. We therefore have in borax a particularly elegant example of a hydrogen-bonded structure and of the characteristically directed configuration of hydrogen bonds about water molecules and hydroxyl groups.

Other hydrates

12.28. In $LiClO_4.3H_2O$ octahedral co-ordination of the cations is again found, but it is clear that such a co-ordination is consistent with the H_2O:Li ratio of 3:1 only if every water molecule is common to two $Li(H_2O)_6$ octahedra. This is achieved in the structure by the octahedra sharing opposite faces, as shown in fig. 12.11*a*, and the endless chains thus formed are bound by hydrogen bonds to ClO_4^- ions, situated between them. In $SrCl_2.6H_2O$ it might be expected that independent $Sr(H_2O)_6$ octahedra would be found, giving a structure analogous to that of $MgCl_2.6H_2O$. In fact, however, the arrangement is quite different on account of the large size of the Sr^{2+} ion. Each of these ions is now co-ordinated by no fewer than nine water molecules, six shared with adjacent polyhedra and three unshared, and the co-ordinating polyhedra are linked into chains of the form shown in fig. 12.11*b*. These chains may be compared with those in $LiClO_4.3H_2O$, but it will be seen that they differ in detail in that here the shared water molecules form a triagonal prism about the cations rather than an octahedron; only in this way does space become available to accommodate the additional water molecules.

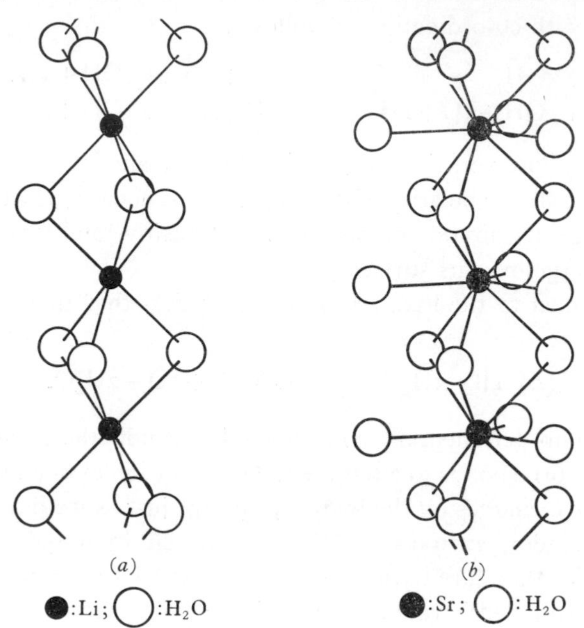

Fig. 12.11. Linkage of cation–water polyhedra: (*a*) in $LiClO_4.3H_2O$; (*b*) in $SrCl_2.6H_2O$.

Cations only partially co-ordinated by water

12.29. When the cation in a hydrate is exceptionally large, or when the number of water molecules present is inadequate to form co-ordinating polyhedra, structures are found in which the cations are co-ordinated both by water molecules and by atoms of the anion. We have already instanced caesium alum as an example of a hydrate with a very large cation, and we may now consider $CaSO_4.2H_2O$ and $CuCl_2.2H_2O$ as examples of compounds containing few water molecules.

Gypsum

12.30. The structure of gypsum, $CaSO_4.2H_2O$, is illustrated in a somewhat idealized form in fig. 12.12. It is essentially a layer structure and the layers are here seen edge-on. Each layer is itself composite, and consists of four sheets containing respectively H_2O, $Ca^{2+}+SO_4^{2-}$,

$Ca^{2+}+SO_4^{2-}$, and H_2O. Within the layers the arrangement is such that each cation is co-ordinated by six oxygen atoms of SO_4 groups and by two water molecules. Each water molecule is linked to one Ca^{2+} ion and (by hydrogen bonds) to two oxygen atoms of two different SO_4 groups, one in its own layer and one in the adjacent layer, the last-mentioned bonds being the only links between the layers. The perfect cleavage arises from the rupture of these hydrogen bonds, and their weakness is revealed by the very much greater coefficient of thermal expansion normal to the sheets than in any other direction.

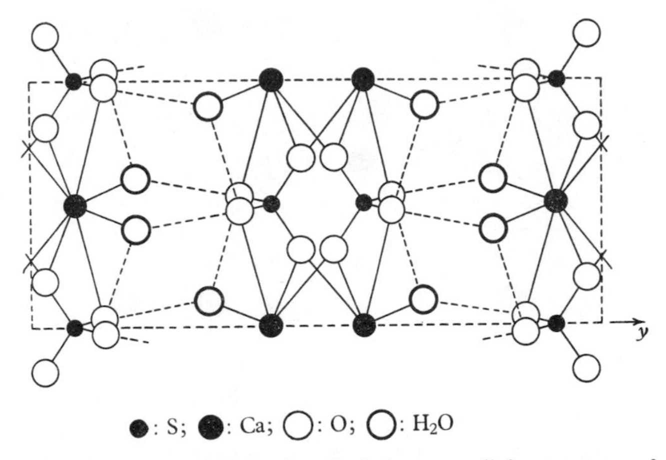

●: S; ●: Ca; ○: O; ○: H_2O

Fig. 12.12. Idealized plan of the unit cell of the monoclinic structure of gypsum, $CaSO_4.2H_2O$, projected on a plane perpendicular to the x axis. The layers (see text) are here seen edge-on, one complete layer in the middle of the diagram and two half-layers at its edges. Hydrogen bonds are represented by broken lines.

Cupric chloride dihydrate

12.31. In $CuCl_2.2H_2O$ each copper atom is co-ordinated by two chlorine atoms and two water molecules arranged in the *trans* configuration at the corners of a square, and these planar $Cu(H_2O)_2Cl_2$ groups are disposed in the crystal in the manner shown in fig. 12.13. The co-ordinating groups are linked together by weak hydrogen bonds of length 3·05 Å between each water molecule and two chlorine atoms of two other groups, and it will be seen from the figure that this system of hydrogen bonds gives the structure coherence in three dimensions. It is interesting to compare this structure with that of the anhydrous chloride $CuCl_2$ (§8.28), in which square planar co-ordination about the copper atom also occurs. In the anhydrous salt, however, each chlorine atom is necessarily shared between two copper atoms, to give a Cl:Cu ratio of 2:1, and the $CuCl_4$ groups are therefore linked into endless bands (fig. 8.11).

We have described the structure of $CuCl_2.2H_2O$ in terms of planar groups of $2H_2O + 2Cl$ co-ordinating the copper atoms; in these groups the Cu–Cl distance is 2·26 Å. It should be noted, however, that each

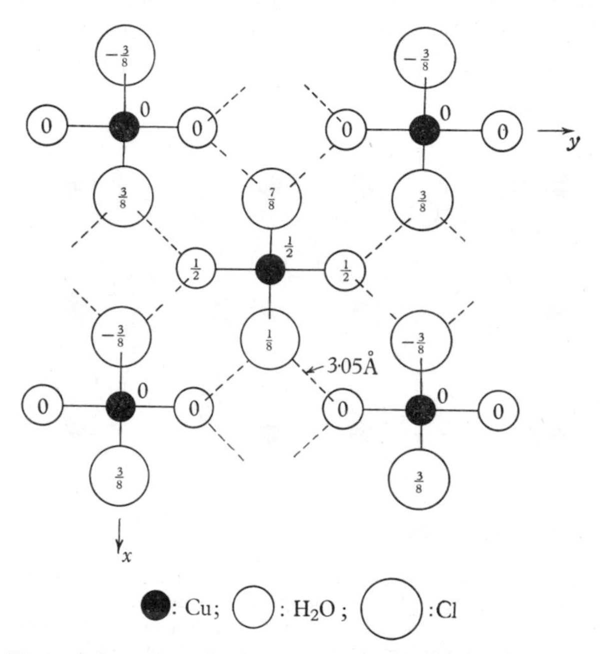

Fig. 12.13. Plan of the unit cell of the orthorhombic structure of cupric chloride dihydrate, $CuCl_2.2H_2O$, projected on a plane perpendicular to the *z* axis. The heights of the atoms are indicated in fractions of the *c* translation and hydrogen bonds are represented by broken lines. The corners of the unit cell coincide with the copper atoms at height 0.

copper atom also has two further chlorine neighbours (namely, one from each of the two groups vertically above and below it) at the considerably greater distance of 2·95 Å, and that if these neighbours are taken into account the co-ordination about the copper atoms is in the form of a distorted octahedron. The structure as a whole is then built up of chains of these octahedra, running parallel to the *z* axis, in which the octahedra are united by sharing two opposite edges. Distorted octahedral co-ordination of this type is found in many structures containing the Cu^{II} atom.

Hydrated acids and acid salts

12.32. The crystal structures of hydrated acids and acid salts are of interest because in them some of the hydrogen ions may be associated with water molecules to form hydroxonium ions H_3O^+. The structure of hydrogen chloride monohydrate, $HCl.H_2O$, is formed by the superposition of a series of puckered sheets, of the type shown in fig. 12.14, in which each chlorine and each oxygen atom is co-ordinated by three atoms of the other kind arranged at the corners of a shallow trigonal

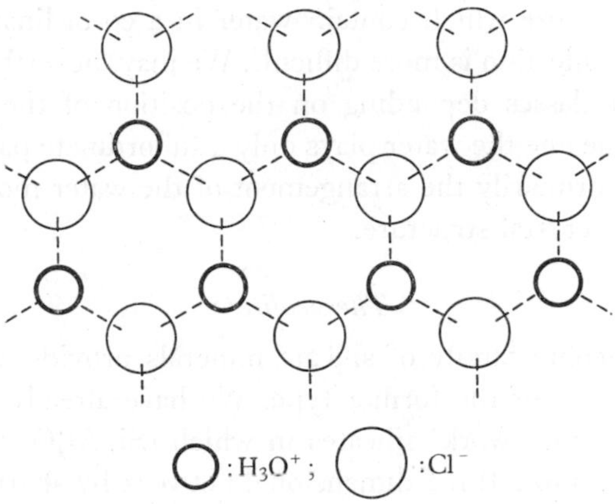

$:H_3O^+$; $:Cl^-$

Fig. 12.14. Idealized plan of a single sheet of the structure of hydrogen chloride monohydrate, $HCl.H_2O$. Hydrogen bonds are represented by broken lines.

pyramid. (In this respect the structure may be compared with that of arsenic (fig. 7.05).) Such an arrangement is clearly not consistent with an assembly of HCl and H_2O molecules, but argues instead the presence of Cl^- and H_3O^+ ions, the latter having a pyramidal structure with seats of positive charge at three corners of a regular tetrahedron. Weak hydrogen bonds, of length 2·95 Å, link each hydroxonium ion to its three chlorine neighbours in the same sheet, and successive sheets are held together only by van der Waals forces. In this structure the presence of H_3O^+ ions has been independently confirmed by both infra-red spectroscopy and nuclear magnetic resonance studies.

Hydroxonium ions are also found in the two hydrates of nitric acid, $HNO_3.H_2O$ and $HNO_3.3H_2O$. The former hydrate again has a structure of puckered sheets in which NO_3^- and H_3O^+ ions are each co-ordinated by three ions of the other kind, the co-ordination about the hydroxonium ions being pyramidal. In $HNO_3.3H_2O$ both water molecules and H_3O^+ ions occur, and the structure as a whole consists of NO_3^- ions, H_2O molecules and H_3O^+ ions linked together in three dimensions by a complex system of hydrogen bonds.

In $Al(OH)_3$, ice and these acid hydrates we have examples of the characteristic spatial distribution of charge on the groups HO^-, H_2O and H_3O^+.

Hydrates containing only structural water

12.33. The hydrates containing structural water are more varied in character than those which contain water in a co-ordinating role, and systematic classification is more difficult. We may, nevertheless, broadly recognize two classes depending on the position of the water in the structure; in the one the water plays only a subordinate part, whereas in the other it is primarily the arrangement of the water molecules which determines the crystal structure.

The zeolites

12.34. The zeolite family of silicate minerals provide an example of hydrate structures of the former type. We have already seen (§11.31) that these are 'framework' silicates in which (Si, Al)O_4 tetrahedra are linked together into a three-dimensional network by sharing all oxygen atoms. This network, however, is far more open than that in other types of framework silicate and can readily accommodate water molecules in its interstices, whereas in, say, the felspars inclusion of water is not possible. The trivial function of water in the zeolites is emphasized by the fact that it can be expelled from the crystal without destruction of the structure and can even be replaced by other neutral molecules, such as those of ammonia, carbon dioxide, iodine and alcohol. The relatively complexity of some of these molecules attests the large size of the cavities available to accommodate them.

The hydrates of the polyacids (§10.24) resemble those of the zeolites in that the water molecules are again situated in the interstices of a very open structure, and it is noteworthy that in these structures, too, the water may be readily removed.

Clathrate hydrates

12.35. In hydrates containing a large proportion of water it is clear that the water molecules must be present in more than a purely interstitial capacity. Unfortunately very few such compounds have been studied, but one structure of particular interest may be described.

It has long been known that a considerable number of gases can be crystallized in the form of highly hydrated solids; among these gases are ammonia, chlorine, sulphur dioxide, argon, krypton and many others. The structure of one such hydrate, that of chlorine of approximate composition $Cl_2.8H_2O$, has been determined in detail. In this structure the cubic unit cell contains forty-six water molecules. At the centre of the cell twenty of these molecules are arranged at the corners of a regular

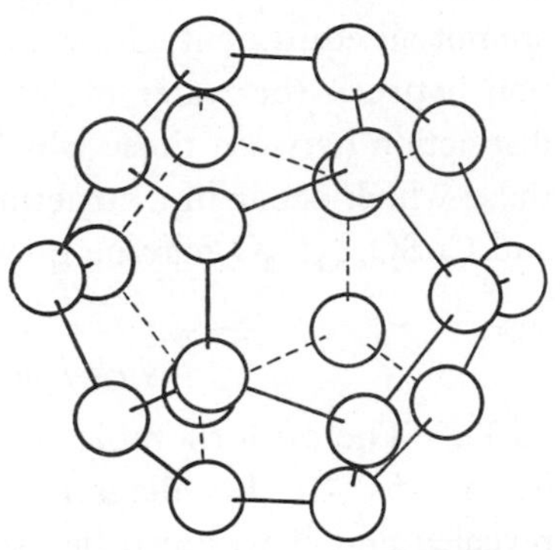

Fig. 12.15. Clinographic projection showing the arrangement of some of the water molecules in the clathrate structure of chlorine hydrate, $Cl_2.7\frac{2}{3}H_2O$.

pentagonal dodecahedron of the form shown in fig. 12.15, and each molecule is bound to its three neighbours by hydrogen bonds of length 2·75 Å (very nearly the same as in ice) directed along the edges of the dodecahedron. These edges are inclined to one another at almost exactly the tetrahedral angle, and each water molecule can therefore participate in one further hydrogen bond extending radially outwards from the polyhedron. Some of these bonds unite the water molecules of the dodecahedron under consideration to those of eight other such dodecahedra arranged about the corners of the cell, thus accounting for forty of the forty-six water molecules in the cell. The remaining six molecules lie in interstices between groups of four dodecahedra, to which they are bound by the as yet unused hydrogen bonds. In this way a structure is achieved in which each water molecule is tetrahedrally co-ordinated by four others and is bound to them by hydrogen bonds to give a three-dimensional framework; the structure so far described may therefore be regarded as that of an imaginary form of ice of low density. The chlorine molecules are situated in six large interstices (of diameter 5·9 Å) in this framework, giving a crystal of composition $Cl_2.7\frac{2}{3}H_2O$, and the existence of analogous hydrates of the inert gases makes it clear that these molecules are held by purely mechanical constraints. Crystals of this type, in which molecules are imprisoned within a framework of molecules of another kind, are termed *clathrate structures*. We shall discuss further examples of such structure in the chapter devoted to organic compounds.

Hydrates containing co-ordinating and structural water

12.36. Many hydrated salts contain an odd number of water molecules, and often there are chemical grounds for believing that these molecules are not all equivalent. In such cases we generally find that the distinction between the water molecules corresponds in the structure to the distinction between those which are present in a co-ordinating role and those which occur in a structural capacity. The hydrates $NiSO_4.7H_2O$ and $CuSO_4.5H_2O$ may be quoted as examples of this.

Nickel sulphate heptahydrate

12.37. The orthorhombic structure of $NiSO_4.7H_2O$ is represented in fig. 12.16. In this diagram the tetrahedral SO_4 groups are clearly revealed and it will also be seen that each nickel atom is octahedrally co-ordinated by six water molecules. In addition to these six water molecules, however, there is also a seventh, marked *Bg*, which is linked to no cation but only to an oxygen atom of an SO_4 group and to other water molecules, a distinction which we can emphasize by writing the formula as $Ni(H_2O)_6SO_4.H_2O$. Not all of the six water molecules co-ordinating any one nickel atom are the same. Four of these molecules (*Aa*, *Ac*, *Ad* and *Af*), which may be described as of Type *A*, make two contacts external to the $Ni(H_2O)_6$ group, and these molecules have their three bonds roughly in a plane. The other two water molecules (*Bb* and *Be*), which are at opposite corners of the octahedron and which may be called Type *B*, make three external contacts, and the four bonds to these molecules are arranged tetrahedrally, as are those to the isolated water molecule *Bg*. The environment of each atom in the structure is represented purely schematically in fig. 12.17, in which the 'polarity' of the bonds to each water molecule is also shown. It will be seen from this diagram that although the detailed co-ordination round all the water molecules of one type is not the same (for example, *Af* differs in its co-ordination from *Ac* and *Ad*) every one of these molecules is co-ordinated by either three or four neighbours in a manner consistent with its characteristic charge distribution. The structure is therefore a particularly elegant example of the principles underlying the formation of hydrated salts, and illustrates very clearly the way in which water is bound in these structures and may appear either co-ordinating the cations or situated in interstices where it is not directly bonded to any cations at all. While a rough description of such structures may often be

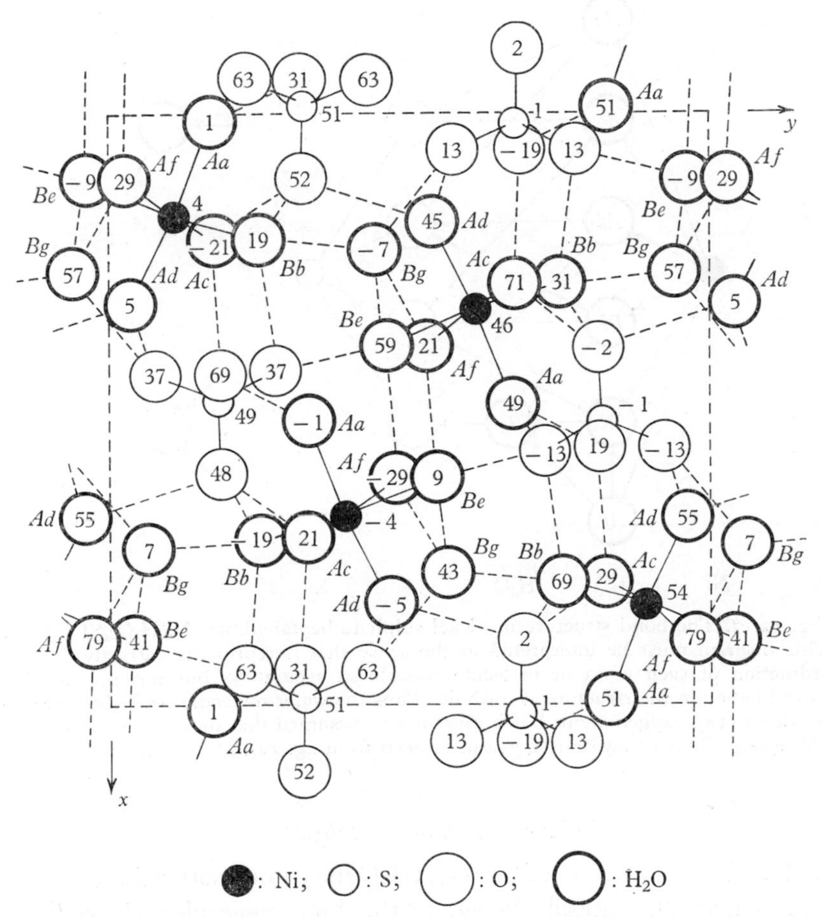

Fig. 12.16. Plan of the unit cell of the orthorhombic structure of nickel sulphate heptahydrate, $Ni(H_2O)_6SO_4.H_2O$, projected on a plane perpendicular to the z axis. The heights of atoms are indicated in units of $c/100$ and hydrogen bonds are represented by broken lines. The heights marked have been expressed in such a way as to emphasize the co-ordination round the nickel and sulphur atoms, but in considering the distribution of the hydrogen bonds it must be remembered that an atom at a height, say, -7 lies outside the cell but is repeated within the cell at a height 93. Some or all of the neighbours of any given oxygen atom or water molecule may therefore be not those marked but others displaced above or below them by a distance c.

given in terms of the packing of the anions, complex or simple, and of the co-ordinating polyhedra of water molecules round the cations, it is clear from this example that the packing must be carried out in such a way that the somewhat exacting demands of the tetrahedral form of the water molecule are satisfied.

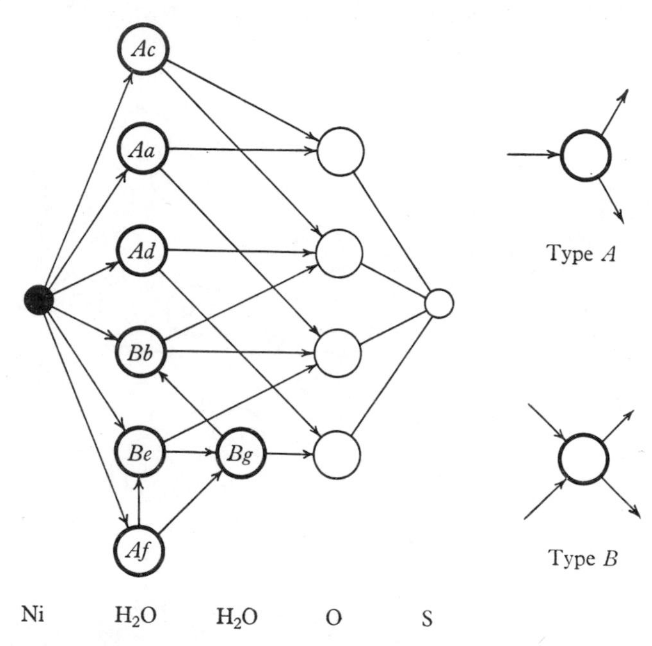

Fig. 12.17. The bond structure in nickel sulphate heptahydrate, $Ni(H_2O)_6SO_4.H_2O$. This diagram must be interpreted in the sense that it represents correctly the co-ordination of each atom or molecule considered separately, but not the detailed co-ordination in the structure as a whole. Thus the water molecule *Ac* is shown to be bonded to two oxygen atoms but it must not be assumed that they belong to a single SO_4 group; in fact they do not, as can be seen from fig. 12.16.

Cupric sulphate pentahydrate

12.38. In the structure of $CuSO_4.5H_2O$ the co-ordination found is that represented schematically in fig. 12.18. Four molecules of structural water (*Aa*, *Ab*, *Ac* and *Ad*) co-ordinate the copper atoms, each of which is, however, also co-ordinated by two oxygen atoms of two SO_4 groups and is therefore octahedrally surrounded by neighbours. The fifth water molecule (*Be*) is co-ordinated only by other water molecules and oxygen atoms. Again it will be seen that the water molecules are of two types, 3- and 4-co-ordinated, respectively, and that the polar character of these molecules is reflected in the distribution and polarity of the bonds about them.

In the two structures just described the 'odd' structural water molecule is linked to other water molecules and to an oxygen atom of the SO_4 groups. It is interesting to note, however, that halides rarely crystallize with an odd number of water molecules because hydrogen bonds to halogen atoms (except fluorine) are not, in general, formed and structural water cannot therefore be held in the crystal.

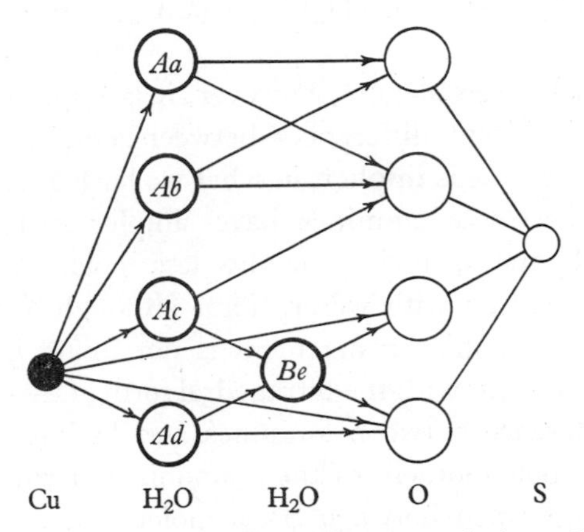

Fig. 12.18. The bond structure in cupric sulphate pentahydrate, $Cu(H_2O)_4SO_4.H_2O$. For the interpretation of this diagram see the legend of fig. 12.17.

AMMINES

12.39. The ammonia molecule resembles that of water in being small and polar, and we accordingly find that there are numerous ammines (or ammoniates), analogous to hydrated salts, in which the primary function of the ammonia is to co-ordinate the cation and to increase its effective size. The great majority of the ammines of which the structures have so far been investigated are of the general type $A(NH_3)_6X_2$, where A is one of the divalent elements Mg, Ca, Zn, Cd, Mn, Fe, Co, Ni, and X is a large univalent anion, being most commonly Cl^-, Br^- or I^- but occasionally a complex group such as ClO_4^-, PF_6^-, SO_3F^-, etc. In all these compounds, with very few exceptions, the structure is one which may be described very simply as a fluorite arrangement of the ions $[A(NH_3)_6]^{2+}$ and X^-, and in each case the ammonia molecules are found in regular octahedral co-ordination about the metallic cation. The complex cyanides $Co(NH_3)_6.Co(CN)_6$ and $Co(NH_3)_6.Cr(CN)_6$

have the same structure as $Ni(H_2O)_6.SnCl_6$ (§ 12.25), again with the cation octahedrally co-ordinated by ammonia, and the close analogy between the roles of the water and ammonia molecules is emphasized even more clearly by certain structures in which these molecules can replace each other statistically. Thus the compounds $Co(NH_3)_5H_2O.Co(CN)_6$ and $Co(NH_3)_4(H_2O)_2.Co(CN)_6$ have the same structure as $Co(NH_3)_6.Co(CN)_6$, but with water and ammonia molecules distributed at random in the co-ordinating polyhedra. Similarly, the salts $Co(NH_3)_6.(ClO_4)_3$ and $Co(NH_3)_5H_2O.(ClO_4)_3$ are structurally isomorphous.

In spite of such resemblances, however, it is important to note that there are also significant differences between ammines and hydrates. One of these differences is implicit in what we have already said, for we find that in many cases ammines have simpler structures than the corresponding hydrates, owing to the less polar character of the ammonia in comparison with water. Thus although $Mg(NH_3)_6Cl_2$ has the fluorite structure, this arrangement is not found in $Mg(H_2O)_6Cl_2$ because it is inconsistent with the tetrahedral form of the water molecule.

A second difference between ammines and hydrates, again arising from the small dipole moment of the ammonia molecule, is that strong bonds cannot be formed between these molecules: they are therefore found in ammines only in a co-ordinating and never in a structural capacity. For this reason the ammine counterparts of hydrates with an odd number of water molecules do not exist, and it is interesting to note, as an example of this point, that cupric sulphate forms only the hydrated ammine $Cu(NH_3)_4SO_4.H_2O$ and not the compound $Cu(NH_3)_4SO_4.NH_3$.

A third point of difference between ammines and hydrates is connected with their stability. It might be expected on general grounds that hydrates would be more stable than ammines since the dipole moment of water is greater than that of ammonia. This is certainly true of the salts of the *A* sub-group metals. When we turn to consider those of the transition metals, however, we find quite the reverse state of affairs, some of these ammines being notable for their exceptional stability. In part this difference may be ascribed to the greater polarizability of ammonia, for not only dipole moment but also polarizability of the co-ordinating molecules will determine the strength of the bonds by which they are held. Thus the low polarizing power of the relatively large ions of the *A* sub-group metals will favour the formation of hydrates, while the smaller and more strongly polarizing ions of the transition metals will be more readily co-ordinated by ammonia. It is doubtful, however, whether such an explanation is adequate, and it seems more likely that in the ammines of the transition metals the ammonia molecules are bound by essentially covalent forces. Evidence for this is available in the crystal structures of a number of these compounds. In $Pd(NH_3)_4Cl_2.H_2O$, for example, each palladium atom is co-ordinated by four NH_3 molecules disposed about it at the corners of a square, and the $[Pd(NH_3)_4]^{2+}$ and Cl^- ions form a structure analogous to that of K_2PtCl_4 (§ 10.19), with the water molecules in the interstices. Such co-ordination is clearly not consistent with an ionic picture of the Pd–NH_3 bond but corresponds to the characteristic spatial distribution of the dsp^2 hybrid bonds about the palladium atom.

13 ALLOY SYSTEMS

INTRODUCTION

13.01. The structures of the metallic elements have already been described in chapters 5 and 7. We now turn to consider systems in which these elements are found in association as alloys.

Alloy systems have been known to man since the Bronze Age. It is, however, only in recent times that they have been the subject of systematic studies, and in these studies no tool has proved more powerful than the technique of crystal structure analysis. Indeed, the extension of our knowledge and understanding of the properties of intermetallic systems to which it has given rise is one of the greatest achievements of crystal chemistry. Prior to the application of X-ray methods, the investigation of the properties of alloy systems was confined principally to observations of their behaviour in the liquid state, and the behaviour of the metal as a solid could be determined only by inference from these observations. Transitions in the solid state and the effect of mechanical or heat treatment could not, of course, be observed in this way, and for information on these properties the microscope and other purely physical methods had to be invoked. Even so, these methods were all more or less indirect, and it is only since the application of X-ray analysis that it has been possible to investigate directly in the solid state, under the precise conditions which are of technical interest and without damage to the specimen, the exact positions of all the atoms in the structure, and so to refer to their ultimate cause the physical and chemical properties of the alloy.

It is not surprising that the application of such a powerful method of investigation should have led, on the experimental side, to a vast extension of our knowledge of the properties of alloy systems. Even more important, however, is the fact that it has also laid the foundations of the modern theory of the metallic state, for, as we have seen in chapter 5, the basic concept on which this theory is based is that of the periodic field in a crystal structure. The development of metallurgy in the past has been hampered by attempts to make metal systems conform to the laws of chemical combination established by observations on bodies in which forces of an entirely different character are operative. Alloys differ profoundly in many of their properties from other chemical compounds. They can generally be formed by no more elaborate synthesis than the simple melting together of their constituents in the appropriate proportions. In marked contrast to other chemical compounds, they still preserve, at least qualitatively, the general character of the elements of which they are composed. In spite of these differences, however, it has been common practice to regard many intermetallic systems as chemical compounds and to attempt to assign to them formulae based on ideas of valency derived from ionic and covalent compounds. Somewhat inconsistently, however, the essentially identical forces operating in the structure of a metallic element have never been regarded as chemical in nature at all. The extent to which it is justifiable to speak of chemical combination in alloy systems will be discussed more fully later, but we may say at once that if we are so to interpret these systems it must be in terms of a wider and far less rigid picture of chemical combination than that of classical chemistry. Compounds of variable composition are often found in metal systems, and it is no longer necessary for the constituents to be present in simple stoichiometric proportions.

In spite of what has been said above, it is nevertheless important to realize that the properties of intermetallic systems are by no means completely characterized by their ideal crystal structures alone. Such systems, more than any others, often display the defects defined as 'faults' in chapter 9, and frequently it is to these faults that many of the technically most important properties, such, for example, as hardness and strength, are to be ascribed. Faults of this type are on a scale intermediate between atomic dimensions and those accessible to microscopic study, and therefore constitute a peculiarly difficult field of investigation. They have, nevertheless, been the subject of much work in recent years but it would be inappropriate here to give an account of this work, the interest of which is primarily physical rather than chemical.

The classification of alloys

13.02. The number of alloy systems to which X-ray methods have been applied is now sufficiently large for many of the principles of metal structural chemistry to have emerged. As always, we shall here make no attempt to review the whole of this wide field but will content ourselves with illustrating these general principles by means of a limited number of appropriately chosen examples. A general classification of binary alloys, which alone we shall consider, can conveniently be based on the

Table 13.01. *The classification of the metallic elements in alloy systems*

True metals											*B* sub-group elements				
T_1		T_2									B_1		B_2		
Li	Be														
Na	Mg											Al	Si		S
K	Ca	Sc	Ti	V	Cr	Mn	Fe	Co	Ni	Cu	Zn	Ga	Ge	As	Se
Rb	Sr	Y	Zr	Nb	Mo	Tc	Ru	Rh	Pd	Ag	Cd	In	Sn	Sb	Te
Cs	Ba	La	Hf	Ta	W	Re	Os	Ir	Pt	Au	Hg	Tl	Pb	Bi	Po

The lanthanide and actinide elements belong to class T_2.

classification of the metallic elements shown in table 13.01. The elements of classes T_1 and T_2 (the subdivision is discussed later) are what we may term the 'true' metals, and are characterized by the fact that with few exceptions they have one of the three typical metal structures, namely, hexagonal and cubic close packing and the cubic body-centred arrangement. They include the elements of the first two *A* sub-groups, the transition elements (excluding zinc, cadmium and mercury), and the elements of the lanthanide and actinide series. The elements of classes B_1 and B_2 are those of the *B* sub-groups. As we have seen, they are characterized by less closely packed structures in which the co-ordination often reveals a tendency towards covalent bonding.

On the basis of this division of the metallic elements into two main groups we may recognize binary alloys of the following three classes: (1) *Alloys of two true metals*; (2) *alloys of a true metal and a B sub-group metal*; (3) *alloys of two sub-group metals.*

To these may be added a fourth class: (4) *the interstitial structures*, which resemble alloy systems in many of their properties, although they are structures in which one of the components is a non-metallic element.

In general terms we may say that in passing through the first three of these classes the properties of the alloys become progressively less metallic and the intermetallic systems show increasingly marked resemblances to true chemical compounds. Few other generalizations can be made concerning alloy systems as a whole, and the characteristic features of those of each class are therefore best discussed separately.

ALLOYS OF TWO TRUE METALS

The copper–gold system

13.03. A typical example of an alloy of this class, and one which has been extensively investigated, is the system copper–gold. These two metals are chemically closely related, have a similar electronic configuration and have the same crystal structure. The atomic radii are not very different, being 1·28 and 1·44 Å, respectively.

When the system is studied at a high temperature, or, more conveniently, in the form of a quenched specimen, it is found that complete solid solution between copper and gold takes place at all compositions.

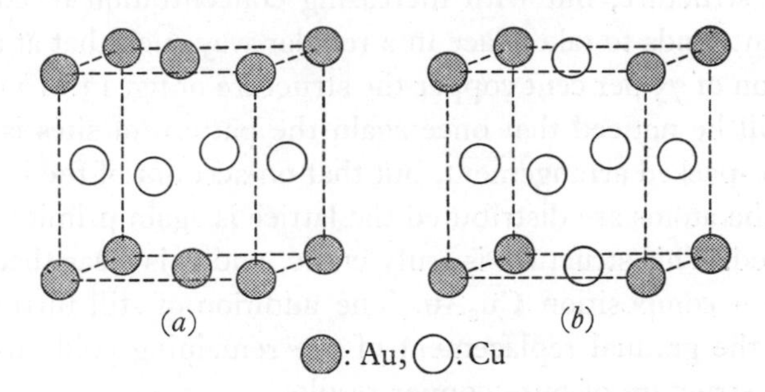

Fig. 13.01. Clinographic projections of the unit cells of the ordered phases in the copper–gold system: (*a*) CuAu; (*b*) Cu_3Au.

If we start, say, with a specimen of pure gold and add to it progressively more and more copper the atoms of this latter element replace those of gold at random at the sites of the cubic face-centred cell, until ultimately the structure of pure copper results. The statistical replacement of gold atoms by the smaller atoms of copper occasions a slight reduction in the cell side, which varies nearly linearly with composition, but otherwise no alteration in the structure occurs.

13.04. The structure of the same system when carefully annealed shows a quite different behaviour. Once again we may consider the addition of progressively more and more copper to initially pure gold. At first, a random replacement of the gold by the copper atoms takes place as before, but when a sufficient quantity of copper has been added it is found that the distribution is no longer statistical, but that instead the two kinds of atoms tend to occupy definite geometrical positions relative to one another. This segregation is complete when the composition corresponds to equal atomic proportions of the two elements, and the structure is then that illustrated in fig. 13.01 *a*. A moment's consideration will reveal that the sites occupied are still those of a cubic close-packed structure, but that, on account of the ordered instead of random distribution of the atoms, the lattice is no longer face-centred but primitive. The segregation of the atoms into layers, and the difference in atomic radius between copper and gold, result in a small departure from cubic symmetry, and the unit cell is actually tetragonal but pseudo-cubic with an axial ratio $c:a = 0{\cdot}932$.

The addition of more copper, followed by careful annealing, results initially in the random replacement of some of the gold atoms of the tetragonal structure, but with increasing concentration of copper this replacement tends to take place in a regular way such that at an atomic composition of 75 per cent copper the structure of fig. 13.01 *b* is formed. Here it will be noticed that once again the pattern of sites is that of a cubic close-packed arrangement, but that on account of the regular way in which the atoms are distributed the lattice is again primitive and not face-centred. This structure is truly cubic, and it is clear that it corresponds to a composition Cu_3Au. The addition of still further copper results in the gradual replacement of the remaining gold atoms, until finally the structure of pure copper results.

13.05. Structures such as those described above for CuAu and Cu_3Au in the ordered state, in which the pattern of sites is that of the parent random solid solution but in which the two kinds of atoms are distributed over these sites in some regular way, are termed *superstructures*. (The term 'superlattice' is also used, but is objectionable for the reasons given in §8.27.) We defer until later any discussion of whether or not these structures are to be regarded as definite chemical compounds, but here we may remark that physically they often show pronounced differences from the corresponding disordered structure. Long before the application of X-ray structure analysis to the problem, it was known that the electrical resistance of quenced copper–gold alloys showed a smooth variation with composition whereas that of the annealed alloys displayed two sharp minima corresponding to the compositions CuAu and Cu_3Au, and more recent work has shown that the resistance of the superstructure at these compositions is less than half that of the corresponding disordered alloy. Similarly, changes occur in many other properties. Thus the elastic constants of the alloy Cu_3Au are profoundly dependent on the heat treatment which it has received, while again the tetragonal structure of CuAu is nearly as soft as pure copper whereas the disordered phase is hard and brittle. In all such cases it is clear that the composition alone is quite inadequate to characterize the alloy.

The iron–aluminium system

13.06. Another system which illustrates the transition from the random arrangement of the solid solution to the ordered structure of the superlattice is that of iron and aluminium. Strictly speaking, this system should be considered under our second group of alloys since we have classed aluminium as a *B* sub-group metal. We have already emphasized, however, that the position of aluminium is somewhat anomalous in that it also behaves in many respects as a true metal, and it is therefore not out of place to consider this system here.

Compared with the copper–gold system, that of iron–aluminium is somewhat more complex in that the two constituents have different crystal structures, iron being cubic body-centred at ordinary temperatures and aluminium cubic close packed. Although the physical properties of the iron–aluminium alloys have not been investigated as completely as those of the copper–gold system, the purely structural features of the order–disorder transformation have been observed in great detail in the range 0–50 atomic per cent of aluminium and may be described in terms of fig. 13.02. This diagram illustrates eight unit cells of the body-centred structure of iron in which different sites, although of course structurally precisely equivalent, are distinguished by different symbols.

If small quantities of aluminium are added to initially pure iron the aluminium atoms replace those of iron quite at random, and in the quenched alloy this obtains throughout the range 0–25 atomic per cent of aluminium. At the composition corresponding to the formulae Fe_3Al the probability of finding any one of the sites *a*, *b*, *c* or *d* occupied by an aluminium atom is, therefore, $\frac{1}{4}$. This condition is represented in fig. 13.03 *a*, where the probability of finding any given site occupied by an aluminium atom is shown as a function of the composition. The purely statistical arrangement of the quenched alloy from 0 to 25 atomic per cent of aluminium is represented by the straight line *a*, *b*, *c*, *d*, which shows that all the four sites *a*, *b*, *c* and *d* are equally favoured. In this arrangement the unit cell is still, of course, one-eighth of the volume shown in fig. 13.02, and the lattice remains body-centred.

In the range 25–50 atomic per cent of aluminium the arrangement of atoms is entirely different. The aluminium atoms are no longer distributed at random over all the sites, but now occupy only the positions *b* and *d*. There is a zero probability of finding an aluminium

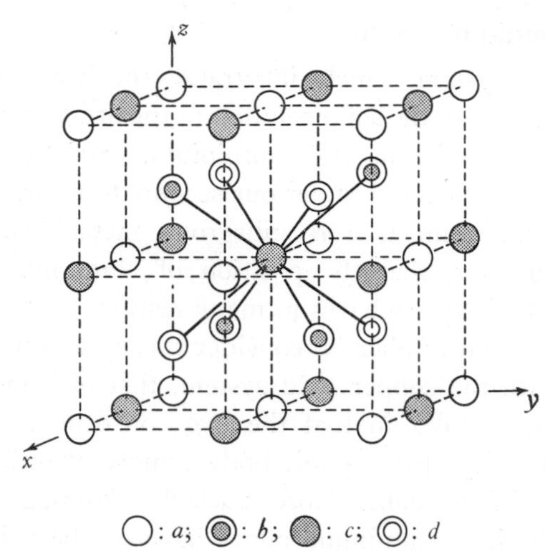

Fig. 13.02. Clinographic projection of the structure of certain phases in the iron–aluminium system (see text).

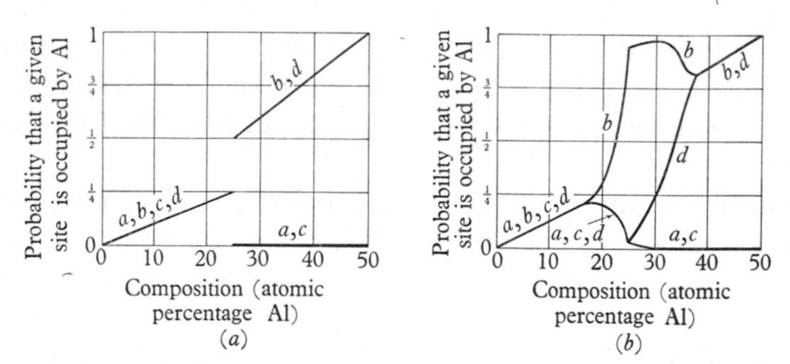

Fig. 13.03. The distribution of atoms over the sites *a*, *b*, *c* and *d* of fig. 13.02 in the iron–aluminium system: (*a*) in the quenched alloy; (*b*) in the annealed alloy.

atom in the sites *a* and *c*, which are always occupied by iron atoms, and a correspondingly greater probability of finding an aluminium atom at *b* or *d*. At a composition near Fe_3Al there is a probability of $\frac{1}{2}$ of finding an aluminium atom in these positions, just one-half of these sites, on the average, being occupied by aluminium and one-half by iron atoms. As the concentration of aluminium increases, so also does this probability, until at the composition FeAl every *b* and *d* site is occupied by aluminium, every *a* and *c* site by iron. The unit cell is still one-eighth of the volume shown in fig. 13.02, but the lattice is no longer body-centred since the sites are not now structurally equivalent. At the composition FeAl the structure is, in fact, that of caesium chloride.

13.07. When the annealed alloy is investigated a different state of affairs is observed. At small concentrations of aluminium the distribution of these atoms is still statistical, but beyond about 18 atomic per cent the aluminium atoms show a marked tendency to favour the *b* positions and to forsake the positions *a*, *c* and *d*. This is represented in fig. 13.03*b*, from which it will be seen that at the composition Fe_3Al almost every *b* site is occupied by aluminium while all the other positions contain iron atoms. With this arrangement, fig. 13.02 now represents only one unit cell. Beyond 25 atomic per cent the *b* sites remain almost fully occupied by aluminium and the new aluminium atoms enter the *d* positions, and from about 30 per cent onwards the concentration in these sites grows rapidly, in part at the expense of the *b* positions, in the way shown by the curve *d* in fig. 13.03*b*. Beyond about 38 per cent the *b* and *d* sites are equally occupied by aluminium and the state of the annealed alloy is then the same as that of the quenched system.

Other systems

13.08. Numerous other binary alloys of two true metals have been investigated by X-ray methods. While these systems necessarily differ in detail many of them show common characteristic features in their structures, some of which have been illustrated in the two examples discussed in detail above. Foremost among these is the tendency to form solid solutions, sometimes embracing the whole range of composition and sometimes separated by intermediate phases of distinctive crystal structure. The extent of this solid solution is determined primarily by the relative radii of the atoms concerned, and it is found that if these radii differ by not more than about 15 per cent conditions are favourable for the formation of a wide range of solid solution whereas if the difference is greater solid solution is correspondingly restricted. This size factor of 15 per cent is not, of course, an absolute criterion and is in any case temperature dependent. The mutual solubility of two metals is generally found to increase with temperature, so that systems in which the size factor is near the limit may behave differently at different temperatures. We have had an example of this in the copper–gold system, where the radius difference is 13 per cent: at high temperatures solid solution occurs at all compositions; at low temperatures the range of solid solution is limited and specimens of composition outside this range crystallize as new phases. In silver–gold, on the other hand, the difference in radius is so small that complete solid solution obtains under all conditions. A system will be capable of existing in both ordered and disordered states only if the difference in radius is neither so small that the disordered solid solution alone is stable nor so large that only the ordered phase can exist.

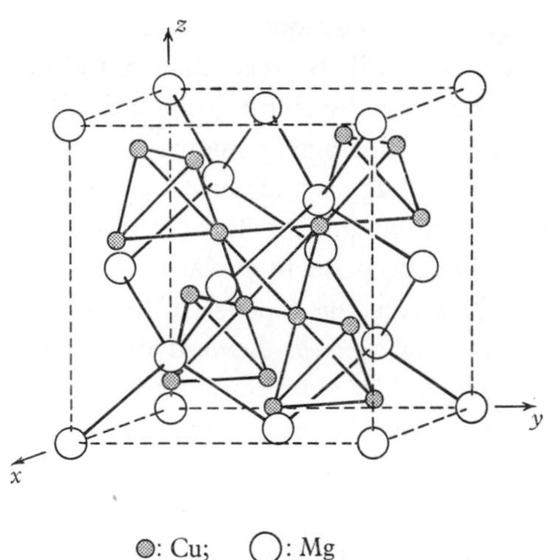

Fig. 13.04. Clinographic projection of the unit cell of the cubic structure of the Laves phase $MgCu_2$.

The intermediate phases found in alloys of two true metals have a number of different structures, but many of these are common to more than one system. In addition to the phases already considered in the copper–gold and iron–aluminium systems we shall describe two, one because of its common occurrence and one because of its geometrical interest.

Laves phases

13.09. A large number of alloys (nearly 100 are known) at a composition MN_2 display one or other of three closely related structures, often termed Laves phases. The simplest and most common of these, of which $MgCu_2$ is an example, is illustrated in fig. 13.04. If we consider first the arrangement of the smaller copper atoms we see that they are disposed at the corners of tetrahedra, and that these tetrahedra are linked into a three-dimensional framework by sharing each copper atom with an adjacent tetrahedron; the distribution of the copper atoms is therefore geometrically the same as that of the oxygen atoms in β-cristobalite (fig. 8.07). The cavities in this framework of copper atoms are occupied by the larger atoms of magnesium, and they, by themselves, are arranged as are the carbon atoms in diamond (fig. 4.02). In the resulting structure each copper atom has 6Mg+6Cu neighbours and each magnesium atom 4Mg+12Cu neighbours. The structures of the other two Laves phases, of which $MgZn_2$ and $MgNi_2$ are examples, are more complex, but in them, too, the smaller N atoms lie at the corners of tetrahedra and these tetrahedra are again united into three-dimensional frameworks, the interstices of which accommodate the larger atoms M.

The majority of Laves phases occur in alloys containing two true metals. This, however, is not always the case, and the structures are found also in $BiAu_2$, KBi_2, $PbAu_2$ and a number of other systems containing metals of the B sub-groups. It is difficult to account for the appearance of the same structures in systems so different in character as, say, KNa_2 and KBi_2, and it seems that the occurrence of these phases is determined solely by geometrical considerations. In the MN_2 structure of fig. 13.04 it is readily seen that the closest packing of the two types of atom will result if their radii are in the ratio $r_M/r_N = \sqrt{3}/\sqrt{2} = 1{\cdot}22$, and the one characteristic common to all the many Laves phases known is that the ratio is never far from this value. Small atoms, such as that of beryllium, can occur only as the N component, and large atoms, such as strontium, only as the M component. Atoms of intermediate size, for example, bismuth, may be found as the N component in association with a larger atom (as in KBi_2) or as the M component when associated with a smaller atom (as in $BiAu_2$).

The WAl_{12} *structure*

13.10. The structure of WAl_{12} is of interest as an example of a phase containing a large excess of one component, and also for geometrical reasons. In the cubic close packing of equal spheres of radius a, any one sphere is co-ordinated by twelve neighbours lying at the corners of a polyhedron of the form shown in fig. 13.05*a*. If, however, we now imagine the central sphere to be removed it becomes possible, by displacing the spheres in such a way as to reduce the distances A–A, B–B, C–C, etc., to achieve a more closely packed arrangement of the remaining spheres in which they now lie at the twelve corners of a regular icosahedron (fig. 13.05*b*). The cavity in the centre of this polyhedron is smaller than before, and can accommodate a sphere of radius only about $0{\cdot}9a$.

Co-ordination of the type just described is found in the WAl_{12} structure. Groups of twelve aluminium atoms lying at the corners of a nearly regular icosahedron surround a smaller atom of tungsten situated at its centre, and the complexes so formed are disposed at the corners and centre of a cubic unit cell in the manner shown in fig. 13.05*b*. Each

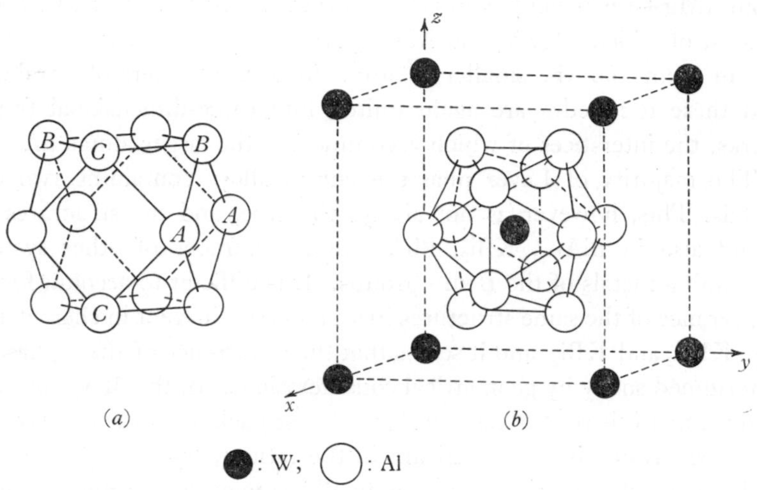

Fig. 13.05. (*a*) Clinographic projection of the polyhedron of atoms co-ordinating any one atom in the cubic close-packed structure. (*b*) Clinographic projection of the unit cell of the cubic structure of WAl_{12}. Tungsten atoms lie at the corners and centre of the cell and are surrounded by twelve aluminium atoms lying at the corners of a regular icosahedron, only one of which is shown. This icosahedron can be regarded as derived from that shown in (*a*) by displacing the atoms in such a way as to reduce the distances *A*–*A*, *B*–*B*, *C*–*C*, etc.

tungsten atom is therefore co-ordinated only by aluminium whereas aluminium atoms have both tungsten and aluminium neighbours. The same arrangement is found in $MoAl_{12}$, and many other structures, although of greater complexity, are also characterized by an icosahedral co-ordination of larger about smaller atoms. In this connexion it is of particular interest to note that this co-ordination is found also in the structures of the α and β forms of elementary manganese, the abnormal properties of which have already been mentioned (§7.20). The icosahedral co-ordination in these structures implies that atoms of different radii are present and confirms the view that atoms occur in the crystal in more than one electronic state.

THE ORDER–DISORDER TRANSFORMATION

13.11. Transitions between ordered and disordered phases, such as those described above in the copper–gold and iron–aluminium systems, have been observed also in many other alloys, both directly by crystal structure analysis and indirectly by inference from changes in physical properties. As a result of these studies it is now clear that the states of order and disorder are but two extreme conditions, and that by appropriate heat treatment an alloy can in general be made to assume intermediate states between these extremes in which the distribution of the atoms is neither the completely regular arrangement of the superstructure nor yet the entirely disordered arrangement of the solid solution. Thus we have already mentioned that the electrical resistance of the ordered phases Cu_3Au and CuAu is less than half that of the disordered phases of the same composition. As these phases are heated the superstructure transforms to the random arrangement of the solid solution, but the transition is not abrupt and the corresponding rise in resistance takes place gradually over a temperature range of the order of 100 °C. Conversely, on cooling the solid solution from high temperatures the superstructure can be gradually regenerated.

Even more direct evidence for the existence of states of intermediate order is available in the copper–gold system at the composition CuAu. In this phase, as we have seen, the disordered state is cubic whereas the superstructure is tetragonal. If the alloy is quenched after being maintained for some time at a temperature T it is found that the axial ratio depends on the exact value of this temperature. When T is appreciably higher than 392 °C the quenched alloy has the completely disordered cubic structure with axial ratio unity, and when T is lower than 320 °C the completely ordered tetragonal superstructure with an axial ratio of 0·932 is obtained. If, however, T lies between these limits a tetragonal structure is formed in which the axial ratio is significantly larger, as the following figures show:

T (°C)	320	380	392	> 392
$c:a$	0·932	0·939	0·947	1

It is clear from these data that in the temperature range 320–392 °C an intermediate degree of order prevails and that as T is raised there is a progressive transition from the tetragonal superstructure towards the cubic arrangement of the disordered solid solution.

Superstructure theory

13.12. The observations described in the preceding paragraphs suggest that at any given temperature an alloy is in a state of thermal equilibrium between the ordered and disordered conditions. Since the fully ordered state is in general the one stable at the lower temperatures we may assume that this state is intrinsically the more stable. As the temperature is raised, however, the energy of thermal agitation seeks to promote a state of disorder, and the degree of disorder actually achieved will be determined both by the temperature and by some parameter of the system representing the difference in energy between the ordered and disordered states. To fix our ideas let us consider the two-dimensional binary alloy of composition MN with the ordered structure shown in fig. 13.06, and let us describe the sites occupied by M and N atoms as m and n sites, respectively. In the state of perfect order the probability p that a given m site, chosen at random, is occupied by an atom M is unity and the probability $(1-p)$ that it is occupied by an atom N is zero. In a state of partial disorder, however, some N atoms will be in m sites and p will be less than unity, and in the state of complete disorder m and n sites are equally likely to be occupied by M or N atoms so that the probabilities p and $(1-p)$ are both $\frac{1}{2}$. We may therefore conveniently adopt as our definition of the degree of order the quantity s defined by the relation*

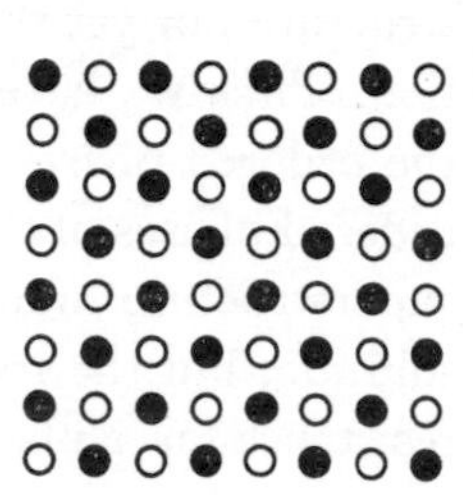

Fig. 13.06. Ordered structure of a two-dimensional binary alloy of composition MN.

$$s = 2p-1,$$

which varies from unity in the ordered state to zero in the state of disorder.

* This relation applies only to a system of composition MN. In the more general case of an alloy of any composition we may adopt the definition $s = (p-r)/(1-r)$, where r is the fraction of atoms of type M; when $r = \frac{1}{2}$ this expression becomes $2p-1$.

Let us now assume that the energy required to interchange a pair of M and N atoms in the structure, thus increasing the degree of disorder, is V. Then at any temperature the degree of order which prevails will be determined in terms of this quantity by the Boltzmann theorem. If V is constant this at once leads to a relation between s and T of the form shown by the broken line in fig. 13.07. As the temperature rises, more and more atoms are shuffled into 'wrong' sites till at very high temperatures the curve approaches $s = 0$ and the arrangement becomes quite random. It is clear, however, that V cannot be regarded as constant,

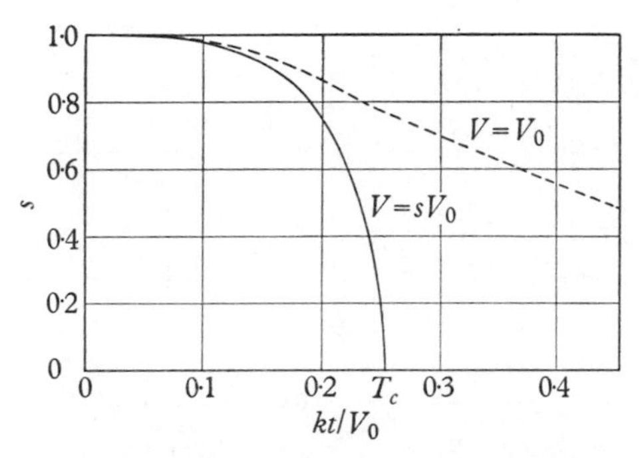

Fig. 13.07. The degree of order in a binary alloy of composition MN as a function of temperature.

for, with increasing disorder, the distinction between 'right' and 'wrong' sites becomes less and less significant, and disappears completely when s is zero. V must therefore vary from a maximum value V_0 when $s = 1$ to zero when $s = 0$, and we may make the simple assumption that this variation is linear, so that $V = sV_0$. The effect of this is to give a relation between the degree of order and temperature of the form shown by the full line in fig. 13.07, and it will be seen that it differs from that previously discussed in that it represents a very much more rapid collapse of the ordered state and the complete disappearance of any trace of order at a characteristic *critical temperature* T_c. Once the disordering process has started the system becomes 'demoralized', its ability to resist the onset of greater disorder is reduced, and finally the degree of order falls catastrophically to zero. In this respect the order–disorder transition is a co-operative phenomenon, and the rapid destruction of order at the critical temperature may be compared with the loss of ferromagnetism in a ferromagnetic material at the Curie point or with the abrupt onset of free rotation of ions or molecules in certain crystals (§9.29). It is a characteristic of such phenomena that the behaviour of the atoms or molecules is influenced by that of their neighbours. Once the process has started the constraints inhibiting further change are lessened, the transformation accelerates, and proceeds rapidly to completion.

13.13. The prediction of a critical temperature is an important feature of the theory of the order–disorder transformation, for it is to be expected that at this temperature many of the physical properties of the alloy will display sharp changes. In particular, anomalies in the specific

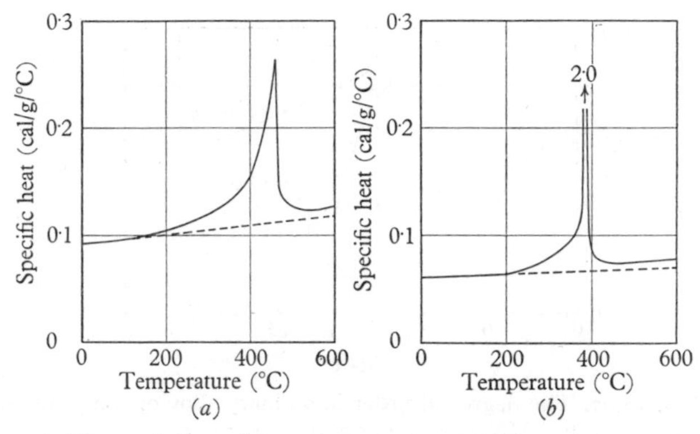

Fig. 13.08. The variation with temperature, on heating, of the specific heat of initially ordered phases: (*a*) CuZn; (*b*) Cu_3Au.

heat are to be expected, for as the temperature of the alloy is raised extra thermal energy must be supplied to break down the ordered structure of the crystal. The specific heat will therefore rise sharply near the critical temperature and then, all order being destroyed, fall again to its normal value. Just such a behaviour is in fact observed in β-brass of composition CuZn (fig. 13.08*a*). In binary systems in which the two components are not present in equal atomic proportions the theory predicts a different relation between s and T: instead of a rapid but continuous decrease to zero at the critical temperature the degree of order suffers a sudden discontinuous change to zero from some finite value. In this case the energy associated with the transformation must be regarded as a latent heat rather than as an anomalously high specific heat. Figure 13.08*b* shows the results of measurements of the specific heat of an annealed alloy of composition Cu_3Au as a function of temperature. Below 250 °C the specific heat is nearly constant. At this temperature it begins to rise sharply and at about 395 °C reaches a value of about 2 cal/g/°C, a value so large that it may be assumed that if the measurements had been made infinitely slowly the specific heat would have been infinite. Above the critical temperature the specific heat falls abruptly but to a value somewhat higher than that at low temperatures. The reason for this difference is discussed below (§ 13.15).

The rate of attainment of equilibrium

13.14. In the above treatment we have assumed implicitly that the alloy is at every temperature in the state of equilibrium appropriate to that temperature and to the value of the parameter V_0. This, however, is in practice not necessarily the case, for it may happen that under a given set of conditions the rate of attainment of equilibrium is so slow that an equilibrium state is never in fact achieved. That this is true is, indeed, clear from observations on those systems in which a disordered state can be preserved at low temperatures by quenching, although the superstructure is then intrinsically more stable. Naturally, a discussion of the rate of attainment of equilibrium is of the greatest importance in considering the effect of heat treatment.

The rate at which a given degree of order is established in any system at a given temperature is determined by that temperature and by the activation energy W associated with the order–disorder transition. This energy is a measure of the height of the potential barrier which must be surmounted before two atoms can change places, and is not to be confused with V_0, the difference in energy of the pair of atoms in their initial and final states. If V_0 is large the ordered structure will at low temperatures be very much more stable than the disordered arrangement, but if W is also large the rate at which equilibrium is attained may, nevertheless, be very slow. These considerations enable us to understand why the order–disorder transformation in crystal structures is relatively rare and why it is in practice confined (with few exceptions) to intermetallic systems. If a given crystal is to be capable of existence in both the ordered and disordered conditions within a range of temperature limited by its melting point it is necessary that V_0 should be neither very large nor very small and that W should be small enough to permit equilibrium to be established in a reasonable period of time. It is possible to imagine, for example, a partially disordered sodium chloride structure, but the energy required to interchange a pair of sodium and chlorine ions would be so large that any appreciable degree of disorder would be possible only at very high temperatures far above the melting point; on crystallizing from the melt, therefore, sodium chloride appears in the fully ordered state. At the other extreme we have a system such as silver–gold in which the two atoms are so nearly the same in size and character that V_0 is very small and the alloy occurs, even at the lowest temperatures, only in the disordered condition.

Short-range order

13.15. The order–disorder transformations which we have described above are concerned with what may be termed 'long-range order', i.e. order which extends throughout the crystal and determines whether a given site is occupied by a 'right' or a 'wrong' atom. It is, however,

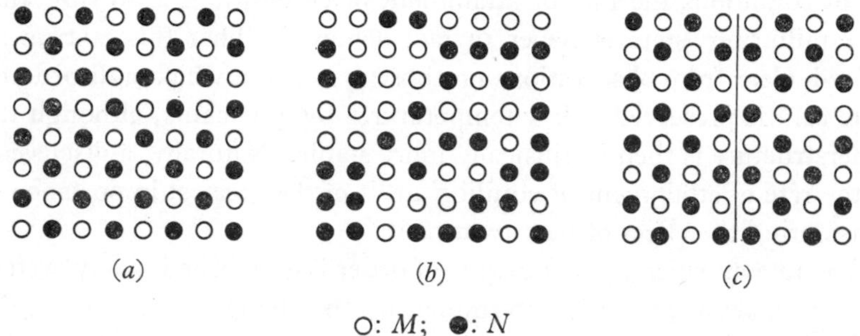

Fig. 13.09. Long- and short-range order in a two-dimensional binary alloy of composition *MN*: (*a*) the fully ordered structure ($s = 1$); (*b*, *c*) structures with no long-range order ($s = 0$). The structure (*c*) differs from (*b*) in that it displays a high degree of short-range order and consists of two anti-phase domains in each of which long-range order is perfect.

possible to adopt a different viewpoint and to consider instead a 'short-range order' which determines not the number of 'right' atoms but the number of 'right pairs' of atoms. The distinction between these two definitions of order may be illustrated in terms of the two-dimensional example of an alloy of composition *MN* shown in fig. 13.09. In (*a*) the alloy is represented in a fully ordered condition with $s = 1$, and in (*b*) in a completely disordered state with $s = 0$. In the latter condition the probability that any given site is occupied by an *M* or *N* atom is $\frac{1}{2}$. Now let us consider the arrangement shown at (*c*). Here, again, the probability that any given site is occupied by a 'right' atom is $\frac{1}{2}$, so that in terms of our definition of long-range order the system is as completely disordered as (*b*). It is clear, however, that it differs from (*b*) in that in it the great majority of the atoms have 'right' neighbours and that a high degree of local order prevails. Within the two domains of which the crystal is composed both long- and short-range order are perfect, and it is only because these domains are disposed in an anti-phase relationship that there is no long-range order in the crystal as a whole. A crystal with short-range order can therefore be regarded as an assembly of small domains with long-range order, the pattern of sites being coherent throughout the whole crystal but the sequence of atoms changing as the domain boundaries are crossed.

It is clear on general grounds that short-range order is a more fundamental concept than long-range order since the principal atomic interactions in a crystal are those between close neighbours. We may define the short-range order σ of a structure in the following way. Consider a crystal containing n atoms, each of which is co-ordinated by z neighbours. In this crystal there will be $\frac{1}{2}nz$ bonds. Let us assume that a fraction q of these bonds are between unlike atoms and that in the completely ordered crystal q has the value q_m; in general q_m will be large and in some cases it may be unity, as in the structure of fig. 13.09*a*. Let us further assume that in the completely disordered structure (with no short-range order) q has the value q_r. Then we define σ by the relation

$$\sigma = \frac{q-q_r}{q_m-q_r},$$

so that $\sigma = 1$ when $q = q_m$ and $\sigma = 0$ when $q = q_r$.

On the basis of this definition we can derive a relationship between short-range order and temperature, using arguments similar to those already employed in our discussion of long-range order. The form of variation thus deduced resembles that between s and T at low temperatures, but predicts that at the critical temperature (when long-range order vanishes) σ does not fall to zero but to a finite residual value which decreases only slowly as the temperature is further increased. Thus even at high temperatures there is a more than random number of MN atom pairs, and although they are unable to link up to give coherent long-range order they are, nevertheless, able to form small domains within which short-range order is present. It is the existence of this short-range order above the critical temperature, and its subsequent gradual dissipation, which accounts for the fact that the specific heat of Cu_3Au above the critical temperature is higher than at low temperatures (fig. 13.08).

Short-range order cannot be as readily detected by X-ray methods as long-range order, and the number of systems in which it has been studied is therefore somewhat limited. It is, however, of particular interest to note that short-range order can exist in a structure in which long-range order is never found. In the silver–gold system at the composition AgAu in the fully disordered state each gold atom has on the average six silver and six gold neighbours. Half the bonds are therefore between unlike atoms and q_r is $\frac{1}{2}$. If the alloy could exist in the fully ordered state, and assuming that it would then have the same ordered structure as CuAu, each gold atom would have eight silver and four gold neighbours and q_m would be $\frac{2}{3}$. In actual fact, at a temperature of 300 °C, the average number of silver neighbours about any gold atom is 6·48. The degree of short-range order is therefore finite and is given by

$$\sigma = \frac{6{\cdot}48/12 - \frac{1}{2}}{\frac{2}{3} - \frac{1}{2}} = 0{\cdot}24.$$

In a similar way the degree of short-range order at the composition Ag_3Au is found to be 0·15.

ALLOYS OF A TRUE METAL AND A *B* SUB-GROUP ELEMENT

13.16. Compared with the alloys of two true metals, those of a true metal with an element of the *B* sub-groups show a much greater diversity of character. We are now dealing with systems in which the two components may be comparable in electronegativity or may differ widely in this respect, and in which not only metallic but also covalent forces may be operative. Systems in which the two elements differ little in electronegativity tend to show a wide range of solid solution, and generally resemble the alloys of the true metals. When the electronegativity difference is large, however, solid solution is restricted and definite chemical compounds, or at least distinct phases with geometrically ordered structures, are likely to be formed.

The influence of the electronegativity of the true metals and of the partially covalent character of the *B* sub-group elements in determining the properties of an alloy system makes it convenient to subdivide both of these classes of element in the way shown in table 13.01. The true metals of division T_1 are the strongly electropositive elements immediately following the inert gases, while those of the division T_2 are the transition metals (excluding Zn, Cd and Hg), characterized by an incompletely filled *d* level. The *B* sub-group elements are divided in a less precise way into the more metallic (B_1) and the less metallic (B_2).

The dimorphism of tin makes it exceptional, and this element can be regarded as belonging to both of these groups.

On the basis of this classification of the metallic elements we may consider four types of alloy containing a true metal and a *B* sub-group element, namely T_1–B_1, T_1–B_2, T_2–B_1 and T_2–B_2. We discuss these in turn, but systems of the type T_2–B_1 have been more widely studied than the others and it is therefore convenient to consider them first. For reasons which will appear later, they may be termed *electron compounds*.

Systems of the type T_2–B_1: electron compounds

13.17. In alloys of the type T_2–B_1, we have in association one of the less strongly electropositive true metals and one of the less strongly electronegative or more 'metallic' elements of the *B* sub-groups. It is therefore not surprising that these systems resemble in a number of respects the alloys of two true metals. As in the latter systems, solid solution of each component in the other is usually found near the two extreme compositions if the size factor is favourable. Now, however, we are dealing with two metals whose structures are generally different, so that the solid solution is necessarily restricted and intermediate phases must also occur. Moreover, in these systems it is normally found that the extent of the solid solution is very far from reciprocal, even when the size factor is favourable, as the following figures show:

System *T*–*B*	Solubility of *B* in *T* (atomic %)	Solubility of *T* in *B* (atomic %)
Cu–Al	20	2·5
Ag–Zn	40	6·5
Ag–Cd	42	4

Hume-Rothery has pointed out, in a principle usually termed the *relative valence effect*, that, if one of the components of the system is copper, silver or gold, these and many other observations can be summarized in the statement that a metal of lower valency will dissolve more of one of higher valency than vice versa. In general terms it is not difficult to see the reason for this. In any structure in which there is a tendency towards covalent binding the electron deficiency resulting from the substitution of an element of low valency is much more likely to lead to a breakdown of the structure than the electron excess produced by the substitution of a metal of high valency.

We shall consider the properties of electron compound by first describing in detail one system which displays a number of features characteristic of many of these structures. Some other systems will then be briefly discussed. Among these are a number, such as brass, of considerable technical importance.

The silver–cadmium system

13.18. The silver–cadmium system has been investigated in considerable detail and the results of these studies are summarized in the equilibrium diagram of fig. 13.10, which refers to room temperature. It will

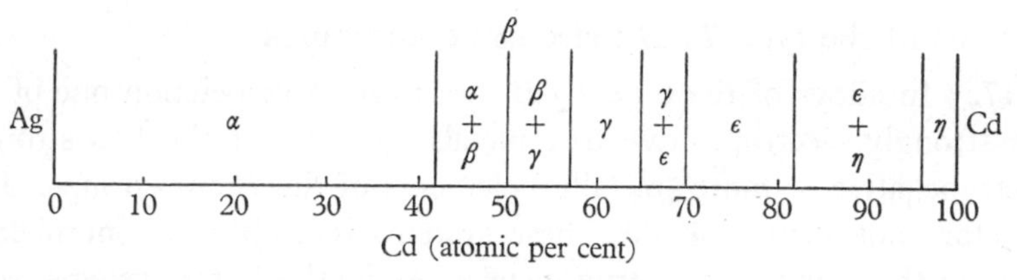

Fig. 13.10. The equilibrium diagram of the silver–cadmium system at room temperature.

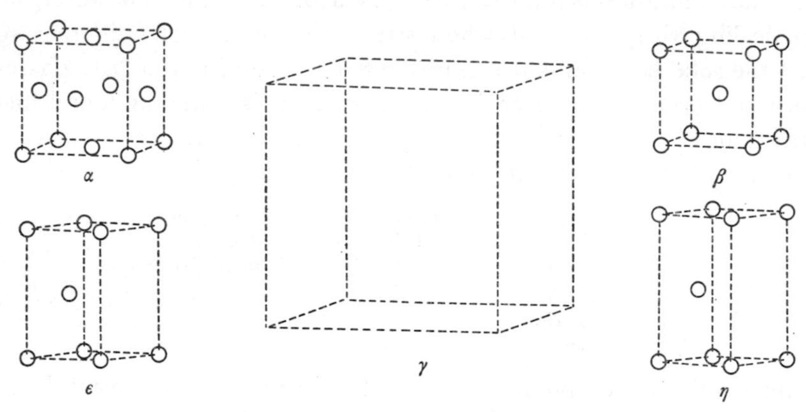

Fig. 13.11. Clinographic projections of the unit cells of the structures of the phases in the silver–cadmium system.

be seen that there are three intermediate phases β, γ and ϵ in addition to the α and η phases of the parent elements.

Pure silver has the cubic close-packed structure α (fig. 13.11). This phase is capable of accommodating up to 42 atomic per cent of cadmium in solid solution by the purely random replacement of silver atoms. The sites occupied are still those of the cubic face-centred structure, no change in which occurs except a progressive and approximately linear variation of cell size with composition. At 42 per cent, however, the α phase appears to be saturated, and further cadmium can be taken up only by the formation of an entirely new phase. This new β phase rapidly grows at the expense of the α phase and becomes homogeneous at 50 per cent cadmium, when it has the simple caesium chloride structure corresponding to the composition AgCd. It must be emphasized, however, that it is the body-centred pattern rather than the actual caesium chloride structure which is characteristic of the β phase, for by suitable heat treatment the phase can be obtained in the disordered form, while in some systems it occurs at a composition which does not admit of an ordered structure.

In the silver–cadmium system the β phase is capable of taking up very little excess of either component in solid solution: it therefore appears only as a line on the equilibrium diagram. When the concentration of cadmium is increased beyond 50 per cent a new γ phase develops and becomes homogeneous at a composition of about 57 per cent. This phase, which is distinguished by a characteristic brittleness and hardness, has a complex cubic structure with a cell containing 52 atoms. It is, however, quite simply related to the cubic body-centred arrangement. If we take 27 unit cells of this structure they can be stacked together to form a larger cube containing 54 atoms. The γ structure is derived from this larger cube by removing two atoms per cell (those at the corners and that at the centre) and displacing the others to fill the interstices thus created. In the particular system under consideration the γ phase can accommodate in solid solution a considerable excess of either component, and on either side of the 'ideal' composition there is therefore a relatively wide range of a single homogeneous phase extending from 57 to 65 per cent of cadmium. Throughout this range there is a progressive change in the lattice dimensions but no alteration in the pattern of sites occupied.

Beyond 65 per cent of cadmium the γ phase becomes saturated, and a new ϵ phase makes its appearance. This phase is simply a hexagonal close-packed arrangement in which silver and cadmium atoms must clearly be distributed at random, since all the sites of such a structure are geometrically precisely equivalent. In this case the phase can occur only in a disordered condition and no superstructure exists. The ϵ phase grows at the expense of the γ phase between 65 and 70 per cent of cadmium and becomes homogeneous at the latter composition. The range of homogeneity extends from 70 to 82 per cent of cadmium, when a new η phase appears. This phase is the structure of elementary cadmium, which we have already described as being closely related to hexagonal close packing and differing from it only in having a somewhat larger axial ratio (§7.19). The η phase is capable of taking up only a small amount of silver in solid solution and the ϵ and η phases therefore co-exist throughout the range 82–96 per cent of cadmium. Only from 96 to 100 per cent of cadmium is the η phase homogeneous.

Other systems

13.19. The silver–cadmium system has been discussed at some length because its behaviour is typical of that of many other alloys containing a transition metal and an element from one of the earlier B sub-groups. While the initial and final phases are necessarily determined by the particular elements involved, the intermediate β, γ and ϵ phases are of very general occurrence in a wide range of such systems. Although these three phases do not necessarily all appear in any given system, and although other phases are often found in addition, one or more of these characteristic phases have been observed in almost every T_2–B_1 system investigated, as is shown by the data of table 13.02, where the compositions of most of those so far observed are recorded. The significance of the composition will be discussed below, but at this point we may note that all three of the β, γ and ϵ phases are found in the systems Cu–Zn, Cu–Si, Cu–Sn, Ag–Zn, Ag–Cd, Au–Zn and Au–Cd, and that two are found in many of the others. In a few cases the simple β phase is replaced by a more complex μ phase in which the atoms are distributed at random in the appropriate proportions over the 20 sites per unit cell of the β manganese structure (§13.10), or alternatively by a ζ phase in which the atoms are arranged as in hexagonal close packing; and sometimes two, or all three, of these phases may occur in the same system but at different temperatures. The μ and ζ phases are indicated in table 13.02 only when the simple β phase is not found under any conditions.

The composition of the phases

13.20. The attribution of formulae to the several phases presents difficulties. Especially is this the case when a wide region of homogeneity exists embracing a number of possible simple compositions. In the silver–cadmium system the very restricted homogeneity range of the β phase and the caesium chloride structure which it possesses at once justify the formula AgCd. The γ phase, however, has a range of

Table 13.02. *β, γ and ε phases in some T_2–B_1 systems*

β	γ	ε	β	γ	ε
CuZn	Cu_5Zn_8	$CuZn_3$	AuZn	Au_5Zn_8	$AuZn_3$
—	Cu_5Cd_8	$CuCd_3$	AuCd	Au_5Cd_8	$AuCd_3$
—	Cu_5Hg_8	—	Au_3Al*	—	Au_5Al_3
Cu_3Al	Cu_9Al_4	—	Au_3In†	Au_9In_4	—
Cu_3Ga	Cu_9Ga_4	—	Au_5Sn†	—	Au_3Sn
Cu_3In	Cu_9In_4	—	—	Mn_5Zn_{21}	—
Cu_5Si	$Cu_{31}Si_8$	Cu_3Si	—	Fe_5Zn_{21}	—
Cu_5Ge†	—	Cu_3Ge	FeAl	—	—
Cu_5Sn	$Cu_{31}Sn_8$	Cu_3Sn	$CoZn_3$*	Co_5Zn_{21}	—
AgZn	Ag_5Zn_8	$AgZn_3$	CoAl	—	—
AgCd	Ag_5Cd_8	$AgCd_3$	—	Ni_5Zn_{21}	—
AgHg*	Ag_5Hg_8	—	—	Ni_5Cd_{21}	—
Ag_3Al	—	Ag_5Al_3	NiAl	—	—
Ag_3Ga†	—	—	NiIn	—	—
Ag_3In	Ag_9In_4	—	—	Rh_5Zn_{21}	—
Ag_5Sn†	—	Ag_3Sn	—	Pd_5Zn_{21}	—
			PdIn	—	—
			—	Pt_5Zn_{21}	—

* Occurs only as the μ phase. † Occurs only as the ζ phase.

solid solution extending from 57 to 65 atomic per cent of cadmium, and this embraces such simple formulae as Ag_2Cd_3, Ag_3Cd_4, Ag_3Cd_5, etc. It is only the X-ray evidence, which demands a unit cell containing 52 atoms, that indicates that none of these is admissible and that the 'ideal' formula must be regarded as Ag_5Cd_8. With the ε phase the position is still more difficult, for the purely statistical distribution of the atoms over the sites of the hexagonal close-packed structure can give no *a priori* reason for preferring any one of the several possible formulae included in the wide range of solid solution.

Even with the β and γ phases the position is not always as simple as in the case just discussed, and in fact it appears that the pattern of sites occupied, and not the actual distribution of the two kinds of atom, is the only significant feature of the structure, for a given phase may appear in different systems at widely different compositions. Thus in the copper–tin system the β phase appears at a composition of about 17 atomic per cent of tin, corresponding approximately to the formula Cu_5Sn. Such a composition is clearly not consistent with the caesium chloride structure, but it is found that the phase is actually a body-centred cubic arrangement with the atoms distributed at random in these proportions.

A similar state of affairs arises with the γ phases, which again occur at widely different compositions in different systems and often at compositions of surprising complexity. In some cases an ideal formula may be determined by structure analysis, as with γ brass. The original analysis of the γ structure revealed that the 52 sites of the unit cell were divided into four groups of equivalent positions containing 8, 8, 12 and 24 atoms, respectively, and that two of these groups consisted of 8+12 copper atoms and the remaining two of 8+24 zinc atoms. In this case there is an ordered distribution of atoms in the structure and the ideal composition is Cu_5Zn_8. Solid solution can, of course, occur within certain limits on either side of this composition by the random replacement of some atoms by atoms of the other kind. In the chemically closely analogous copper–cadmium system the structure of the γ phase is entirely different. The sites occupied remain, as always, the same, but now 8+8 positions contain 16 copper atoms while the remaining 12+24 positions are occupied by 32 cadmium and 4 copper atoms distributed quite at random. The copper–cadmium system also differs from that of copper–zinc in showing, in addition to the β, γ and ϵ phases, others of considerable complexity. Such an example illustrates that in alloy systems the chemical properties of the elements concerned play little part in determining the structures which obtain, and that chemical analogy does not necessarily lead to the formation of analogous phases. What, then, are the factors involved?

Hume-Rothery's rule

13.21. The answer to this question was provided by Hume-Rothery, who in 1926 made the empirical observation that the widespread occurrence of the β, γ and ϵ phases in chemically dissimilar systems and at widely differing compositions is determined, not by the chemical properties of the elements concerned, or by any arguments based on valency concepts, but solely by the relative number of valency electrons and atoms in the crystal structure. This generalization, usually known as Hume-Rothery's rule, is illustrated by the data of table 13.03, from which it will be seen that the β, γ and ϵ phases are characterized by electron : atom ratios of 3:2, 21:13 and 7:4, respectively.* In each case this ratio alone determines the structure, and the relative number of atoms and the particular atoms by which the electrons are contributed appear to be of

* These ratios may be readily memorized if they are expressed 21:14, 21:13 and 21:12.

Table 13.03. *Electron:atom ratios in some T_2–B_1 systems*

Phase	Composition	Electrons	Atoms	Ratio
β	CuZn	$1+2 = 3$	2	3:2
β	$CoZn_3$	$0+3\times2 = 6$	4	3:2
β	Cu_3Al	$3+3 = 6$	4	3:2
β	FeAl	$0+3 = 3$	2	3:2
β	Cu_5Sn	$5+4 = 9$	6	3:2
γ	Cu_5Zn_8	$5+8\times2 = 21$	13	21:13
γ	Fe_5Zn_{21}	$0+21\times2 = 42$	26	21:13
γ	Cu_9Al_4	$9+4\times3 = 21$	13	21:13
γ	$Cu_{31}Sn_8$	$31+8\times4 = 63$	39	21:13
ϵ	$CuZn_3$	$1+3\times2 = 7$	4	7:4
ϵ	Ag_5Al_3	$5+3\times3 = 14$	8	7:4
ϵ	Cu_3Sn	$3+4 = 7$	4	7:4

little importance. It is for this reason that Bernal proposed the description 'electron compounds' for such systems.

It will be seen from table 13.03 that the characteristic electron:atom ratios for systems containing metals from the eighth group of the Periodic Table are obtained only if these elements are assumed to make no electron contribution to the structure. This appears to be associated with the ease with which the valency electrons in these elements can be absorbed into the incompletely filled *d* level of the penultimate shell; thus nickel in the ground state has the configuration 2, 8, $3s^2 3p^6 3d^8$, $4s^2$, but in the alloy behaves as if it had the configuration 2, 8, $3s^2 3p^6 3d^{10}$ of zero valency. Confirmatory evidence for this behaviour comes from the large contraction in volume which accompanies the formation of these electron compounds from their parent elements, and also, in some cases, from their magnetic properties.

All the phases shown in table 13.03 will be seen to obey Hume-Rothery's rule. In cases where the composition of any phase has been determined on structural grounds it has, with few exceptions, been found to be consistent with the rule. In the majority of cases, however, the formula quoted is that deliberately chosen to give the appropriate electron:atom ratio. In the copper–tin system, for example, the γ phase has a purely statistical distribution of atoms, so that structurally no particular composition can be preferred. The range of homogeneity is in this case very narrow, but once the possibility of a formula as complex as $Cu_{31}Sn_8$ is admitted, several of no greater complexity could doubtless be found; this particular formula is chosen because it gives the appropriate electron:atom ratio of 21:13. In such cases, however, it is significant that the composition satisfying Hume-Rothery's rule is found always to lie within the range of homogeneity of the phase, while purely physical measurements reveal characteristic properties at this composition.

The μ phase occurs at the same electron:atom ratio as the β phase.

The theoretical basis of Hume-Rothery's rule

13.22. The Hume-Rothery rule was originally advanced as a purely empirical generalization without theoretical foundation, but the recent developments of metal theory have thrown much light on its physical significance and have led to an interpretation of many of the properties of electron compounds. We have seen in chapter 5 that the energy of an electron in a metal cannot assume any value, but only values lying in a series of discrete zones separated, in general, by regions of forbidden energy. Energy values lying near the middle of the permitted zones are approximately those of free electrons, but this is not true of energy values close to the zone boundaries. For such electrons the energies just below the forbidden values are abnormally depressed and those just above these values abnormally raised. It follows that in general a system in which the occupied Brillouin zones are just filled, without overlap into higher zones, will be one of special stability, and that an alloy system in which overlap occurs will tend to be unstable and will seek if possible to alter the zonal configuration by a phase change.

In terms of these ideas the characteristic electron:atom ratio of 21:13 for the γ phase has been satisfactorily explained. In this structure the Brillouin zones corresponding to the 52 atoms of the unit cell can accommodate 90 electrons without overlap. Further electrons could be received only in states of considerably higher energy. The actual number of electrons per unit cell being 84, the zone is very nearly completely filled, and the structure is therefore stable. Similar arguments applied to the β and ϵ phases give electron:atom ratios of 1·48 and 1·73, very close to the observed values of 3:2 and 7:4, respectively. It will, nevertheless be realized that no particular significance is to be attached to the precise value of these ratios since the phases are generally stable over an appreciable range of composition and, therefore, of electron:atom ratio. What is significant, however, is that these ratios always correspond to a structure in which the Brillouin zone is nearly full.

13.23. The question of the range of solid solution over which any given phase may be expected to be stable may also be discussed in terms of the theory. Thus if we consider the α phase of the copper–zinc system in which zinc is taken up without change of copper structure, it is clear that every zinc atom so accommodated involves an increase of one in the number of valency electrons in the phase. A limit to the extent of solid solution may therefore be expected to occur at an electron:atom ratio at which overlap into a higher zone impends, and this ratio (calculated to be 1·36) will be characteristic of the α phase and independent of the solute. With metals of higher valency as solute the range of solid solution will be correspondingly restricted, and will in fact be approximately inversely proportional to the valency. The data of table 13.04 show that this generalization is very roughly true. The range of solid solution of the various metals in copper decreases rapidly with increasing valency even where the relative sizes are favourable, but the electron:atom ratio in the saturated phase is approximately constant and in very satisfactory agreement with the theoretical value.

Table 13.04. *Solubility of metals of different valencies in copper*

System	Composition of saturated phase (atomic per cent)	Electron:atom ratio of saturated phase
Cu–Zn	38·4	1·38
Cu–Al	20·4	1·41
Cu–Ga	20·3	1·41
Cu–Si	14·0	1·42
Cu–Ge	12·0	1·36
Cu–Sn	9·3	1·28

The range of solid solution in the γ phases throws further light on the same problem. In the majority of these the limit of solid solution on the side of the component of higher valency occurs at an electron:atom ratio of 1·70 (corresponding to 88 electrons in the Brillouin zone) and beyond this point a new phase develops. There are, however, some γ phases, such as those of copper–aluminium or copper–gallium, in which this ratio is appreciably exceeded, and it is found in these cases that the phase has a defect structure in which some of the 52 sites normally occupied are vacant. This defect structure begins to develop when the electron:atom ratio reaches the value of 1·70, and thereafter atoms drop out of the unit cell in such a way that the total number of electrons *per cell* remains constant at 88; in the copper–gallium system as many as five vacant sites per cell are created before the phase becomes saturated. A similar defect structure has been observed in the β phase of the nickel–aluminium system, where again vacant sites are created as the composition changes, but again in such a way that the number of electrons per cell remains constant. We must therefore conclude from these examples that in the distinctive phases of the electron compounds the total number of electrons in the cell is a more fundamental characteristic than the electron:atom ratio. In phases with no defect structure the two quantities are directly related and either may be used; but if a defect structure occurs only the former is significant.

Systems of the type T_2–B_2

13.24. The electron compounds described above occur primarily in systems containing a transition metal and an element from one of the earlier *B* sub-groups. When the *B* sub-group element belongs to one of the later groups, or when any *B* sub-group metal is combined with an *A* group metal, a tendency towards the formation of more or less well defined chemical compounds is apparent. With the increasingly covalent character of the *B* sub-group element the range of solid solution becomes more restricted and definite structures of varying degrees of complexity, entirely different from those of the parent constituents, are formed. The metallic character, however, is by no means completely lost and we find that such systems still resemble alloys in being more or less metallic in appearance, in having a relatively high electrical conductivity and in showing variable composition.

The nickel arsenide and related structures

13.25. Many T_2–B_2 systems, at the appropriate composition, display either the nickel arsenide structure or one geometrically closely related to it. This structure has already been described (§8.12 and fig. 8.02) and it will be remembered that it is characterized by the asymmetry of the co-ordination: each of the more metallic *T* atoms is octahedrally co-ordinated by six atoms *B* of the less metallic component, whereas each *B* atom is surrounded by six neighbours *T* at the corners of a trigonal prism. For our present purposes, however, it is convenient to describe the structure in somewhat different terms, for it can alternatively be regarded as an expanded hexagonal close-packed array of *B* atoms with the *T* atoms disposed in its octahedral interstices. This interstitial picture explains the wide range of composition over which the structure can exist as a single phase, for on the one hand we may have a defect crystal in which the interstitial sites are only partially occupied and on the other a crystal in which not only the octahedral but also other interstices are filled. In the former case, if just one-half of the octahedrally co-ordinated interstices are occupied we have a structure of composition MN_2, geometrically identical with that of cadmium iodide (fig. 8.09) and typified among the systems now under consideration by $CoTe_2$, $NiTe_2$ and others. In the latter case, if octahedral and trigonal interstices are fully occupied we obtain a structure of composition M_2N, of which Ni_2Ge is an example.

The widespread occurrence of the nickel arsenide and related structures in T_2–B_2 systems is illustrated by the data in table 13.05. Where the composition is given in the form MN it is to be understood that the simple nickel arsenide structure normally obtains, but that in general a range of solid solution on each side of the equiatomic composition is found. In a few cases, however, the range of homogeneity does not include the simple composition, and phases occur only with an excess of one component. The copper–tin system is an example of this, for the phase shown as 'CuSn' does not exist at this composition but only with an excess of interstitial copper. A similar state of affairs obtains in the other systems marked with an asterisk in the table.

Table 13.05. *The nickel arsenide and related structures in some T_2–B_2 systems*

	S	Se	Te	As	Sb	Bi	Ge	Sn
Ti	TiS_2	$TiSe_2$	$TiTe_2$*	—	—	—	—	—
V	VS	{VSe, VSe_2}	VTe	—	—	—	—	—
Cr	CrS	CrSe	CrTe	CrAs†	CrSb	—	—	—
Mn	—	—	MnTe	MnAs†	MnSb	MnBi	Mn_2Ge	Mn_2Sn
Fe	FeS	FeSe	FeTe	FeAs†	FeSb	—	Fe_2Ge	FeSn*
Co	CoS	CoSe	{CoTe, $CoTe_2$}	CoAs†	CoSb	—	Co_2Ge	CoSn*
Ni	NiS	NiSe	{NiTe, $NiTe_2$}	NiAs	NiSb	NiBi	{NiGe†, Ni_2Ge}	NiSn*
Pd	—	—	{PdTe, $PdTe_2$}	—	PdSb†	—	PdGe†	PdSn†
Pt	PtS_2	$PtSe_2$	{PtTe, $PtTe_2$}	—	PtSb	—	PtGe†	PtSn
Cu	—	—	—	—	—	—	—	CuSn*

* In these systems the homogeneity range does not include the equiatomic composition shown.

† $B31$ structure (see text).

The systems shown as M_2N and MN_2 in table 13.05 have the Ni_2Ge and cadmium iodide structures, respectively, again with a range of homogeneity on each side of the ideal composition. When two compositions are shown for the same system this range of homogeneity extends continuously from one to the other. Thus CoTe has the nickel arsenide structure, but if more and more tellurium is added (or, more correctly, if more and more cobalt is removed) a solid solution is obtained in which some of the octahedrally co-ordinated sites are vacant. When the composition reaches 66·7 atomic per cent of the tellurium just half these sites are empty and the structure is that of cadmium iodide. The close relationship of this structure to the original nickel arsenide arrangement is revealed by the fact that throughout this wide range of homogeneity the cell dimensions change by less than 3 per cent.

In our original discussion of the nickel arsenide structure we explained that although each T atom is co-ordinated primarily by six B atoms at a distance d_1, say, it also has two other T neighbours at a distance d_2 vertically above and below it. In the less metallic systems d_2 may exceed d_1 by as much as 0·4 Å or more, but as the systems become progressively more metallic the structure shrinks in the vertical direction and the eight neighbours become more and more nearly equidistant. It follows that the axial ratio $c:a$ is directly related to the metallic character of the structure, and it is interesting to note how close is this correlation in the systems now under discussion. The ranges of axial ratios in a number of T_2–B_2 systems with the nickel arsenide structure are shown in table 13.06 for different individual B sub-group elements, and it will be seen that axial ratios fall from values as high as 1·75 in systems containing the most strongly electronegative elements of group 6B to values

Table 13.06. *The axial ratios of the nickel arsenide structure in some T_2–B_2 systems*

B sub-group element	Electro-negativity	Range of $c:a$ ratios
S	2·5	1·54–1·75
Se	2·4	1·46–1·75
Te	2·1	1·37–1·66
As	2·0	1·39–1·53
Sb	1·9	1·25–1·40
Bi	1·9	1·31–1·42
Ge	1·8	1·25–1·27
Sn	1·8	1·21–1·40

as low as 1·21 in systems in which the more electropositive elements of group 4*B* are present.

13.26. A limited number of T_2–B_2 systems at the equiatomic composition *MN* crystallize not with the nickel arsenide structure but with the so-called '*B* 31' arrangement. This structure is of lower symmetry than that of nickel arsenide but is in fact closely related to it, and can be derived from it by a small displacement of the atoms. The systems marked with an obelisk in table 13.05 show the *B*31 structure.

The pyrites structure

13.27. A number of systems containing a transition metal in association with a *B* sub-group element from one of the later groups crystallize at a composition MN_2 with the pyrites structures, already described (§ 8.38 and fig. 8.15). Some of these systems are shown in table 13.07, and it will be noted that among them are systems which crystallize at a composition *MN* with the nickel arsenide structure. There is, however, no simple geometrical relationship between these two structures such as that which exists between the cadmium iodide and nickel arsenide arrangements, and a homogeneous phase extending continuously between the compositions *MN* and MN_2 is therefore no longer possible. The factors which determine the appearance of the cadmium iodide structure in some systems and of the pyrites structure in others are not known.

Table 13.07. *The pyrites structure in some* T_2–B_2 *systems*

	S	Se	Te	As	Sb
Mn	MnS_2	—	$MnTe_2$	—	—
Fe	FeS_2	—	—	—	—
Co	CoS_2	$CoSe_2$	—	—	—
Ni	NiS_2	$NiSe_2$	—	—	—
Ru	RuS_2	$RuSe_2$	$RuTe_2$	—	—
Pd	—	—	—	$PdAs_2$	$PdSb_2$
Os	OsS_2	$OsSe_2$	$OsTe_2$	—	—
Pt	—	—	—	$PtAs_2$	$PtSb_2$

A third structure, that of molybdenum sulphide (§ 8.37 and fig. 8.14), is found in the phases MoS_2, $MoSe_2$, $MoTe_2$, WS_2, WSe_2 and WTe_2.

Systems of the type T_1–B_1

13.28. In systems of this type we are concerned with two elements of widely different electronegativity, and we accordingly find that in general solid solution is restricted and that the systems show a series of phases of different crystal structure. Many of these phases are characteristic of only a limited number of systems, but two structures, both corresponding to the equiatomic composition *MN*, are of common occurrence.

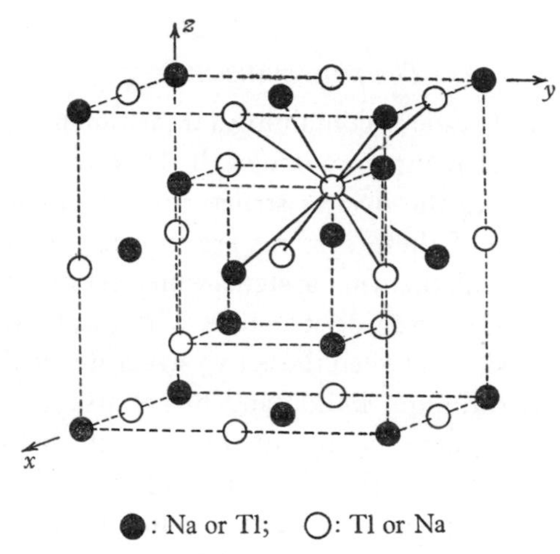

Fig. 13.12. Clinographic projection of the unit cell of the cubic structure of the Zintl phase NaTl. If the distinction between the two types of atoms is ignored the structure can be described in terms of the smaller cubic body-centred cell shown at the centre.

The caesium chloride structure

13.29. The simpler of these two structures is the caesium chloride arrangement, found in the phases LiHg, LiTl, MgTl, CaTl and SrTl. This is, of course, also the structure of the β phase in the silver–cadmium system and in other electron compounds (fig. 13.11), and for this reason the systems just mentioned are sometimes quoted as exceptions to Hume-Rothery's rule. Apart from this geometrical resemblance, however, these systems have little in common with the electron compounds, and it seems preferable to regard the Hume-Rothery rule as applicable only to alloys of the T_2–B_1 type.

The sodium thallide structure

13.30. The second structure common to a number of T_1–B_1 systems is that of sodium thallide, sometimes called the *Zintl phase*. This structure (fig. 13.12) is closely related to that of caesium chloride in that the pattern of sites occupied forms a cubic body-centred lattice. The distribution of the atoms, however, is such that each atom has four neighbours of each kind, and the true cell is therefore the larger unit shown, containing sixteen instead of only two atoms. Some phases in which the sodium thallide structure occurs are LiZn, LiCd, LiAl, LiGa, LiIn, NaIn and NaTl. It is a characteristic feature of all of these phases that in them the alkali metal atom appears to have a radius considerably smaller than in the structure of the element (even when allowance is made for the change in co-ordination number), suggesting that this atom is present in a partially ionized condition and that forces other than purely metallic bonds are operative in the structure.

The $Mg_{17}Al_{12}$ *structure*

13.31. Another phase in which the atoms appear to have abnormal radii is found in the magnesium–aluminium system at a composition of approximately Mg_3Al_2. The structure of this phase is based on that of α-manganese, in which there are 58 atoms in the unit cell divided into three sets of 24, 24 and 10. The true composition of the phase must therefore be $Mg_{17}Al_{12}$. A detailed consideration of the interatomic distances reveals that these differ from the values to be expected on the basis of the normal radii. Thus the aluminium atoms are all found to be grouped in pairs, with a separation considerably smaller than twice the aluminium radius. We may regard these atoms as bound into diatomic molecules by covalent bonds, for which, however, two further electrons per pair of atoms are required. These electrons may be derived by the ionization of magnesium atoms, in which case we shall have in each unit cell 12 negatively charged molecules, $[Al_2]^{2-}$, 24 positively charged magnesium ions and 10 neutral magnesium atoms:

$$[Al_2]_{12}{}^{2-}\,Mg_{24}{}^{+}\,Mg_{10}.$$

In confirmation of this picture it is found that the distance between the aluminium and the 24 magnesium atoms of the first kind is substantially smaller than that between the aluminium and the remaining 10 magnesium atoms.

Systems of the type T_1–B_2

13.32. The systems containing a strongly electropositive atom in association with an element from one of the later B sub-groups display few characteristically metallic properties. From a metallurgical point of view such systems are very simple, for they normally show only one intermediate phase and this occurs at a simple composition on each side of which little or no solid solution is possible. The phase is generally very stable and behaves essentially as a normal valency compound. This is confirmed by the fact that the intermediate phases occur in different systems at different compositions, corresponding to the normal valencies of the elements concerned, and that they crystallize with the simple structures characteristic of chemical compounds of these compositions. The data in table 13.08 illustrate this point. The sulphides of the alkaline earth metals are undoubtedly ionic compounds, but even when the difference in electronegativity between the two constituent elements is less the influence of chemical binding is still apparent, not only in the crystal structures but also in the physical properties of the phases. Thus the phase Mg_3Sb_2 has a melting point over 500 °C higher than either magnesium or antimony, and all the phases, even including such seemingly metallic systems as Mg_2Sn, are poor conductors of electricity.

Table 13.08. *The structures of some phases in T_1–B_2 systems*

	S	Se	Te	As	Sb	Bi	Ge	Sn
Li	—	Li_2Se^b	Li_2Te^b	Li_3As^a	Li_3Sb^a	Li_3Bi^a	—	—
Na	—	Na_2Se^b	Na_2Te^b	Na_3As^a	Na_3Sb^a	—	—	—
Be	BeS^e	$BeSe^e$	$BeTe^e$	—	—	—	—	—
Mg	MgS^d	$MgSe^d$	$MgTe^f$	$Mg_3As_2{}^c$	$Mg_3Sb_2{}^c$	$Mg_3Bi_2{}^c$	Mg_2Ge^b	Mg_2Sn^b
Ca	CaS^d	$CaSe^d$	$CaTe^d$	—	—	—	—	—
Sr	SrS^d	$SrSe^d$	$SrTe^d$	—	—	—	—	—
Ba	BaS^d	$BaSe^d$	$BaTe^d$	—	—	—	—	—

(*a*) Anti-BiF_3 structure.* (*b*) Anti-fluorite structure. (*c*) Anti-La_2O_3 or anti-Mn_2O_3 structure.* (*d*) Sodium chloride structure. (*e*) Zincblende structure. (*f*) Wurtzite structure.

* These structures are not described in this book.

ALLOYS OF TWO *B* SUB-GROUP ELEMENTS

13.33. In systems containing two B sub-group metals the partially covalent character of these elements leads to a still more pronounced tendency towards the formation of definite chemical compounds. When the metals both belong to the earlier groups the generally metallic character of the systems persists, but any considerable extent of solid solution obtains only between metals of the same group and even then only when the radii are very nearly equal. Thus cadmium and mercury form a much wider range of solid solution than cadmium and zinc, while cadmium and tin are practically immiscible. When both elements belong to the later *B* sub-groups the sodium chloride structure is common, as in the compounds SnSb, SnTe, PbSe, PbTe.

Compounds with the zincblende or wurtzite structure are commonly formed when one element belongs to the *n*th and the other to the $(8-n)$th *B* sub-groups. In these structures each atom is bound to four neighbours by purely covalent bonds and it is necessary that the number of valency electrons available for the formation of these bonds should average four per atom. It is not, however, necessary that these should be contributed equally by the atoms of the two types, and the condition is therefore satisfied if these atoms are related in the manner indicated. Some *B–B* compounds with the zincblende and wurtzite structures are shown in table 13.09. These structures are not, of course, confined to systems of the type *B–B*; we have encountered numerous examples of their occurrence in compounds of quite other kinds.

Table 13.09. *The zincblende and wurtzite structures in some B–B systems*

Zincblende structure

BeS	ZnS	CdS	HgS	AlAs	GaAs	—
BeSe	ZnSe	CdSe	HgSe	AlSb	GaSb	InSb
BeTe	ZnTe	CdTe	HgTe			

Wurtzite structure

ZnS	CdS
—	CdSe

13.34. This completes our discussion of the three principal classes into which we have divided the alloy systems. As a summary, a condensed survey of the chief structural characteristics of these classes is given in table 13.10. The remaining class of alloys to be considered is that embracing the interstitial structures. These systems, however, differ materially from the alloy systems so far discussed, and a description of them is accordingly deferred until later in the chapter (§ 13.37).

Table 13.10. *The classification of binary intermetallic systems*

See table 13.01 for the classification of the elements.

	True metals T_1	True metals T_2	*B* sub-group elements B_1	*B* sub-group elements B_2
T_1	Wide range of solid solution. Superstructures Laves phases		CsCl and NaTl structures	NaCl, ZnS, anti-CaF_2 and anti-BiF_3 structures
T_2			Electron compounds	NiAs and related structures FeS_2 and MoS_2 structures
B_1	See above		Solid solution if chemically similar and of comparable size	ZnS structures
B_2			See above	NaCl structure

THE CHEMISTRY OF METAL SYSTEMS

13.35. The structural properties of the metal systems described above naturally challenge consideration from a purely chemical point of view, for occasionally the results of structure analysis cannot readily be reconciled with accepted chemical principles and demand a re-orientation of the chemical picture of the metallic state. It has often been the practice of the metallurgist to seek to represent the several phases in an alloy system by simple chemical formulae corresponding to idealized stoichiometric compositions, departures from which were interpreted as solid solution in the ideal phase of excess of one or other of the components. Such formulae inevitably tend to convey the impression, implicitly if not explicitly, that these phases are to be regarded as definite chemical compounds, and it is necessary to consider carefully to what extent such a viewpoint is justifiable, and even whether there are valid grounds for attributing to the phases any formulae at all.

In our description of the metal structures we have, of course, included many in which covalent or ionic binding predominates. Such systems may often quite properly be regarded as chemical compounds, but it is not in such systems that the chief interest lies, and here we shall discuss only the systems of more pronouncedly metallic character, including the electron compounds. It was early pointed out by Hume-Rothery that if formulae are assigned to such phases they must not be expected to obey the ordinary chemical valencies of the metals concerned. In fact, the satisfaction of chemical valencies is the negation of metallic properties since in chemical combination the loosely bound electrons of the metal atoms are bonded in the stable groups of the ionic and covalent linkages. If free valency electrons are to be left over to form a truly metallic phase the valency relations must necessarily be different from those obtaining in definite chemical compounds. Chemical combination in the generally accepted sense is confined to covalent and ionic structures, and alloy systems are sharply distinguished from such compounds by the indefiniteness of their composition, the readiness with which they are synthesized and their general qualitative resemblance to their component metals. If such systems are to be regarded as compounds it must be in terms of a wider conception of chemical combination. The extent to which such a wider conception is desirable is discussed below.

Even though metal systems cannot be regarded as ordinary chemical compounds, it does not follow that it is unjustifiable to represent such systems by definite formulae. In covalent and ionic structures stoichiometric proportions are normally necessitated by rigid electrostatic demands, but in a crystal structure simple proportions may be necessary to conform to purely geometrical requirements. If atoms of two different radii are to form a structure, a simple stable arrangement is possible only when the atoms are present in some simple proportions. It is this purely geometrical effect which justifies definite formulae for the phases of many true metal systems, but it must be clearly understood that such formulae are primarily geometrical in their origin and do not represent chemical combination in its generally accepted sense. In most cases the geometrical demands are not very exacting, and a greater or lesser degree of tolerance, giving rise to a corresponding indefiniteness of composition, is permitted.

13.36. To meet the difficulties presented by metal systems, various wider definitions of chemical combination in terms of crystal structure have been proposed. For example, it has been suggested that we should regard an ideal chemical compound as one in which structurally equivalent positions are occupied by chemically identical atoms, and an ideal solid solution as a structure in which all atoms are structurally equivalent. It is clear that such a definition of chemical combination embraces all the generally accepted compounds, but it is not without objection when applied to metal systems. Thus, to take only one example, the β phase in the silver–cadmium system already discussed has, in its ordered state, the simple caesium chloride structure and must therefore be regarded as a definite compound. In the state of disorder, however, the two kinds of atom are distributed at random over the structurally equivalent sites of a body-centred cubic lattice, and the phase then satisfies the definition of an ideal solid solution. It is difficult to view the distinction between solid solution and chemical compound as nothing more than the relatively trivial distinction between these two states, and when it is remembered that not only the two extremes but every intermediate condition of order can be realized, the difficulty is even greater.

The electron compounds present special difficulties of their own, for with them geometrical demands do not generally determine the structure, and many phases occur only in the completely disordered state. In such cases, and especially when an extended range of solid solution occurs, it is possible to assign many formulae, a choice between which can be made only on structural grounds in terms of Hume-Rothery's rule. The β phase in the system copper–aluminium has a considerable range of solid solution on either side of the composition Cu_3Al assigned to it, and clearly a phase of this composition cannot have the caesium chloride structure but must occur in the disordered state. From a structural point of view, therefore, any formula within the homogeneity range would be equally acceptable, and it is only the demands of the electron: atom ratio which determine the ideal composition of the phase. A similar difficulty arises with the μ phase of the silver–aluminium system. This system has a vanishingly narrow homogeneity range coinciding with the composition Ag_3Al, and this composition has in fact long been attributed to the phase on purely chemical grounds. The structure, however, cannot give any confirmation of this or of any other formula, for the arrangement is that of β-manganese with 20 atoms in the unit cell divided into two equivalent groups of 12 and 8. It is clearly impossible to distribute two kinds of atoms regularly in such a structure in any other ratio than 3:2, and the phase Ag_3Al is therefore a completely disordered one. It is only from Hume-Rothery's rule that confirmation of the experimentally determined composition is obtained.

Exactly similar arguments apply to the γ phases. Here the extent of the solid solution has often led to the assignation of incorrect formulae, especially since the composition satisfying Hume-Rothery's rule is relatively complex so that a composition differing but little from it can often be expressed by a simpler formula. Disordered structures are again common, and even when an ordered arrangement obtains the distribution of the atoms in closely related structures is not necessarily the same. We have already seen that in the phases Cu_5Zn_8 and Cu_5Cd_8 the disposition of the atoms is quite different although the sites occupied are the same in both cases.

A further difficulty which arises in many alloy systems is that the range of stability of a phase may not in fact embrace the composition corresponding to its ideal formula. Thus in the aluminium–chromium system a phase exists with a relatively simple tetragonal structure, the unit cell of which contains ideally $4Cr+2Al$. This phase is capable of taking up in solid solution an excess of chromium and is found, say, at a composition Cr_3Al. The range of solid solution does not, however, include the ideal composition Cr_2Al, and if an attempt is made to crystallize an alloy of this composition an inhomogeneous mixture of two phases is formed. Further examples of the same phenomenon are the phases of idealized composition $CuAl_2$ and $NaPb_3$, both of which exist in the systems concerned but in neither case at the simple composition corresponding to the ideal formula.

These examples serve to show that in metal systems chemical analogy has no necessary counterpart in analogy of structure, and that in general if a formula is assigned to an intermetallic phase it must be interpreted as representing composition alone without any implication of chemical combination. Even when the composition is known, however, the properties of the alloy are by no means completely specified. It may exist in an ordered or in a disordered state, or in a condition intermediate between these extremes, and its properties may, moreover, be profoundly dependent not only on structure but also on physical characteristics such as grain size, texture and the presence of 'faults'.

INTERSTITIAL STRUCTURES

13.37. The interstitial structures comprise the compounds of certain metallic elements, notably the transition metals and those of the lanthanide and actinide series, with the four non-metallic elements hydrogen, boron, carbon and nitrogen. In chapter 8 we discussed the structures of a number of hydrides, borides, carbides and nitrides of the most electropositive metals, and these we found to be typical salt-like compounds with a definite composition and with physical properties entirely different from those of the constituent elements: they are generally transparent to light and poor conductors of electricity. The systems now to be considered are strikingly different. They resemble metals in their opacity and characteristic lustre, they are good electrical conductors and, in common with alloy systems, they often show an indeterminacy of composition and a sequence of distinct phases. They differ from intermetallic systems, however, in that they are brittle and extremely refractory, and have a hardness often approaching if not exceeding that of diamond. By virtue of these properties some of these systems are of considerable technical importance.

The structures of the interstitial systems, as their name implies, can all be described in geometrical terms as an array of metal atoms with the smaller atoms of the non-metallic element disposed in the interstices, and it is found that they are formed only if the ratio of the radii of the atoms is less than about 0·59. The arrangement of the metal atoms alone is usually that of one of the three structures common to the true metals, i.e. cubic or hexagonal close packed with a co-ordination number 12 or cubic body centred with a co-ordination number 8, but occasionally these atoms are arranged on a simple hexagonal lattice.

The distribution of the non-metal atoms in the interstices of the metal array is found always to be such that these atoms are in contact with their metal neighbours. Subject to this condition, the non-metal atoms occupy the largest spaces in the structure in which this contact can be attained, and so achieve the highest possible co-ordination. Thus, if the metal atoms are arranged as in cubic close packing the non-metal atoms, if sufficiently large, occupy the 6-co-ordinated interstices at the centres of the sides of the unit cell; if all these sites are occupied the structure is that of sodium chloride and corresponds to the composition MX. If the non-metal atom is too small for this type of co-ordination it occupies instead the smaller 4-co-ordinated sites in the centres of the eight cubelets into which the unit cell can be divided. If all these sites are occupied the structure is that of fluorite and corresponds to the composition MX_2. If one-half of the sites are symmetrically filled, the zincblende structure of composition MX results. In an analogous way a series of structures, including those of nickel arsenide and wurtzite, can be developed by filling the 6- or 4-co-ordinated interstices in a hexagonal close-packed array of metal atoms.

Classification

13.38. These observations provide the basis for a classification of interstitial structures, originally proposed by Hägg, in which we consider as of one type all those structures produced by the occupation of interstices of a given kind in a given type of metal array, regardless of the number of these interstices which are in actual fact so occupied. The classification is therefore expressed in terms of (1) the nature of the metal array, and (2) the co-ordination number of the non-metal atoms. The justification for such a classification lies in the observation that interstitial structures never occur in which more than one type of interstice is occupied, so that these structures never contain more atoms of the non-metal than are required completely to fill the equivalent positions of one kind. On the other hand a deficiency of non-metal atoms, with only some of the equivalent positions occupied, is possible.

Table 13.11. *The classification of interstitial structures*

C.N.	Arrangement of metal atoms	C.N. of non-metal	Type symbol	Condition for occurrence $r_X:r_M$
12	12*a* Cubic close packed	6	12*a*, 6*	> 0·41
		4	12*a*, 4†	> 0·23
	12*b* Hexagonal close packed	6	12*b*, 6§	> 0·41
		4	12*b*, 4‡	> 0·23
8	8*a* Cubic body centred	4	8*a*, 4	> 0·29
	8*b* Simple hexagonal $c:a = 1$	6	8*b*, 6	> 0·53

* Cf. sodium chloride. † Cf. fluorite and zincblende. § Cf. nickel arsenide. ‡ Cf. wurtzite.

The classification of interstitial structures just described is summarized in table 13.11, where the radius ratio conditioning the appearance of each structure type is also given. In table 13.12 the structures of a number of interstitial compounds are shown. A study of these data reveals that structures based on a close-packed arrangement of metal atoms are far more common than those in which the metal is 8-co-ordinated. With structures of composition MX a cubic close packing of metal atoms predominates but with M_2X compounds hexagonal close packing is of more frequent occurrence. The appearance of both these structures at different compositions in a single system, as in Ta–C or Cr–N, shows that the arrangement of the metal atoms is not necessarily that which obtains in the structure of the element. The occurrence of the structure type 12*a*, 6 at the composition M_4X in the systems Mn–N and Fe–N reveals that interstitial structures may exist with only a fraction of the equivalent sites occupied by non-metal atoms.

Table 13.12. *The structures of some interstitial phases*

System	$r_X:r_M$	M_4X	M_2X	MX	MX_2
Zr–H	0·29	12*a*, 4	12*b*, 4	12*a*, 4	—
Ta–H	0·32	—	12*b*, 4	8*a*, 4	—
Ti–H	0·32	—	12*b*, 4	12*a*, 4	12*a*, 4
Pd–H	0·34	—	12*a*, 4	—	—
La–N	0·37	—	—	12*a*, 6	—
Ce–N	0·38	—	—	12*a*, 6	—
Pr–N	0·38	—	—	12*a*, 6	—
Nd–N	0·38	—	—	12*a*, 6	—
Th–N	0·39	—	—	12*a*, 6	—
Nd–C	0·42	—	—	12*a*, 6	—
Th–C	0·43	—	—	12*a*, 6	—
Zr–N	0·44	—	—	12*a*, 6	—
Sc–N	0·44	—	—	12*a*, 6	—
U–N	0·45	—	—	12*a*, 6	—
Zr–C	0·48	—	—	12*a*, 6	—
Nb–N	0·48	—	—	12*a*, 6	—
Ti–N	0·48	—	—	12*a*, 6	—
Hf–C	0·49	—	—	12*a*, 6	—
U–C	0·50	—	—	12*a*, 6	—
W–N	0·50	—	12*a*, 6	—	—
Mo–N	0·50	—	12*a*, 6	8*b*, 6	—
V–N	0·52	—	—	12*a*, 6	—
Nb–C	0·53	—	—	12*a*, 6	—
Ti–C	0·53	—	—	12*a*, 6	—
Ta–C	0·53	—	12*b*, 6	12*a*, 6	—
Mn–N	0·53	12*a*, 6	12*b*, 6	—	—
W–C	0·55	—	12*b*, 6	8*b*, 6	—
Cr–N	0·55	—	12*b*, 6	12*a*, 6	—
Mo–C	0·56	—	12*b*, 6	—	—
Fe–N	0·56	12*a*, 6	12*b*, 6	—	—
V–C	0·58	—	12*b*, 6	12*a*, 6	—

Bonding in interstitial structures

13.39. Although the geometrical picture of interstitial structures presented above provides a convenient basis for their classification, and is, indeed, the origin of the very name by which they are known, it must not be assumed that these structures can be regarded simply as interstitial solid solutions in the crystal structure of the parent metal. There are many reasons for this view, among which the following may be mentioned:

(1) The structures are formed only by the transition, lanthanide and actinide elements, and not by other metals of comparable atomic radius and electronegativity.

(2) The arrangement of the metal atoms is often different from that

in the structure of the element, and may be different in different phases of the same system.

(3) The phases are hard and brittle whereas the metals are usually soft and malleable.

(4) The structures have very high melting points, in some cases over 1000 °C higher than that of the parent metal.

(5) In spite of the hardness and refractory character of the phases, the metal–metal distance in them is generally greater than in the structure of the elementary metal.

It is clear from these properties that intermetallic forces alone cannot be responsible for the cohesion of interstitial structures, and that in addition strong forces must operate between the metal and non-metal atoms. The nature of these forces has been considered in some detail in the particular case of the numerous carbides and nitrides of composition MX. It will be seen from table 13.12 that the great majority of these crystallize with the sodium chloride structure (12*a*, 6), in which, of course, each atom is octahedrally co-ordinated by six of the other kind. If this octahedral co-ordination is to be ascribed to covalent bonds from the non-metal atoms the difficulty arises that the covalency of these elements is normally not greater than four. We must therefore regard the six bonds as fractional bonds formed by resonance, and this can be achieved in two ways. The first alternative is that the orbitals involved are the three $2p$ orbitals of the light atom. These are mutually at right angles and by resonance would give six bonds, each of number $\frac{1}{2}$. A second alternative is that one of the p orbitals is hybridized with the s orbital to give two sp bonds oppositely directed, with two further bonds, due to the remaining unhybridized orbitals, in a plane perpendicular to them. Again, resonance would give six equivalent bonds, but now each of number $\frac{2}{3}$. The former arrangement is to be expected in interstitial structures of the more electropositive metals, the latter when the metal is less electropositive. There is some evidence that this is so. The metal–non-metal distance is always greater than the sum of the normal covalent radii, implying a bond of number less than unity, but the difference is greatest in those systems containing the most electropositive metals.

It is not sufficient, however, to consider only the bonding of the non-metal atoms in the interstitial structures. The argument outlined above explains why these atoms are octahedrally co-ordinated, but the same co-ordination could be achieved in a 12*b*, 6 structure, with the metal atoms arranged as in hexagonal rather than cubic close packing. In fact, the majority of metals forming carbides and nitrides with the sodium chloride structure have a hexagonal close-packed structure in the elementary state, and we have to enquire why this changes to a cubic close-packed array in the interstitial compound. The answer to this question appears to be that although the co-ordination about the non-metal atoms is in both cases octahedral this is not so for that about the metal atoms. If the metal atoms are arranged as in cubic close packing, giving the sodium chloride structure, they also are octahedrally co-ordinated. If, however, the metal atoms are in a hexagonal close-packed array the resulting interstitial compound has the nickel arsenide structure with the metal atoms co-ordinated by six non-metal neighbours at the corners of a trigonal prism. The preference for octahedral co-ordination about the metal atom as well as about that of the non-metal suggests that the metal orbitals involved are the d^2sp^3 hybrids (table 4.01), and it is noteworthy that the interstitial structures are formed by just those metals in which such hybridization is commonly found. In terms of this picture the exceptional stability of the interstitial compounds with the sodium chloride structure is to be ascribed to the fact that this structure satisfies simultaneously the requirements of the bond configuration of both metal and non-metal atoms.

The structure of steel

13.40. The interstitial structures of by far the greatest technical importance are those which occur in the iron–carbon system, and the application of X-ray analysis to this system has resulted in a great extension of our understanding of the properties of carbon steels, and in a considerable simplification in the description of their behaviour. We cannot give here a detailed account of all the work in this field but certain features are of general interest and may be briefly discussed.

The crystal chemistry of the iron–carbon system is especially complex on account of the relatively small size of the iron atom, resulting in a carbon:iron radius ratio of about 0·60, which is so close to the critical value 0·59 discussed above that both interstitial structures and structures of greater complexity may be expected. Added to this is the further complication that iron is dimorphous. Below about 910 °C, and from about 1400 °C to the melting point, the structure is cubic body centred, and is known as α iron. Between these two temperatures a cubic close-packed structure, termed γ iron, is formed. The ferromagnetism of iron is found only in the α phase, but is not a property related to atomic arrangement alone and vanishes at the Curie point of 766 °C without change of structure. Between this temperature and 910 °C, and again above 1400 °C, the α phase is not ferromagnetic.

13.41. Above about 900 °C all carbon steels form a non-magnetic interstitial solid solution of carbon in γ iron, called *austenite*. The amount of carbon which can be taken up in solid solution is, however, limited to about 1·7 per cent by weight, equivalent to 7·5 atomic per cent, and it is therefore clear that the number of carbon atoms is insufficient for them to be arranged in any regular way and that they must be distributed at random in the interstices of the structures. If more carbon is present the excess occurs as graphite, and in this case the system is usually described as 'cast iron'. It is convenient to restrict the term 'steel' to those systems in which the carbon content does not exceed 7·5 atomic per cent.

The presence of carbon in solid solution stabilizes the γ iron structure and inhibits the transition to α iron. On slow cooling, therefore, austenite remains stable below the normal transition temperature of 910 °C, and it is only at about 700 °C that the solid solution breaks down and transforms into a mixture of *ferrite* and *cementite*. Ferrite is an interstitial solution of carbon in α iron, but the amount of carbon which can be taken up in solution is very small and is limited to about 0·3 atomic per cent. The excess of carbon is therefore thrown out of solution and appears in the definite compound cementite, Fe_3C, with a complex orthorhombic structure. The solid is no longer homogeneous, and the characteristic appearance, due to the separation of ferrite and cementite, gives rise to the name *pearlite*. In this condition the steel is very soft.

If cementite is reheated it transforms into austenite and graphite, and there is reason for believing that even below 700 °C cementite is, strictly speaking, metastable and that the stable system is graphite in equilibrium with ferrite. In practice, however, the rate of transformation is so slow that cementite behaves essentially as a stable phase.

When steel is cooled rapidly, by quenching, the transformation from austenite to pearlite takes place much more slowly, and at a temperature of 200 °C a period of the order of hours is required for the change to occur. The transformation may even be suppressed altogether by the addition of other elements, so that non-magnetic austenite can then be preserved in a stable state at room temperature. This, however, is not possible with carbon steels (although a limited amount of 'retained austenite' is, indeed, found at low temperatures), and below about 150 °C austenite transforms rapidly at a rate which *increases* with decreasing temperature and becomes almost instantaneous at room temperature. The mechanism of the transformation, however, is entirely different from before, and now not pearlite but *martensite* is formed. This is a supersaturated interstitial solid solution of carbon in α iron, the unit cell of which is deformed by the large carbon content and is tetragonal instead of cubic body-centred. The axial ratio depends upon the carbon content and has a maximum value of 1·07 at a composition of about 7 atomic per cent carbon. It is important to note that martensite is stable only below about 150 °C, and that in the neighbourhood of this temperature its rate of formation from quenched austenite is far greater than that at which austenite is transformed to pearlite at somewhat higher temperatures.

Quenched steel containing martensite is hard and brittle. A softer, tougher material may, however, be obtained by reheating to temperatures in the range 200–300 °C. At these temperatures martensite is unstable, and in the tempering process breaks down into ferrite and cementite. Although of the same composition as pearlite, the solid thus formed is generally of a coarser microcrystalline texture and is termed *sorbite*. The rate of transition from martensite to sorbite increases rapidly with temperature, and careful control of the tempering process is therefore essential if steel of the desired quality is to be produced.

13.42. The relation of martensite to austenite, on the one hand, and to the mixture of ferrite and cementite, on the other, may be readily understood if the decomposition of austenite is regarded as involving two separate processes: (1) the transformation from an interstitial solid solution of carbon in γ iron to a solid solution in α iron, and (2) the rejection of the carbon in the α iron. If the first process takes place alone the carbon is retained in a supersaturated solid solution in α iron and tetragonal martensite is formed. If the second process now follows carbon is precipitated from the martensite and taken up as cementite, so that sorbite results. This is exactly what occurs at low temperatures, when martensite is first formed by quenching and is subsequently converted to sorbite by tempering, because at these temperatures the rate of precipitation or carbon from solution is very slow compared with the rate at which the γ- $\rightarrow$ α-iron transformation takes place. At higher temperatures, however, this is no longer the case and now the second process immediately follows the first, with the result that austenite is directly converted to ferrite and cementite. The relationships of the various phases in the iron–carbon system and the effect of heat treatment on them are summarized in fig. 13.13.

Austenite (solid solution of C in γ-Fe)
|
Slow cooling to 700 °C
↓
Pearlite (mixture of ferrite (solid solution of C in α-Fe) and cementite, Fe_3C)
|
Reheat
↓
Austenite + graphite

Austenite
|
Quench to 150 °C
↓
Martensite (supersaturated solid solution of C in α-Fe) + 'retained austenite'
|
Temper at 200–300 °C
↓
Sorbite (mixture of ferrite and cementite)

Fig. 13.13. Phase relationships in the iron–carbon system.

13.43. It is perhaps hardly necessary to add that the practical metallurgy of steels, and especially of the alloy steels, is vastly more complex than the above greatly simplified account might suggest. Not the least important of the reasons for this is that the mechanical and other physical properties of steel are not only dependent on the number and nature of the phases present but are also profoundly sensitive to the microstructure of the alloy, particularly in respect of the size of the crystal grains present and their mutual orientation. It is because all these factors can be varied, to a large extent independently, by the appropriate control of composition, heat treatment and mechanical working that steels can be prepared with an almost infinite variety of physical properties and that they constitute by far the most versatile and by far the most important of all the alloy systems known to man. It is interesting to reflect that the characteristics of the iron–carbon system are quite unique, and in a sense accidental in origin, and to ponder how our way of life would be altered if steel did not exist.

14 ORGANIC STRUCTURES

INTRODUCTION

14.01. We have already described a considerable number of structures which are molecular in the sense that more or less well defined molecules, of finite or infinite extent, can be recognized as discrete entities in the crystal. These molecules are bound within themselves by forces of one kind and to one another by forces of another, and usually weaker, type. By far the most important molecular structures, however, are those which embrace the whole field of organic chemistry, and it is to them that this chapter is devoted. In organic structures the molecules are usually relatively stable units in which covalent bonds are operative, but they are normally held together only by forces of a weaker kind. Hence organic compounds are broadly distinguished as a class by their softness and fusibility and by the fact that they generally melt without decomposition.

The application of the methods of crystal structure analysis in the field of organic chemistry has in recent times been one of the major achievements of the X-ray crystallographer. Initially, however, progress was slow, and it was not until some twenty years after the discovery of X-ray diffraction that this tool was systematically applied to the study of organic compounds. In part this was due, no doubt, to the practical problems presented by such compounds, the analysis of which is technically more difficult than that of inorganic structures, owing, on the one hand, to the scarcity of good crystals and, on the other, to the closely comparable scattering powers of the atoms of carbon, nitrogen and oxygen and to the very small scattering power of hydrogen. In part, however, it was also due to the fact that the field appeared potentially less fruitful. Before the development of crystal structure analysis little or nothing was known about the arrangement of the atoms in inorganic compounds in the solid state; the power of the new technique was therefore rapidly appreciated as the significance of the results to which it gave rise became apparent. In organic chemistry, however, the position was quite different. The organic chemist had long had at his disposal the classical techniques of stereochemistry, and these, supplemented by spectroscopic and other physical methods, had already thrown much light on the nature of organic compounds. Moreover, he had the further advantage that his studies were not confined to the solid state; organic molecules preserve their identity in the melt, and sometimes even in the vapour, so that molecular configurations determined in these phases more accessible to experiment could be assumed to exist also in the solid. The structures of large numbers of organic molecules were therefore known with reasonable certainty long before the methods of crystal structure analysis were applied in this field, and the results of the early analyses were primarily verifications of molecular configurations already established by other means. Today this is no longer true. Refinements in the methods of X-ray analysis have made possible the location of hydrogen atoms and the successful solution of structures of ever increasing complexity, and in some instances molecular configurations have been established for the first time in this way after all other techniques had failed.

14.02. It must not be assumed from what has just been said that X-ray methods are of value only in the study of molecules of great complexity. On the contrary, in even the simplest structures these methods have much to contribute. In the first place they add a metrical basis to the structural formula of the chemist by providing a quantitative and very precise measure of bond length, interbond angles and electron distribution in the molecule. We shall give examples of the interpretation of these quantities later, but here we may recall that in covalent compounds we have found bond lengths to be the sum of the appropriate covalent radii. These radii, however, are not constant but are themselves a function of bond number, so that bond number and length are directly related, as is indicated by the data given in table 14.01, in which the lengths of a few bonds of common occurrence in organic compounds are recorded. A detailed study of interatomic distances can thus give immediate evidence of the type of binding in a molecule. Such considerations, moreover, are not confined to bonds of integral number, for if an interatomic distance is observed to depart appreciably from the value given in the table we must assume the existence of a resonance structure to which two or more bond configurations contribute. In complex structures, where chemical evidence alone cannot provide such information, this method of approach is especially fruitful.

A second important contribution of X-ray methods is the light which they throw on the nature of the forces responsible for the coherence of the crystal and on the way in which the molecules are linked together in

Table 14.01. *Some covalent bond lengths*

The bond lengths given are those of 'pure' single, double or triple bonds, as observed in compounds in which the bonds have substantially the number indicated

Bond	Length (Å)	Bond	Length (Å)
C—C	1·54	C—Cl	1·76
C=C	1·33	C—Br	1·91
C≡C	1·20	C—I	2·15
C—O	1·42		
C=O	1·20		
C—OH	1·42	O···O*	2·5–2·9
C—N	1·48	N···O*	2·7–3·1
C=N	1·27	N···N*	3·0–3·3

* Hydrogen bond.

the solid under the influence of these forces. This is a problem which is normally ignored in chemistry, and when the intermolecular forces are purely of the van der Waals type it is indeed true that they have little influence on chemical properties. There are, however, many organic structures, especially of compounds containing reactive groups such as —OH or =NH, in which far stronger hydrogen bonds operate between the molecules and in which the packing of the molecules and the properties of the substance are largely determined by the nature of these intermolecular forces. In such structures the most direct evidence for the existence and disposition of the hydrogen bonds is that provided by the crystal structure analysis. Finally, we must recognize a third type of force, the purely ionic bond, which is found in the salts of organic acids and bases.

The classification of organic structures

14.03. The structures of upwards of a thousand organic compounds have now been elucidated by X-ray analysis. The choice of substances studied has naturally been often determined primarily by the chemical interest of the compounds in question, and we therefore find that some structures of considerable complexity are known whereas many other and simpler bodies have not as yet been investigated. For this reason it is not possible to present a classification of organic structures as systematic as that which has been given for structures of other types. In ionic or metal structures, say, we are concerned with the packing of spherical units under the influence of non-directed interatomic forces, so that geometrical factors are largely responsible for determining the structure adopted; in covalent compounds, although the bonds are now limited in number and characteristically directed, the component units in the structure are still isolated atoms. In organic structures the position is quite different. Here the units of structure are not separate atoms but complete molecules of irregular and often complex shape, and the forces between these molecules may be either relatively weak non-directed van der Waals bonds or considerably stronger hydrogen bonds characteristically disposed in space. Each substance therefore has its own distinctive structure, and it is only in the most general terms that it is possible to point to structural features of widespread occurrence among compounds of a given type. Our discussion of organic compounds must therefore be primarily a descriptive account of some of the structures analysed, in the course of which any general principles shown by these structures will be emphasized. A relatively limited number of quite simple structures will suffice to illustrate these principles, and we shall accordingly make no attempt to discuss the many structures of greater complexity which display no novel structural features. This is not to imply that such structures are of any less importance; on the contrary, it is often in them that the greatest chemical interest resides, and it is certainly among them that we find what are, as yet, the most striking examples of the power of X-ray analysis as a tool for the study of the solid state.

14.04. Even though a strictly systematic classification of organic structures is not possible it is nevertheless essential, as a matter of convenience, that some basis of classification should be adopted. We shall therefore classify structures primarily in terms of molecular shape, and for this purpose we may assign many molecules to one of three somewhat arbitrarily defined groups: (*a*) small or symmetrically shaped molecules, (*b*) long molecules, and (*c*) more or less flat molecules. Such a division is inevitably in some respects artificial, for there are molecules which can be assigned with equal justification to more than one of these groups, and there are other molecules, particularly of complex compounds, which cannot properly be assigned to any of them. Even so, the classification has certain practical advantages and will enable us to point to some characteristic structural features closely associated with molecular shape. The most important of these is that in many structures the molecules tend to assume as closely packed an arrangement as is consistent with this shape, so that molecules of similar shape are often found to be similarly arranged. This, however, is subject to the important qualification that it is true only if the molecules are bound together by van der Waals forces. When hydrogen or other ionic bonds between active groups are responsible for intermolecular cohesion the position of these groups in the molecule and the spatial disposition of the bonds may give rise to structures in which the packing is very different and in which molecular shape plays little part in determining the structural arrangement. Superimposed on the classification in terms of molecular shape, therefore, we have a second classification in terms of the nature of the intermolecular binding.

COMPOUNDS WITH SMALL OR SYMMETRICAL MOLECULES

Apolar intermolecular binding

14.05. A characteristic feature of many structures of this type is an approximately close-packed arrangement of the molecules arising from their simple form and from the symmetrical, undirected distribution of the bonds between them.

Hydrocarbons

14.06. The simplest possible example of such a structure is that of *methane*, CH_4. Above a temperature of about 22 °K this structure is cubic close packed, with the molecules in free rotation. The C–C distance, representing the weak intermolecular van der Waals bonds, is 4·1 Å. It will be remembered that the inert gases also have a cubic close-packed structure, and it is of interest to note that argon and krypton, in which the interatomic distances (3·82 and 4·02 Å, respectively) are very close to those in methane, form solid solutions with this gas. Methane also forms a clathrate-type hydrate of composition $CH_4 . 5\frac{3}{4}H_2O$, with a structure the same as that of the corresponding xenon hydrate and similar to the structure of $Cl_2 . 7\frac{2}{3}H_2O$ already described (§12.35). Below 22 °K the methane molecules are no longer in free rotation and, arising from the change in structure, an anomalous increase in the specific heat is observed on heating through the temperature range 18–22 °K.

14.07. Some other simple hydrocarbons also show approximately close-packed structures. In *ethane*, C_2H_6, the centres of the molecules are arranged as in hexagonal close packing and the axes of the molecules are all parallel to the principal axis. The C–C distance within the molecule is 1·54 Å, exactly the same as the single-bond distance in diamond, and the C–C distance between adjacent molecules is about 3·6 Å. In *acetylene*, C_2H_2, the centres of the molecules are disposed in a cubic close-packed array with the axes of the molecules parallel to the

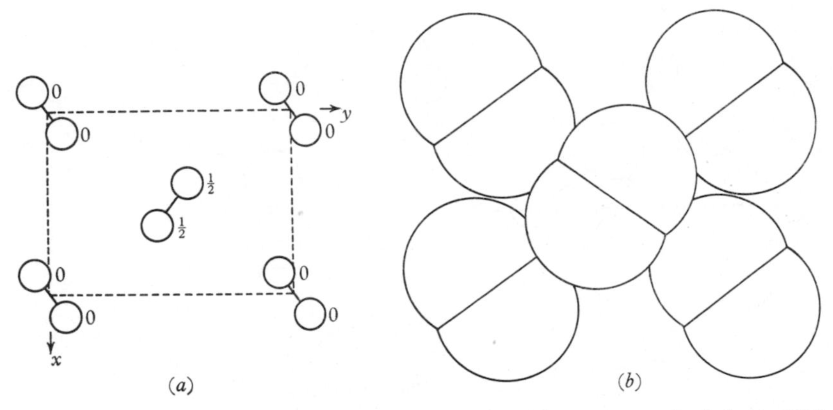

Fig. 14.01. (*a*) Plan of the unit cell of the orthorhombic structure of ethylene, C_2H_4, projected on a plane perpendicular to the *z* axis. (*b*) The same, with the CH_2 groups given their correct van der Waals radii to emphasize the packing of the molecules.

four diagonal axes of the cube, giving an arrangement analogous to that in carbon dioxide (§8.35 and fig. 8.13). Here the C–C distance within the molecule is only 1·20 Å, corresponding to a triple-bond radius of 0·60 Å for the carbon atom. The structure of *ethylene*, C_2H_4, is shown in fig. 14.01. This structure is orthorhombic, but it will be seen that it is, nevertheless, a compact arrangement in which each molecule is surrounded by ten others and in which the mutual disposition of adjacent molecules is such that they fit closely together. The C–C distance within the molecule, corresponding to the double bond, is 1·33 Å. These three structures are of particular importance in that they give values for the single-, double- and triple-bond covalent radius of the carbon atom under circumstances in which the bond number is unambiguously known, and it was in fact the lengths of the C–C bonds in these three hydrocarbons which were used to plot fig. 4.07 and which are quoted in table 14.01.

Other structures

14.08. Even relatively complex molecules may be packed together to give simple structures if they are themselves highly symmetrical in form. The structure of *hexamethylenetetramine*, $C_6H_{12}N_4$, is an example of this and is also of historic interest as the first organic structure to be completely determined by X-ray analysis. The molecule of this compound has the symmetrical configuration shown in fig. 14.02*a*, with six

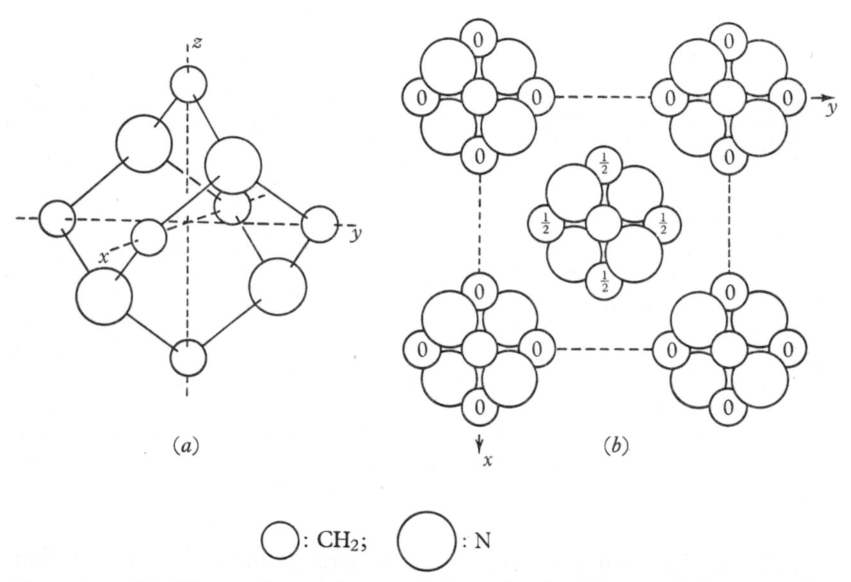

Fig. 14.02. (*a*) Clinographic projection of the molecule of hexamethylenetetramine, $C_6H_{12}N_4$. (*b*) Plan of the unit cell of the cubic structure projected on a plane perpendicular to the *z* axis. The heights of four carbon atoms in each molecule are indicated in fractions of the *c* translation.

CH_2 groups arranged octahedrally and four nitrogen atoms arranged tetrahedrally about a common centre, and, in the crystal, molecules of this form lie at the centre and at the corners of a cubic unit cell (fig. 14.02*b*). Each molecule is surrounded by eight neighbours at the corners of a cube and the closest approach between carbon atoms of adjacent molecules is 3·7 Å. In the hydrocarbon *adamantane* (*s*-tricyclodecane), $C_{10}H_{16}$, the molecule has precisely the same cage-like form as in hexamethylenetetramine (with the nitrogen atoms replaced by CH groups), but the structure of the crystal is different in that the molecules now lie at the corners and face centres of a cubic unit cell to give an even more closely packed arrangement. In *tetraphenylmethane*, $C(C_6H_5)_4$, the central carbon atom is tetrahedrally surrounded by four planar phenyl groups, and the resulting tetrahedral molecules are arranged in the crystal with their centres at the corners and middle of a tetragonal unit cell. This gives rise to a very symmetrical structure (fig. 14.03) in which the four phenyl groups of each molecule fit compactly into the interstices between the phenyl groups of its neighbours.

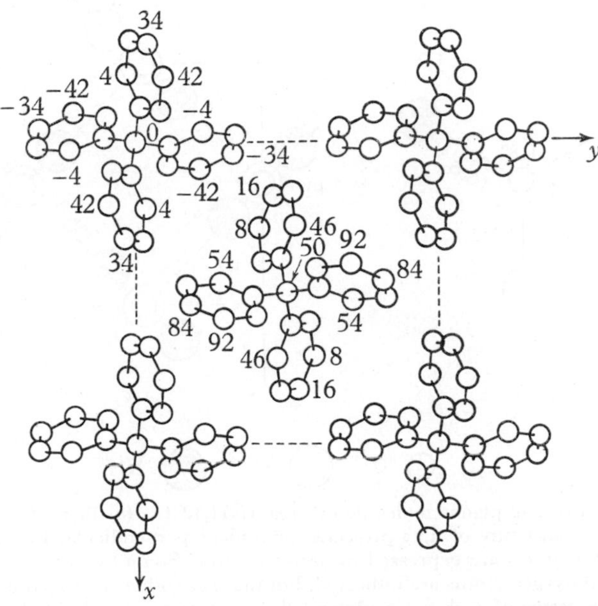

Fig. 14.03. Plan of the unit cell of the tetragonal structure of tetraphenylmethane, $C(C_6H_5)_4$, projected on a plane perpendicular to the z axis. The heights of some of the atoms are indicated in units of $c/100$ to emphasize the form of the molecules and their packing.

We defer a discussion of the structure of the phenyl group itself until we come to consider some other aromatic compounds.

Hydrogen bonded structures

14.09. Simple closely packed structures of the type just described are rarely found in compounds containing reactive groups such as —OH, =O or $—NH_2$. The arrangement of the molecules is now no longer determined by the van der Waals forces between them but rather by the hydrogen bonds between the active groups, and the molecules are therefore so disposed that groups of appropriate polarity are adjacent.

Occasionally this can be achieved in a way which still gives a closely packed arrangement of molecules, but more commonly it gives rise to a more complex and less closely packed structure than would be expected if the hydrogen bonds were not operative. Hydrogen bonds in a crystal structure can generally be readily recognized by the fact that the atoms between which they operate approach more closely than those bound only by van der Waals forces.

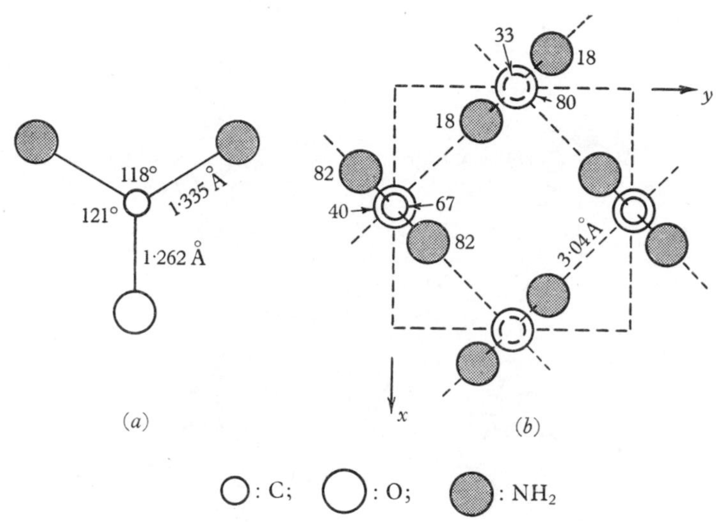

: C; : O; : NH_2

Fig. 14.04. (*a*) The planar molecule of urea, $(NH_2)_2CO$. (*b*) Plan of the unit cell of the tetragonal structure of urea projected on a plane perpendicular to the z axis. The heights of the atoms are expressed in units of $c/100$. Some hydrogen bonds between nitrogen and oxygen atoms are indicated, but there are other hydrogen bonds between the nitrogen atoms of each molecule and the oxygen atom of the molecule vertically above or below it.

Urea and thiourea

14.10. The shape of the planar molecule of urea, $(NH_2)_2CO$, and the arrangement of the molecules in the tetragonal unit cell are shown in fig. 14.04. It will be seen that the molecules are disposed with their axis of symmetry parallel to the z axis of the cell, but that their planes lie in two directions at right angles and that the molecules are pointing alternately up and down. The closest distance of approach between molecules is between oxygen and nitrogen atoms, each oxygen having two nitrogen neighbours (those in the molecules vertically above or below it) at a distance of 2·99 Å and two other nitrogen neighbours (in the two molecules on each side of it) at a distance of 3·04 Å. These distances are much smaller than the sum of the van der Waals radii and represent hydrogen bonds between the NH_2 groups and oxygen atoms, the molecular packing being such that every hydrogen atom is involved in a hydrogen bond. The hydrogen bonds are not, however, as strong as in certain amino acids, in which N···O distances as small as 2·7 Å are observed.

The principal interest of the urea structure is in the detailed form of the molecule observed and in the light which this throws on the nature of the interatomic bonds. The dimensions shown in fig. 14.04*a* are not consistent with the simple structure (I), for the C–O distance of 1·262 Å is considerably longer than the known length of 1·20 Å for a C–O double bond while the C–N distance of 1·335 Å is shorter than the C–N single-bond length of 1·48 Å. We must therefore assume that both these bonds are intermediate in character between a single and a double bond, and this will be the case if the molecule has a resonance structure to which not only configuration (I) but also configurations (II) and (III) make a contribution. Relative contributions from these three configurations of 40, 30 and 30 per cent give the C–O bond 40 per cent and the C–N bond 30 per cent double-bond character and account satisfactorily for the observed interatomic distances.

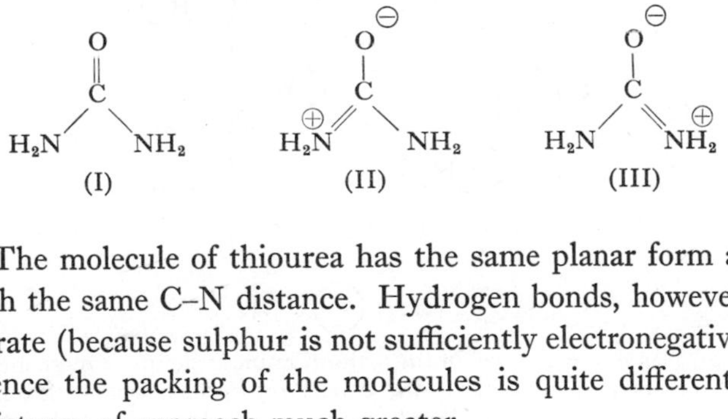

(I)	(II)	(III)
$H_2N{-}C({=}O){-}NH_2$	$H_2\overset{\oplus}{N}{=}C(O^{\ominus}){-}NH_2$	$H_2N{-}C(O^{\ominus}){=}\overset{\oplus}{N}H_2$

14.11. The molecule of thiourea has the same planar form as that of urea, with the same C–N distance. Hydrogen bonds, however, cannot now operate (because sulphur is not sufficiently electronegative) and in consequence the packing of the molecules is quite different and the closest distance of approach much greater.

Methanol

14.12. In urea each molecule is involved in four hydrogen bonds, and these bonds are responsible for the coherence of the structure in three dimensions. In many structures, however, the pattern of hydrogen bonds extends in only one or two dimensions, linking molecules into chains or sheets between which van der Waals forces alone are operative. An example of the former arrangement is provided by methanol (methyl alcohol), CH_3OH, the very simple structure of which is represented in fig. 14.05. The number of hydrogen atoms in the hydroxyl groups is sufficient to form only two hydrogen bonds per molecule, and these are expended in linking the molecules through their hydroxyl groups into endless bands parallel to the z axis of the crystal. These bands are held together laterally only by van der Waals forces, under the influence of

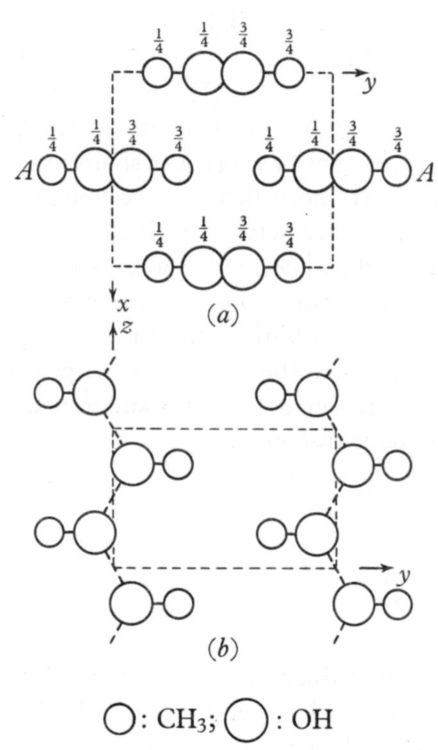

Fig. 14.05. (*a*) Plan of the unit cell of the orthorhombic structure of methanol, CH_3OH, projected on a plane perpendicular to the z axis. (*b*) Vertical section through the unit cell on the plane *A–A*. Hydrogen bonds are represented by inclined broken lines.

which they form an approximately close-packed arrangement with each band symmetrically surrounded by six neighbours. There is evidence that long-chain polymers still survive in the melt, so that on fusion relatively few hydrogen bonds are ruptured and the process is primarily one involving the severance of the van der Waals links. The melting point (−98 °C) is therefore much lower than that of, say, ice (where melting involves the breakdown of the three-dimensional framework of hydrogen bonds), and the latent heat of melting is not abnormally high. On vaporization, however, more bonds must be broken, and in consequence methanol is stable as a liquid over a wide temperature range and shows a large latent heat of evaporation.

Pentaerythritol

14.13. The molecule of pentaerythritol, $C(CH_2OH)_4$, has the symmetrical form shown in fig. 14.06. The central carbon atom is tetrahedrally co-ordinated by four CH_2 groups, and the hydroxyl groups lie at the corners of a square, coplanar with this central carbon atom. Each molecule can take part in eight hydrogen bonds and these are utilized to link the molecules into infinite sheets, as shown in the figure. The

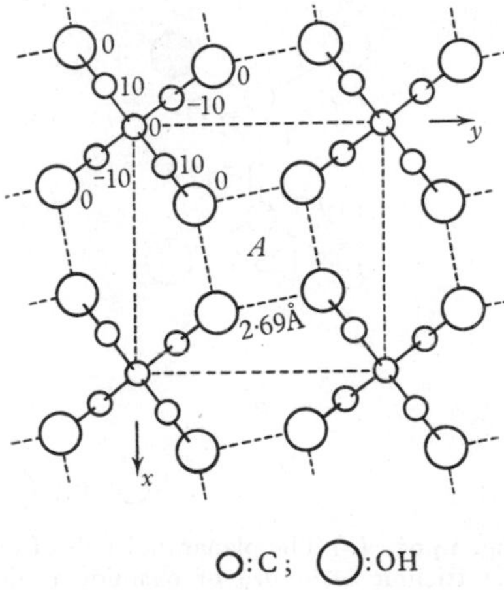

Fig. 14.06. Plan of one layer of the tetragonal structure of pentaerythritol, $C(CH_2OH)_4$, projected on a plane perpendicular to the z axis. The heights of the atoms in one molecule are indicated in units of $c/100$ and hydrogen bonds between hydroxyl groups are represented by broken lines.

sheets lie perpendicular to the z axis of the tetragonal unit cell and the structure as a whole is built up by the superposition of these sheets, their mutual arrangement being such that the centres of the molecules in any one sheet lie vertically above or below the points A in the adjacent sheets. Within a sheet the distance between hydroxyl groups, corresponding to the hydrogen bond, is 2·69 Å, but adjacent sheets are held together only by residual forces and their closest distance of approach is about 3·5 Å. In this particular structure the positions of the hydrogen atoms has been established by neutron diffraction. They are found to lie at a distance of 0·94 Å from the nearest oxygen atom, thus confirming the view already expressed (§ 12.02) that the hydrogen bond is unsymmetrical. The physical properties of pentaerythritol reflect the influence of the hydrogen bonding. The compound readily forms excellent crystals which are hard and brittle and which have a high melting point (253 °C). Parallel to the sheets the crystals show a cleavage comparable in perfection to that of calcite or mica.

Oxamide

14.14. The structure of oxamide, $(CONH_2)_2$, is geometrically closely analogous to that of pentaerythritol. The molecule is planar, with the *trans* configuration, and has the dimensions shown in fig. 14.07*a*.

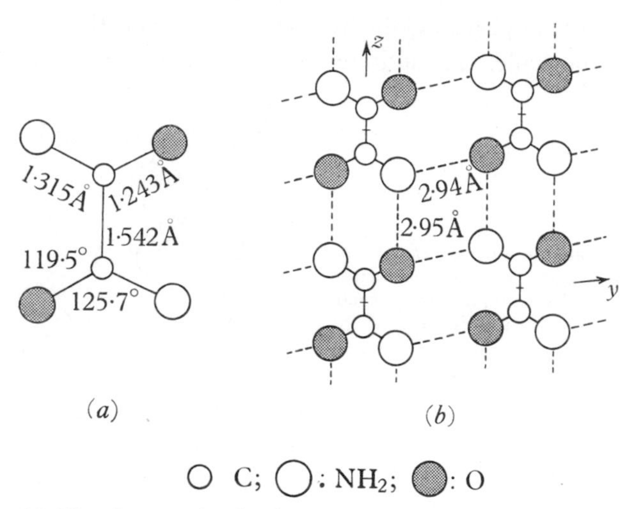

Fig. 14.07. (*a*) The planar molecule of oxamide, $(CONH_2)_2$. (*b*) Plan of one layer of the triclinic structure of oxamide projected on the *y*, *z* plane. Hydrogen bonds between molecules are represented by broken lines. Crosses indicate the corners of the unit cell.

Molecules of this form are linked in the crystal by hydrogen bonds of length approximately 2·95 Å to form two-dimensional sheets (fig. 14.07*b*), and the three-dimensional structure is built up by a superposition of these sheets. As in pentaerythritol, the crystals are very stable (they decomposite at 420 °C without melting) and possess a perfect cleavage parallel to the sheets.

The dimensions of the oxamide molecule call for some comment. The C–C distance of 1·542 Å corresponds closely to that in diamond, but both the C–O and C–N distances are intermediate in length between pure single and double bonds. It must therefore be concluded that the molecule is a resonance structure to which both the configuration (I) and (II) make a contribution.

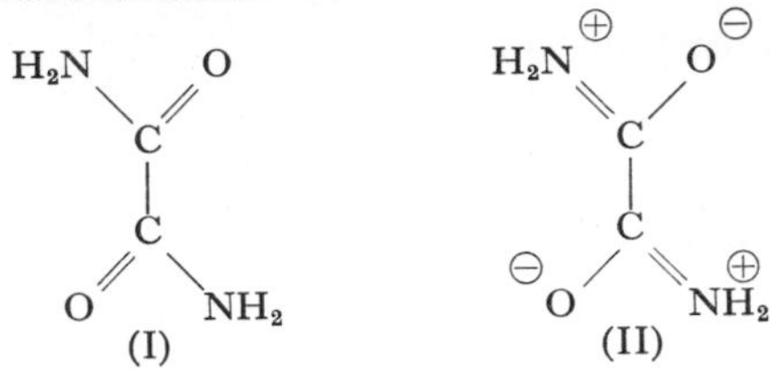

Formamide and methylamine

14.15. A number of other structures are known in which hydrogen bonds link the molecules into sheets. In formamide, $H.CO.NH_2$, the number of hydrogen atoms in the NH_2 groups is sufficient for each

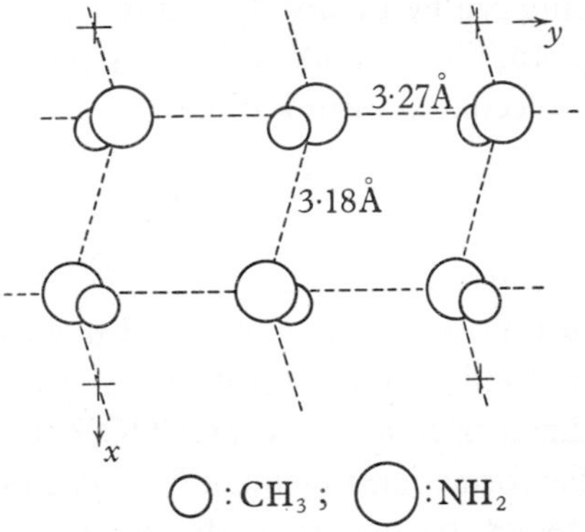

Fig. 14.08. Plan of one layer of the orthorhombic structure of methylamine, $CH_3.NH_2$, projected on a plane perpendicular to the z axis. The nitrogen atoms are nearly coplanar, and hydrogen bonds between them are represented by broken lines. Crosses indicate the corners of the unit cell.

molecule to be involved in only four hydrogen bonds, and these bonds, between NH_2 groups and oxygen atoms, bind the molecules into puckered sheets which are superimposed in the crystal structure. In methylamine, $CH_3.NH_2$, each molecule is again linked by four hydrogen bonds to four neighbours, and in this case the relatively simple arrangement shown in fig. 14.08 results. Nitrogen atoms, lying very nearly in a plane perpendicular to the z axis of the orthorhombic unit cell, and situated nearly at the corners of a square, are joined to their four neighbours by hydrogen bonds of length about 3·2 Å. The molecules to which these nitrogen atoms belong point alternately up and down, and they are therefore bound into sheets consisting of a layer of NH_2 groups sandwiched between two layers of CH_3 groups. The structure as a whole is formed by the superposition of such sheets, which are held together only by residual forces. The hydrogen bond between two nitrogen atoms is weaker than either the O···O or O···N hydrogen bond. This is reflected in the methylamine structure by the length of the bond (3·2 Å compared with 2·69 Å in pentaerythritol).

Carboxylic acids

14.16. It is well known that the carboxylic acids, in common with the alcohols, show a pronounced tendency to polymerization. This tendency finds an immediate explanation in the crystal structures of these compounds, for in every such structure so far investigated it has been found that the molecules are strongly linked to their neighbours by hydrogen bonds between oxygen atoms and hydroxyl groups. In simple monocarboxylic acids of composition *R*.COOH, where the number of active hydrogen atoms is only one per molecule, each molecule can be involved in only two hydrogen bonds, and in this case two possibilities arise: either (*a*) the two bonds are directed to one other molecule to form a dimer:

```
       O···HO
      //      \
R—C            C—R
      \       //
       OH···O
```

or (*b*) the bonds are directed to two other molecules, in which event the molecules may be linked into closed rings or endless chains. In the dimer of *formic acid*, H.COOH, found in the liquid and vapour phases, the former arrangement has been established by electron diffraction. In the crystal, however, dimers are not found, and instead the molecules are linked into infinite chains:

```
              H                   H                   H
              |                   |                   |
              C                   C                   C
            /  \\               /  \\               /  \\
         HO      O           HO      O           HO      O
  ·.       ·:      ·.         ·:      ·.          ·:      ·.
  HO      O          HO      O          HO      O
    \   //             \   //             \   //
      C                  C                  C
      |                  |                  |
      H                  H                  H
```

The O···O distance within the chains is 2·58 Å but the molecules in different chains are united only by van der Waals bonds of length 3·18 Å or more.

14.17. Few organic compounds have been studied by X-ray methods in such detail, or with such precision, as *oxalic acid*, $(COOH)_2$. This substance is polymorphous, occurring in two anhydrous forms and also as a dihydrate, and the crystal structures of all three modifications have been established. The molecule has substantially the same planar *trans* configuration in each case, but the structures differ in the way in which the molecules are linked together by hydrogen bonds. In the two forms of the anhydrous acid the molecules are arranged in the manner represented schematically in fig. 14.09. In the β form each molecule is

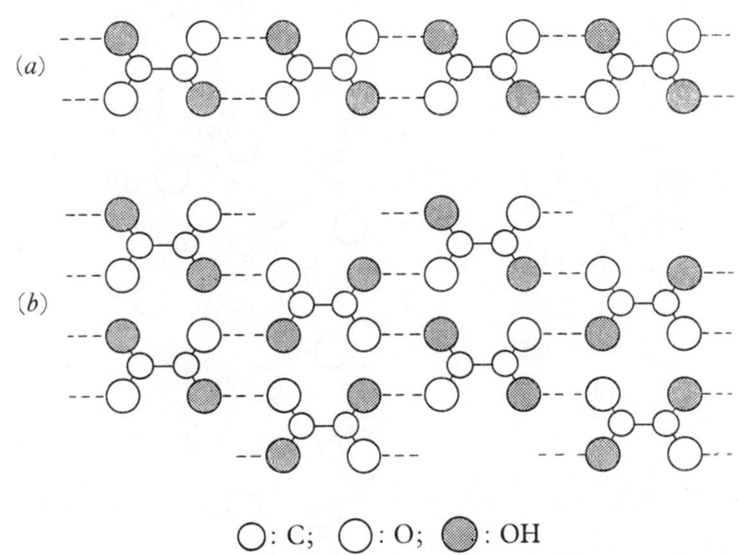

Fig. 14.09. Schematic representation of the hydrogen bonding between molecules in the structure of the two forms of anhydrous oxalic acid, $(COOH)_2$: (*a*) the β form; (*b*) the α form.

bound to two others by hydrogen bonds to form endless chains running parallel to one another in the crystal, analogous to the chains in formic acid. In the α form, on the other hand, each molecule is linked to four different neighbours and the molecules are therefore bound together in two dimensions by hydrogen bonds. The physical properties of the two forms reflect this difference in crystal structure. The α form has a good cleavage parallel to the sheets while the cleavages of the β form are in two planes parallel to the length of the chains, in both cases the cleavages being such that only van der Waals bonds are broken.

The structure of oxalic acid dihydrate, represented in fig. 14.10*b*, merits more detailed discussion. The molecules are planar and centrosymmetrical, and their centres lie at the corners and at the centre of the monoclinic unit cell. The planes of the molecules are, however, inclined to the plane of projection and the molecules thus appear foreshortened in the illustration; nevertheless, their form can be clearly recognized. The molecules are bound together by hydrogen bonds, but it will be

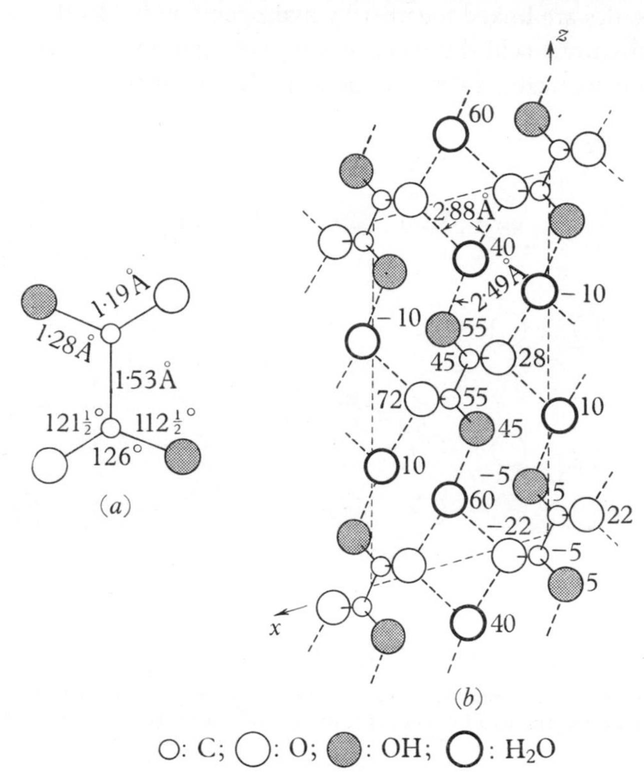

Fig. 14.10. (*a*) The planar molecule of oxalic acid in the structure of oxalic acid dihydrate, $(COOH)_2.2H_2O$. (*b*) Plan of the unit cell of the monoclinic structure, projected on a plane perpendicular to the *y* axis. The heights of the atoms are indicated in units of $b/100$ and hydrogen bonds between oxalic acid and water molecules are represented by broken lines.

seen that this binding is always through the medium of a water molecule, each of which is linked to two oxygen atoms and to one hydroxyl group of three different oxalic acid molecules. The water molecules, by providing more hydrogen atoms, have the effect of multiplying the number of hydrogen bonds in the structure and so enable each oxalic acid molecule to be indirectly linked to six others instead of to a maximum of only four. The structure therefore differs from those of the anhydrous acid in that cohesion in all three dimensions is due to hydrogen bonding. The three hydrogen bonds from each water molecule differ significantly in length and strength, the $H_2O\cdots OH$ and $H_2O\cdots O$ distances being 2·49 and 2·88 Å, respectively.

Of even greater interest than the arrangement of the molecules in the crystal is the detailed configuration of the atoms in the isolated oxalic acid molecule. This is represented in fig. 14.10*a*. We have already remarked that the molecule is planar, and in this connexion it is interesting to note that the observed C–C distance of 1·53 Å does not differ appreciably from the standard single-bond distance found in diamond. The planar form of the molecule cannot therefore be attributed to any appreciable degree of double-bond character in the central C–C bond. The C–O and C–OH distances, 1·19 and 1·28 Å, respectively, differ significantly, and a similar significant difference has been found in the structures of many other carboxylic acids. This must be interpreted as a confirmation that the hydrogen bond is unsymmetrical and that the hydrogen atom in the bond is associated more intimately with the oxygen atom of the hydroxyl group than with the other oxygen atom to which it is linked. It will be noted, however, that the C–OH distance is considerably shorter than the normal C–O single-bond length of 1·42 Å, implying some double-bond character in this link and suggesting that the carboxylic acid radical must be interpreted as a resonance structure to which both the configurations (I) and (II) contribute. A contribution

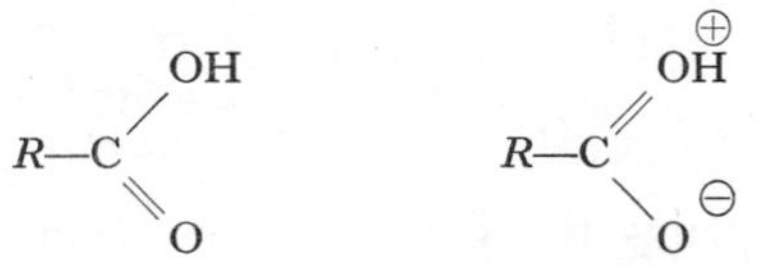

$$R\text{—C}(\text{—OH})(\text{=O}) \qquad R\text{—C}(\text{=}\overset{\oplus}{\text{OH}})(\text{—O}^{\ominus})$$

(I) (II)

from configuration (II) will increase the strength of the hydrogen bond and explain the fact that in carboxylic acids this bond is much stronger than, say, in water.

We have discussed the structure of oxalic acid in some detail because it provides a particularly clear example of the type of information which X-ray methods may be expected to yield when applied to molecular compounds, and especially of the extent to which these methods can furnish information, beyond the scope of direct chemical experiment, on detailed molecular configuration and intermolecular binding. Our arguments reveal how important in such discussions is an exact knowledge of interatomic distances and how important, therefore, are precision structure analyses.

Ionic binding

14.18. The metal salts of organic acids, and certain other compounds such as the substituted ammonium halides, form essentially ionic structures with the component ions held together by electrostatic forces. When the complex ion is small and simple in form these compounds often crystallize with structures analogous to those of some of the simple inorganic ionic compounds already discussed. Thus *mono-methylammonium iodide*, NH_3CH_3I, has an arrangement of $NH_3CH_3^+$ and I^- ions which may be regarded as a distorted sodium chloride structure, and the structure of *tetramethylammonium chloride*, $N(CH_3)_4Cl$, similarly resembles that of caesium chloride. *Sodium oxalate*, $(COO)_2Na_2$, has a relatively simple structure in which each sodium ion is octahedrally co-ordinated by six oxygen atoms of adjacent planar $(COO)_2^{2-}$ ions. In this salt, and in many others of carboxylic acids, it is found that no distinction can be made between the two oxygen atoms of the carboxylate ion; both are at the same distance (1·23 Å) from the carbon atom and both similarly co-ordinate the sodium ions. Thus, in contrast to the acids discussed above, we must assume that in the ions complete resonance takes place between the equivalent configurations

$$-\mathrm{C}(=\mathrm{O})-\mathrm{O}^{\ominus} \quad \text{and} \quad -\mathrm{C}(-\mathrm{O}^{\ominus})=\mathrm{O}$$

This resonance explains the fact that the observed C–O distance is intermediate between the lengths of C–O single and double bonds, and it is to stablization arising from this resonance that the exceptional acid strength of the carboxylate ion must be ascribed.

Closely similar interatomic distances within the oxalate ion have been observed in other oxalate structures. This ion, however, is not always planar, and in the strongly hydrogen-bonded structure of *ammonium oxalate monohydrate*, $(COO)_2(NH_4)_2 . H_2O$, the hydrogen bonds have the effect of twisting the ion about the C–C bond. This illustrates the important point that the structure of a molecule in an organic crystal is determined not only by the intermolecular bonds but also by the forces between molecules. Although molecular configuration can be determined with great precision by X-ray analysis, and although the results of these analyses throw much light on molecular structure, the configuration derived is necessarily that in the particular structure investigated; it may be different in isolated molecules or in the different forms of a compound which is polymorphous.

COMPOUNDS WITH LONG MOLECULES

14.19. The most important compounds of this class are the long-chain paraffins and their related alcohols, esters, ketones, fatty acids and other derivatives. There are, however, also other compounds which contain long molecules and which display structural features characteristic of the class.

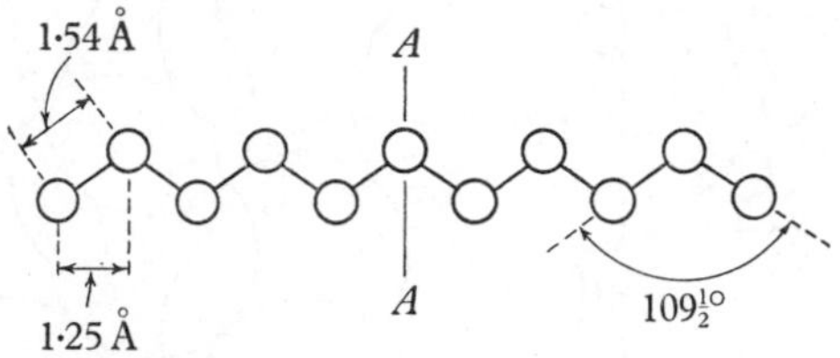

1·54 Å
A
A
1·25 Å
109½°

Fig. 14.11. The structure of the normal paraffin molecule $C_{11}H_{24}$. *A–A* represents a mirror reflexion plane perpendicular to the length of the chain. Paraffin molecules with an even number of carbon atoms have no such symmetry plane.

Apolar intermolecular binding

The paraffins

14.20. The paraffins have been extensively studied by X-ray methods, but the practical difficulties of obtaining single crystals, and the fact that many of these compounds are polymorphous, have as yet prevented any precision structure determination from being made. Certain features, however, are clear and are supported by the results of work on the more readily crystallized long-chain acids and other derivatives. In all the normal paraffins the molecules consist of planar zigzag chains of carbon atoms with a characteristic tetrahedral interbond angle of 109½° and with a C–C single-bond separation of 1·54 Å (fig. 14.11). These molecules are stacked in the crystal with their axes parallel, giving rise to a structure the unit cell of which has one axis very much longer than the other two. As the number of carbon atoms increases, the length of this axis increases linearly (for paraffins which are isostructural) by an amount corresponding to an increment of 1·25 Å in the length of the molecule for each carbon added. This distance is just the projection on to the axis of the length of the C–C bond.

14.21. Some of the above mentioned features of paraffin structures are illustrated by an early study of n-$C_{29}H_{60}$. This paraffin is orthorhombic with cell dimensions

$$a = 7{\cdot}45, \quad b = 4{\cdot}97, \quad c = 77{\cdot}2\ \text{Å}$$

and with the structure represented in fig. 14.12. The carbon chains are all arranged parallel to the z axis of the cell, and are therefore seen end-on, but have their planes mutually inclined in two different directions to give an approximately close-packed arrangement in which each

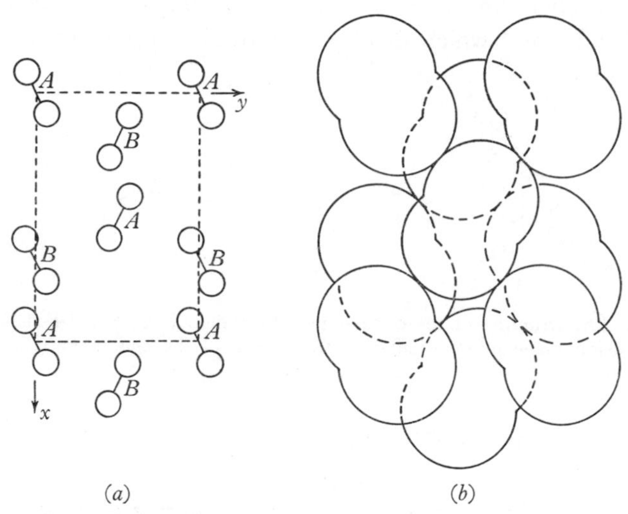

Fig. 14.12. (*a*) Plan of the unit cell of the orthorhombic structure of n-$C_{29}H_{60}$ projected on a plane perpendicular to the z axis. The molecules A and B are mutually displaced vertically through a distance $\frac{1}{2}c$. (*b*) The same, with the CH_2 groups given their correct van der Waals radii to emphasize the packing of the molecules.

chain has six close neighbours. The c translation of the cell embraces the length of two molecules, and those in the upper and lower halves of the cell are distinguished as A and B, respectively. The terminal methyl groups of the A and B molecules are separated vertically by a gap of about 4·0 Å, and it will be seen from the packing diagram that each such group in one layer lies symmetrically above three methyl groups in the layer beneath. In this respect there is an interesting geometrical distinction between the paraffins with an odd and those with an even number of carbon atoms. The symmetry of the orthorhombic structure is such that there are mirror reflexion planes perpendicular to the z axis, separated by a distance $\frac{1}{2}c$, and the arrangement of the molecules must of course satisfy this symmetry. It will be seen from fig. 14.11 that a paraffin chain with an odd number of carbon atoms possesses a plane of symmetry perpendicular to its axis passing through the central atom. Such a chain can therefore be arranged in the orthorhombic cell in the way represented schematically in fig. 14.13*a*, with the molecules astride the planes of symmetry; and this is the arrange-

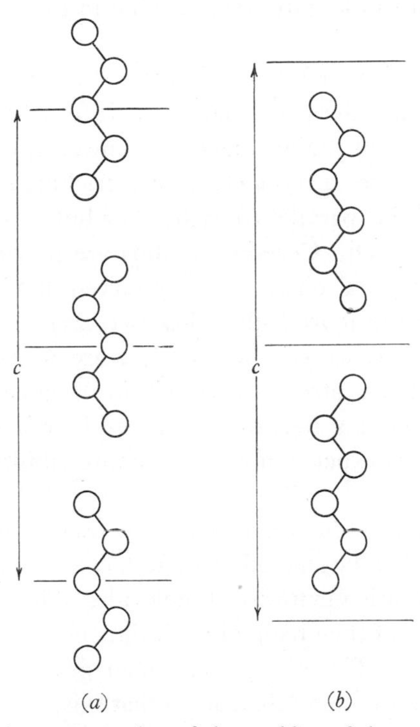

Fig. 14.13. Schematic representation of the packing of the molecules in the crystal structure of the normal paraffin C_nH_{2n+2}: (*a*) for *n* odd; (*b*) for *n* even. The horizontal lines represent mirror reflexion planes.

ment in the structure of *n*-$C_{29}H_{60}$ just described. Paraffin chains with an even number of carbon atoms, however, possess no such plane of symmetry and they must therefore be placed in the unit cell *between* the mirror planes (fig. 14.13*b*). In both cases the repeat distance *c* embraces two molecules, but in the latter case the terminal methyl groups in adjacent layers, being related by the reflexion operation, are less closely packed and fall vertically above one another. This distinction is revealed by a comparison of the cell dimensions of $C_{29}H_{60}$ given above with those of *n*-$C_{30}H_{62}$, namely

$$a = 7{\cdot}45, \quad b = 4{\cdot}97, \quad c = 81{\cdot}60 \text{ Å}.$$

It will be seen that the a and b dimensions are identical but that the c dimension in $C_{30}H_{62}$ is larger, as is of course to be expected. The increment in c of 4·40 Å, however, is greater than can be ascribed to an increase in the length of the molecule, which would account for an increment of only 2·50 Å ($= 2 \times 1{\cdot}25$ Å); the rest of the increment is due to the difference in longitudinal packing in the two paraffins.

14.22. A number of other long-chain *n*-paraffins have been studied in less detail, and all show the same characteristic form of the carbon chain. The detailed features of the arrangement, however, are not always the same, and in some cases a monoclinic or a triclinic structure is found with the carbon chains parallel to each other but inclined obliquely to the base of the unit cell. Certain paraffins are polymorphous, and in some instances structural changes are observed at high temperatures. This point is discussed more fully below (§ 14.23).

The physical properties of the paraffins are in satisfactory accord with the structures. The strong forces within the chains compared with the feeble lateral binding between them result in a very pronounced anisotropy of thermal expansion and of compressibility.

14.23. Some paraffins show a progressive transition with increasing temperature towards a hexagonal structure, corresponding to the close packing of freely rotating cylindrical molecules. This transition may be detected by a study of the change in cell dimensions, for, although the thermal expansion coefficient is very much greater in any direction normal to the axis of the chain than along that axis, there is, nevertheless, a pronounced anisotropy of the lateral expansion as between the x and y crystallographic directions. The sense of this anisotropy is such as to decrease the angle ψ (fig. 14.14) with increasing temperature until it approaches the value of 60° for the hexagonal structure, but in most paraffins this freely rotating form would be reached only at temperatures above the melting point. In some cases, however, the hexagonal structure is actually realized, and completely free molecular rotation is established both by the X-ray investigations and by the optical properties which, above the transition temperature, are those of a uniaxial crystal.

Dicyanoacetylene

14.24. A different type of packing of long molecules is found in dicyanoacetylene, N≡C—C≡C—C≡N. The molecules are strictly linear but are no longer all parallel, being disposed instead approximately in sheets with their axes mutually inclined in two directions nearly at right angles to give the compact arrangement represented schematically in fig. 14.15. The structure as a whole is built up by the superposition of such sheets.

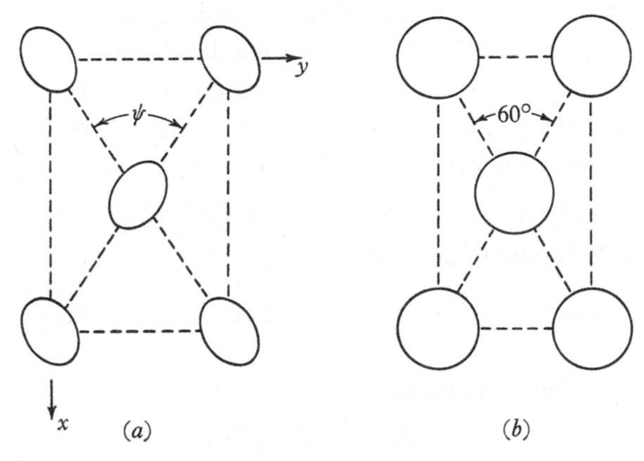

Fig. 14.14. The influence of molecular rotation on the structures of the paraffins: (*a*) the orthorhombic structure; (*b*) the hexagonal structure of close-packed cylindrical molecules.

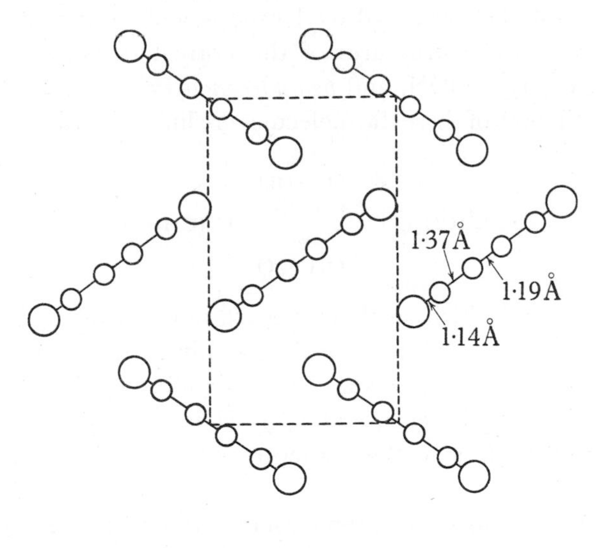

Fig. 14.15. Plan of one sheet of the monoclinic structure of dicyanoacetylene, N≡C—C≡C—C≡N. The molecules lie very nearly in the plane of the sheet and the structure as a whole is formed by the superposition of these sheets.

The molecular dimensions in dicyanoacetylene, also shown in the figure, are of interest. The difference between the different C–C bonds is clearly reflected in the difference in interatomic distances, but it is noteworthy that the 'single bond' distance of 1·37 Å is very considerably shorter than the standard C–C single-bond distance of 1·54 Å. Very nearly the same distance is found also in cyanogen and in cyanoacetylene, and it is clear that in these molecules the simple formulation is an inadequate representation of the configuration. In the particular case of dicyanoacetylene we must assume that there is appreciable resonance between the ordinary configuration

$$\mathrm{N{\equiv}C{-}C{\equiv}C{-}C{\equiv}N} \quad \text{or} \quad \mathrm{:N:::C:C:::C:C:::N:}$$

and the double-bonded configuration

$$\mathrm{N{=}C{=}C{=}C{=}C{=}N} \quad \text{or} \quad \mathrm{:\overset{\oplus}{N}::C::C::C::C::\overset{\ominus}{\underset{\cdot\cdot}{N}}:}$$

Hydrogen bonded structures

Long-chain carboxylic acids

14.25. In long-chain carboxylic acids, as in the simpler carboxylic acids already discussed, the molecules are always found to be linked by hydrogen bonds between hydroxyl groups and oxygen atoms of the carboxyl group. The structures of the normal monocarboxylic acids *lauric acid*, $C_{11}H_{23}.COOH$, and *pentadecanoic acid*, $C_{14}H_{29}.COOH$, are known, and in both of these the molecules are linked in pairs by hydrogen bonds:

$$\begin{array}{ccc} & \mathrm{O\cdots HO} & \\ & {/\!/} \qquad\quad \diagdown & \\ \mathrm{CH_3.(CH_2)_n.C} & & \mathrm{C.(CH_2)_n.CH_3} \\ & \diagdown \qquad\quad {/\!/} & \\ & \mathrm{OH\cdots O} & \end{array}$$

The long 'dimers' which result are packed in the structure in a manner similar to that of the molecules in the paraffins. As in other carboxylic acids, the C–O and C–OH distances are significantly different.

Normal dicarboxylic acids, $HOOC.(CH_2)_n.COOH$, have been more extensively studied, and the structures of several are known. In *succinic acid* ($n = 2$), *adipic acid* ($n = 4$) and *sebacic acid* ($n = 8$), all with an even number of carbon atoms, the molecules are centrosymmetrical and planar, and are linked in the structure into endless chains (fig. 14.16*a*) exactly analogous to those found in the β form of oxalic acid (fig. 14.09*a*). The cell dimensions of the acids $n = 6$ and $n = 16$ show that they also undoubtedly have the same structure.

The acids with n odd show an interesting difference in structure. In these acids the molecule cannot be centrosymmetrical, and if they were planar they could form chains only of the crumpled type shown in fig. 14.16*b*. A more nearly linear chain can be achieved, however, if the carboxyl groups are twisted out of the plane of the carbon atoms by rotation about the terminal C–C bonds, and just such a rotation, of about 30°, is found in the structures of *glutaric acid* ($n = 3$) and *pimelic acid* ($n = 5$). Even so, the chains are not as straight as in the acids with

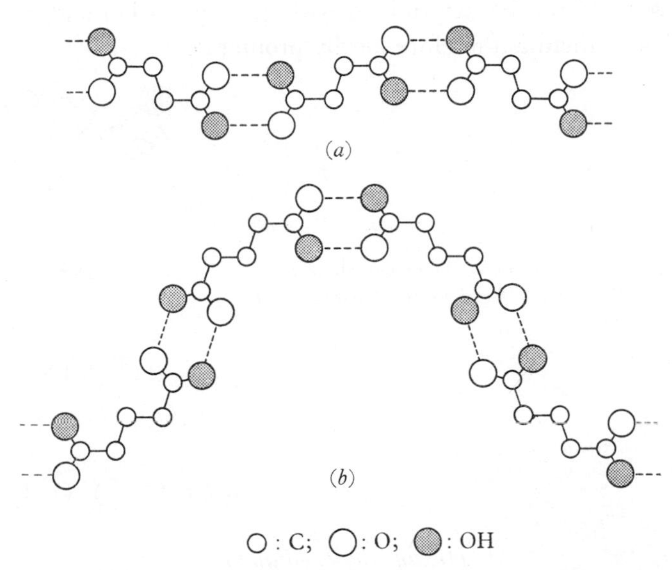

Fig. 14.16. Schematic representation of the association of the planar molecules of normal dicarboxylic acids, $HOOC.(CH_2)_n.COOH$: (*a*) for n even; (*b*) for n odd.

n even and they therefore pack together laterally less tightly. This difference in structure accounts for the well-known alternation in physical properties as between the acids with even and odd numbers of carbon atoms.

14.26. Certain dicarboxylic acids crystallize as hydrates. In *acetylene dicarboxylic acid dihydrate*, $HOOC—C{\equiv}C—COOH.2H_2O$, the molecule has the planar form shown in fig. 14.17, with the carbon atoms in a straight line. The arrangement of these molecules in the crystal very closely resembles that found in oxalic acid dihydrate (§14.17). The water molecules are situated between the carboxyl groups of neighbouring acid molecules, which are thus linked into a three-dimensional array by hydrogen bonds to these water molecules. The lengths of the hydrogen bonds show the same pronounced difference as in oxalic acid, the $H_2O \cdots OH$ and $H_2O \cdots O$ bonds being of length 2·56 and about 2·85 Å, respectively. The C–C distances in the chain may be compared with those in dicyanoacetylene (§ 14.24), and again show that the simple formulation of the molecule is inadequate and that resonance with other configurations must occur. The C–O and C–OH distances in the molecule appear to be nearly equal, but in this particular structure these distances could not be determined with great precision and a detailed discussion of them is therefore hardly profitable.

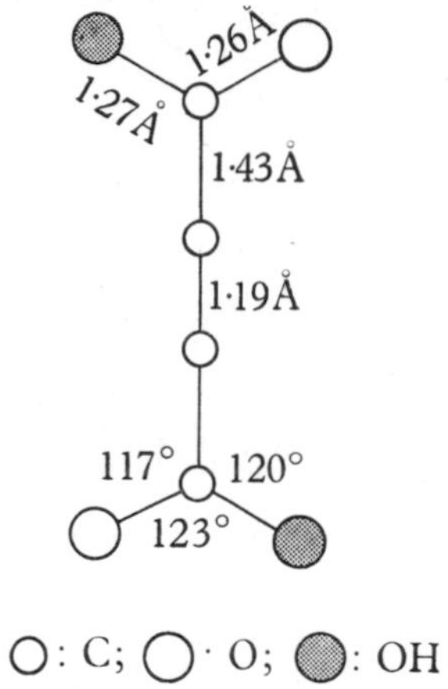

Fig. 14.17. The planar molecule of acetylene dicarboxylic acid, HOOC—C≡C—COOH, in the structure of its dihydrate.

1·26 Å
1·27 Å
1·43 Å
1·19 Å
117°
120°
123°
○: C; ○: O; ●: OH

Hexamethylenediamine

14.27. An elegant example of the dominating influence of hydrogen bonding in crystal structures is provided by hexamethylenediamine, $NH_2.(CH_2)_6.NH_2$. This compound resembles the dicarboxylic acids in having active groups at the ends of an aliphatic carbon chain, but it differs from them in that these active groups now consist of only a single atom (that of nitrogen). The spatial distribution of the hydrogen bonds about these nitrogen atoms requires each to be linked to more than one neighbour, so that a chain structure analogous to that of the dicarboxylic acids is impossible. The arrangement which obtains is shown in fig. 14.18, where the planar centrosymmetrical molecules are clearly revealed. These molecules lie with their axes parallel to the paper, but at two different heights and with their planes alternately inclined in two different directions. The molecules at the same height are bound into infinite sheets parallel to the paper by hydrogen bonds of length

3·21 Å between each nitrogen atom and those of two neighbouring molecules, and the structure as a whole is built up by the superposition of these sheets, in the manner shown, in such a way that the molecules of each sheet lie symmetrically between those of its neighbours.

The length of the N···N hydrogen bond in the hexamethylenediamine structure, very nearly the same as that in methylamine (§14.15), again reveals that this is a relatively weak link. It is also of interest to note that in this structure only half of the available hydrogen atoms take part in hydrogen-bond formation, for each NH_2 group, potentially

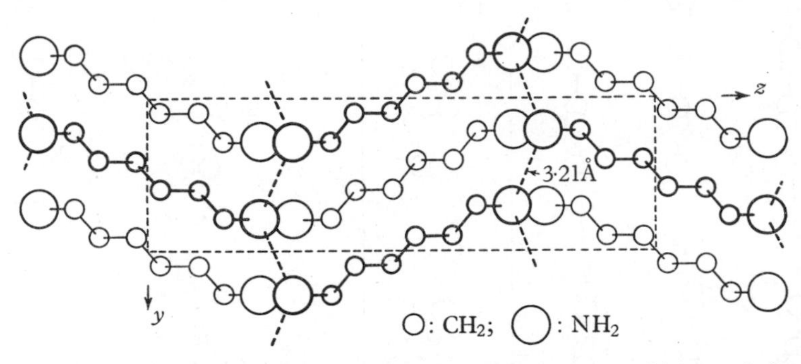

Fig. 14.18. Plan of the unit cell of the orthorhombic structure of hexamethylenediamine, $NH_2.(CH_2)_6.NH_2$, projected on a plane perpendicular to the x axis. The axes of the molecules are parallel to the plane of projection at heights 0 and $\frac{1}{2}a$, the molecules at $\frac{1}{2}a$ being drawn more boldly than the others. Hydrogen bonds between nitrogen atoms, linking the molecules into sheets, are represented by broken lines.

capable of participating in four hydrogen bonds, as in the structures of formamide and methylamine, in fact takes part in only two. This is a not uncommon feature of structures containing N···N bonds and one in respect of which they are sharply distinguished from those in which O···O hydrogen bonds occur. In the latter structures it appears to be a principle, sometimes termed the *principle of maximum hydrogen bonding*, that all the hydrogen atoms in the active groups are employed in hydrogen-bond formation.

Ionic binding

14.28. The structures of some salts of long-chain fatty acids have been determined. In these the anion is far too large to give a simple, symmetrically co-ordinated structure and in consequence arrangements of marked asymmetry result.

Soaps

14.29. The structure of the soap *potassium caprate*,

$$CH_3.(CH_2)_4.COOK,$$

is shown in fig. 14.19. In (*a*) the planar anions are seen edge-on and it will be noted that they lie very nearly in a series of sheets parallel to the

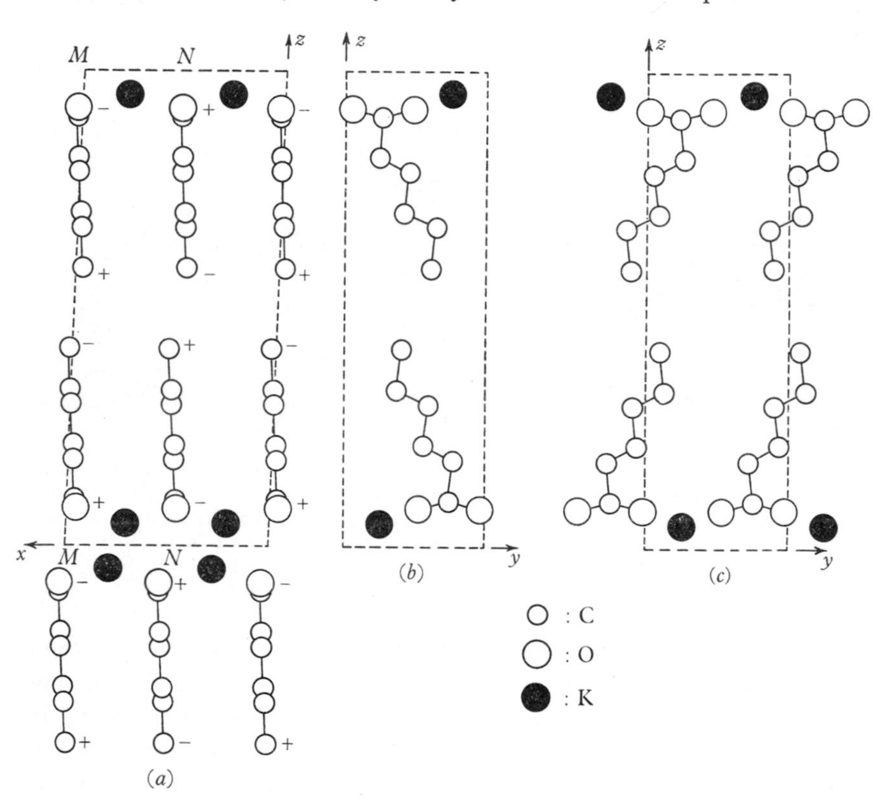

Fig. 14.19. (*a*) Plan of the unit cell of the monoclinic structure of potassium caprate, $CH_3.(CH_2)_4.COOK$, projected on a plane perpendicular to the y axis. The inclination of the individual anions is indicated by the symbols + and −. (*b*, *c*) Sections through the structure on planes parallel to the y and z axes and containing (*b*) *M–M* and (*c*) *N–N*. The potassium ions do not lie in the planes of these sections but are shown in projection on to them.

y, z plane of the monoclinic unit cell. Within any one sheet the anions are arranged with their axes parallel, but in successive sheets the inclination of the anions is reversed so that the structure as a whole is formed by the alternate superposition of sheets of the types shown at (*b*) and (*c*).

Within each sheet the CH_3 groups of neighbouring anions approach one another and are bound together by weak van der Waals forces; the slight inclination of the anions to the y, z plane enables binding between these groups to take place not only within each sheet but also between adjacent sheets. At the other end of the anions the $—CO_2$ groups similarly approach one another, but they are bound very much more strongly in three dimensions by ionic bonds to the potassium ions lying between these groups and between the sheets. A more significant description of the structure, therefore, is to consider it as a stack of thick 'sandwiches' parallel to the x, y plane of the unit cell. Each sandwich consists of two layers of anions strongly united by ionic bonds to the potassium ions which they enclose, but successive sandwiches are held together only by residual bonds between CH_3 groups.

Potassium caproate, $CH_3.(CH_2)_8.COOK$, has an analogous structure, but of course with longer anions, and it is clear from the cell dimensions that similar structures also obtain in other potassium soaps.

Alkylammonium halides

14.30. In the substituted ammonium halides it is the cation which is the complex unit. A number of the *n*-alkyl compounds have been investigated, and their structures, although not all the same, are found to have features in common and to bear a formal resemblance to the structures of the long-chain fatty acids. As is to be expected, the halogen ions are co-ordinated by the $—NH_3$ groups of the cations, and these cations are arranged, usually parallel to one another, in such a way that this co-ordination can be achieved. These features are illustrated by the relatively simple tetragonal structure of *n-propylammonium chloride*, $CH_3(CH_2)_2NH_3Cl$, and of the isomorphous bromide and iodide, represented schematically in fig. 14.20*a*, *b*. It will be seen that each chlorine ion is tetrahedrally co-ordinated by four nitrogen neighbours and that these ions are sandwiched between two sheets of cations arranged with their axes parallel but pointing in opposite directions. Sandwiches of this type run through the top and bottom of the unit cell, with their planes perpendicular to the z axis, and are held together only by van der Waals forces. The tetragonal symmetry of this structure is of interest, for it can be reconciled with the accepted planar configuration of the *n*-aliphatic carbon chain only if it is assumed that the cations effectively acquire cylindrical symmetry by free rotation about the vertical axis. This assumption is confirmed by the fact that at low temperatures (about

-90 °C in the case of the chloride) the three halides have the monoclinic structure shown in fig. 14.20c. The symmetry is now, of course,

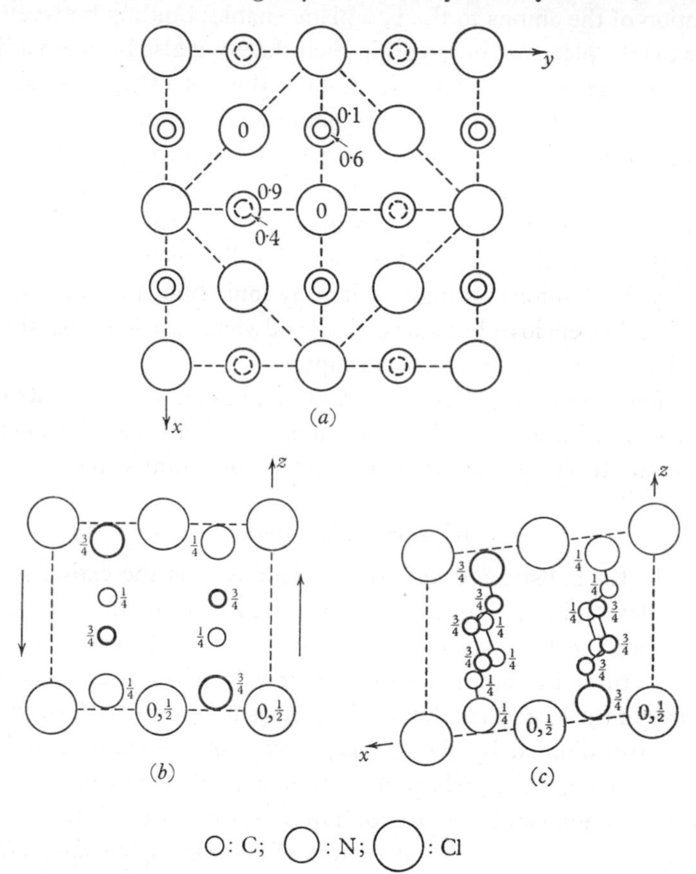

Fig. 14.20. (*a*) Plan of four unit cells of the tetragonal structure of *n*-propylammonium chloride, $CH_3(CH_2)_2NH_3Cl$, projected on a plane perpendicular to the *z* axis. (*b*) Plan of the same structure projected on a vertical diagonal plane and referred to the cell indicated by inclined broken lines in (*a*). (*c*) Plan of the unit cell of the monoclinic structure of *n*-propylammonium chloride projected on a plane perpendicular to the *y* axis. The cell can be regarded as derived from that shown in (*b*) by a shear in the direction of the arrows.

In (*a*) and (*b*) only the terminal atoms of the $CH_3(CH_2)_2NH_3^+$ cations are shown. In all the diagrams the heights of the atoms are expressed as fractions of the appropriate translation perpendicular to the plane of the paper.

lower and is consistent with a static planar cation, but careful comparison of the two structures will reveal that they are, nevertheless, very closely related: if the cations in the tetragonal structure are imagined to be 'frozen' parallel to a vertical diagonal plane of the tetragonal structure, and if this structure is then deformed by a shear in that plane, the monoclinic arrangement results. The transition between the structures is therefore displacive, in the sense described in §9.15, and it is interesting to note that it does not take place abruptly. As soon as the temperature falls to a point at which free rotation is impossible a monoclinic structure is formed, but initially the deformation of the tetragonal arrangement is slight and it is only at an appreciably lower temperature that complete arrest of the cations finally gives rise to the structure represented in fig. 14.20*c*.

COMPOUNDS WITH FLAT MOLECULES

14.31. The number of organic compounds containing molecules more or less flat in shape of which complete structure analyses have been made is greater than that of any other class of molecular compounds. This is primarily due to the chemical interest of the problem of the configuration of the benzene ring, as a result of which numerous extensive and detailed studies of the structures of aromatic compounds have been carried out. There are, however, also a number of non-aromatic compounds in which the molecule is substantially planar.

Apolar intermolecular binding

Hexamethylbenzene and benzene

14.32. For practical reasons the structure of benzene itself was not determined until several of its derivatives had been analysed. The earliest complete structure determination of an aromatic compound was that of hexamethylbenzene, $C_6(CH_3)_6$, and this analysis is of historic interest as the first unequivocal proof of the regular planar character of the benzene ring. All the carbon atoms lie in a plane, and those of the ring are situated at the corners of a regular hexagon of side 1·39 Å. The methyl groups are at a distance of 1·53 Å from these ring carbon atoms, this distance being closely comparable, as is to be expected, with the normal C–C single-bond distance found in diamond and the aliphatic hydrocarbons. The structure of the crystal containing these molecules is triclinic, with only one molecule per cell. The planes of the molecules are, therefore, necessarily parallel to one another, but it is found that they also lie almost exactly in the plane of the x and y axes of the cell. The structure may accordingly be regarded as composed of sheets of molecules parallel to this plane, and the cell dimensions are such that within each sheet each molecule is surrounded by six others at the corners of a very nearly regular hexagon, as shown in fig. 14.21. The structure as a whole is built up in three dimensions by the superposition of identical sheets laterally displaced relative to one another, the closest distance of approach between CH_3 groups, both within a sheet and between sheets, being about 3·9 Å.

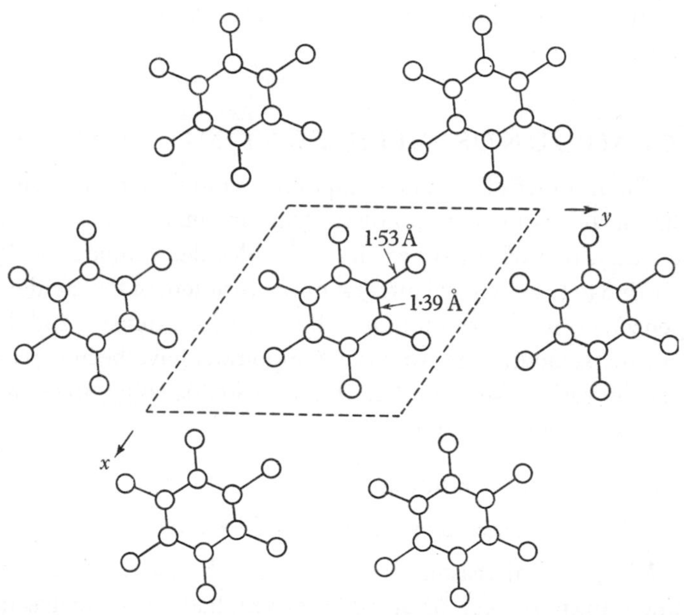

Fig. 14.21. Plan of one sheet of the triclinic structure of hexamethylbenzene, $C_6(CH_3)_6$, projected on the plane of the *x* and *y* axes. The molecules lie very nearly in the plane of the sheets and the structure as a whole is formed by the superposition of these sheets.

The orthorhombic structure of solid benzene is shown in fig. 14.22. The molecules are again planar hexagonal rings of closely the same dimensions as in hexamethylbenzene but their mutual arrangement is quite different. Each molecule has its plane very nearly parallel to the *z* axis of the cell, but the molecules are inclined about this axis in two directions nearly at right angles. The centres of the molecules lie at the corners and at the centres of the faces of the cell, so the structure is a relatively compact arrangement which may be looked upon as a cubic close-packed array deformed to accommodate the aspherical molecules. The closest distance of approach between adjacent molecules is about 3·8 Å.

The structural studies of hexamethylbenzene and of benzene, although less precise than later work on aromatic compounds, were sufficiently accurate to establish the planar character of the benzene ring and the equivalence of all the C–C bonds. These studies therefore provide direct experimental evidence that the benzene molecule is a resonance structure of the type already discussed (§§4.14 and 4.28). The observed C–C distance of 1·39 Å in the ring is intermediate in length between the normal C–C single and double bonds, as is to be expected in a structure to which the two Kekulé configurations make the principal contribution. It is not, however, the arithmetic mean of these bond lengths, showing that although the bond has formally 50 per cent double-bond character it is closer, in length and strength, to a double than to a single bond owing to the stabilization of the molecule which the resonance has brought about (§4.23).

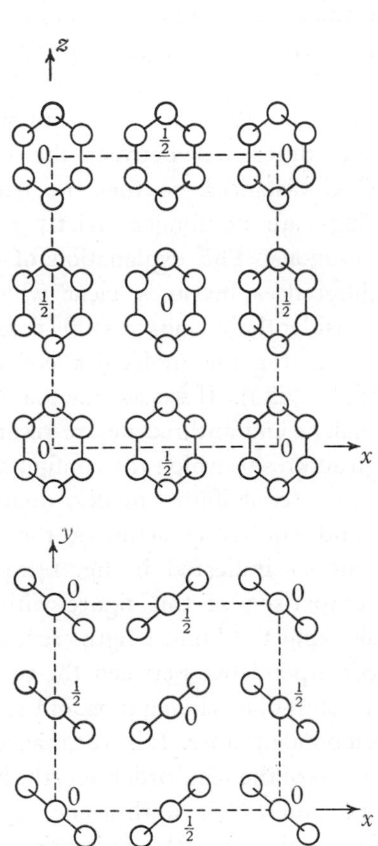

Fig. 14.22. Plans of the unit cell of the orthorhombic structure of benzene, C_6H_6, projected on planes perpendicular (*a*) to the *y* axis and (*b*) to the *z* axis. The heights of the centres of the molecules are expressed as fractions of the appropriate translations perpendicular to the plane of the paper.

Naphthalene and anthracene

14.33. The structures of naphthalene, $C_{10}H_8$, and anthracene, $C_{14}H_{10}$, are closely related and have been determined with great precision. In both structures the molecules are centrosymmetrical and strictly planar, and in naphthalene these molecules are arranged in the monoclinic unit cell as shown in fig. 14.23. It will be seen that the molecules all lie with their long axes nearly parallel to the *z* axis of the cell but that the planes of the molecules are mutually inclined about this axis in two different directions. To this extent the structure resembles that of benzene, but otherwise the arrangement is quite different. In the anthracene structure the molecules are arranged in precisely the same way in a cell with almost the same *a* and *b* translations as in naphthalene but with a greater *c* translation to accommodate the longer molecule. In both structures the closest distance of approach between carbon atoms of different molecules is about 3·7 Å.

Of greater importance than the arrangement of the molecules in the crystal is the configuration of the individual molecules themselves. This is represented in fig. 14.24*a*, from which it will be seen that in both molecules there are small but significant differences between the various C–C distances, so that the benzene rings are no longer strictly regular hexagons. The explanation of these differences becomes clear when we consider the various possible configurations for the molecules concerned (fig. 14.24*b*). If we assume that naphthalene and anthracene are resonance structures to which the configurations shown contribute equally, then each bond effectively acquires the bond number indicated in fig. 14.24*c*. A comparison of this figure with that showing the bond lengths reveals the correspondence between the two, the shortest bonds being those of the highest bond number. It is true that certain bonds of the same order (e.g. the bonds *AB* and *FG* in anthracene) are not identical in length. This arises from the fact that these bonds, although chemically equivalent, are crystallographically distinct and are therefore independently measured in the structure analysis. The differences may be real and may arise from the way in which the molecules are packed in the crystal, but they are in any case small compared with those between the lengths of bonds of different number. The naphthalene and anthracene structures constitute an elegant example not only of the precision with which molecular configuration can be determined by X-ray analysis but also of the chemical significance of the results achieved.

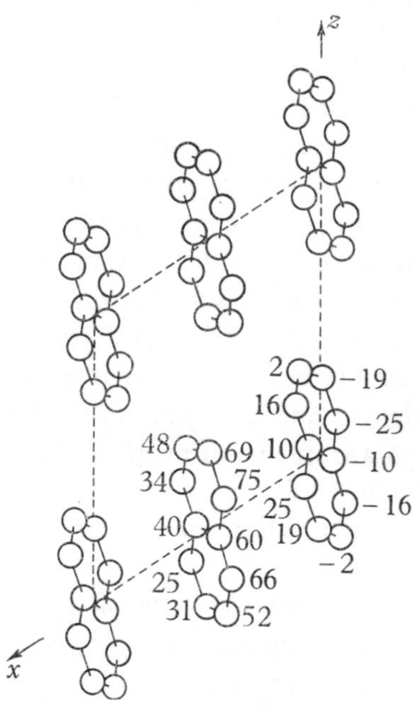

Fig. 14.23. Plan of the unit cell of the monoclinic structure of naphthalene, $C_{10}H_8$, projected on a plane perpendicular to the *y* axis. The heights of the carbon atoms are expressed in units of *b*/100.

Other aromatic compounds

14.34. Many other aromatic compounds have been shown to have planar molecules, and often these are arranged in the crystal in a manner similar to that found in naphthalene and anthracene. A few of the numerous substances of which this is true are shown in fig. 14.25. The cell dimensions of course vary widely from one compound to another to

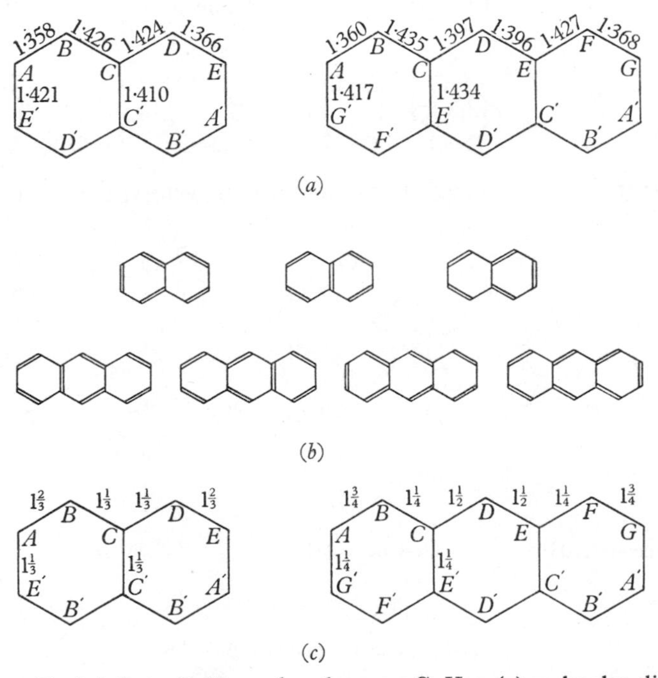

Fig. 14.24. Naphthalene, $C_{10}H_8$, and anthracene, $C_{14}H_{10}$: (*a*) molecular dimensions in Ångström units; (*b*) possible bond configurations; (*c*) bond numbers for the C–C bonds, assuming resonance between the configurations shown in (*b*).

accommodate the molecules of very different shape and size, but in all these structures the molecules are arranged with their planes nearly parallel to a common direction and inclined about this direction in two symmetrically related directions. The planar form of the molecules of *diphenyl* and *p-diphenylbenzene* (*p*-diphenyldiphenyl is planar too) must imply conjugation in these molecules, for if the bonds between benzene rings were single bonds rotation about them would be possible and a molecule with the rings inclined to one another would be expected to be more stable. The observed length of these bonds is appreciably less than 1·54 Å and supports the view that they possess some degree of double-bond character. In the condensed ring compounds, such as *coronene* and *ovalene*, small but significant differences are found between different C–C distances, and these can be satisfactorily interpreted, as in naphthalene and anthracene, in terms of a resonance structure to which the various possible configurations of the molecule all contribute. The number of such configurations increases rapidly with increasing complexity of the compound (it is 20 in coronene and 50 in ovalene) and

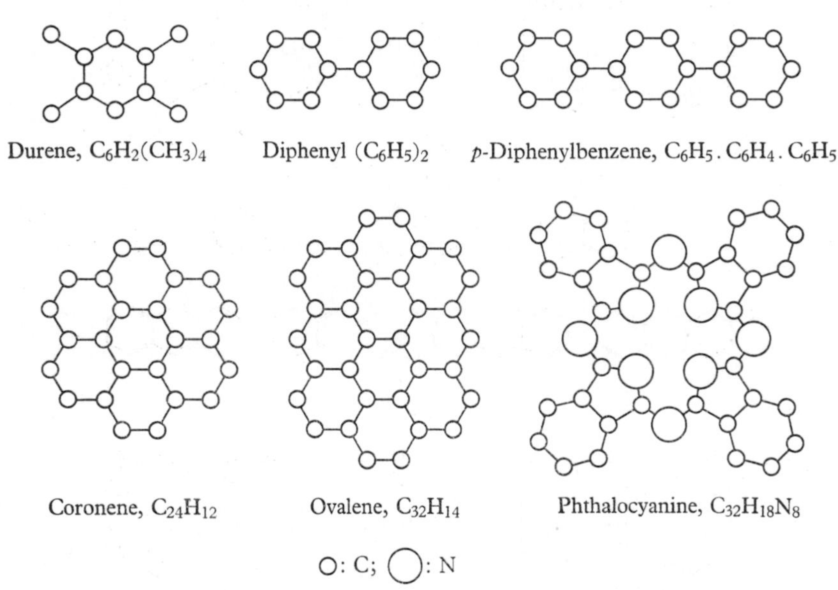

Fig. 14.25. Some planar molecules.

reaches its limit in the structure of graphite (§7.10). Here each sheet may be considered as an infinite condensed-ring molecule with the C–C bonds all equivalent and each of bond number $1\frac{1}{3}$.

14.35. The compound *phthalocyanine*, $C_{32}H_{18}N_8$, is of exceptional interest on chemical grounds in that in combination with the metals Be, Mn, Fe, Co, Ni, Cu and Pt it forms derivatives of composition $C_{32}H_{16}N_8M$ of quite extraordinary stability; the copper compounds, for example, sublimes without decomposition at 580 °C. The structure analysis reveals that the parent metal-free substance and all these derivatives (except that of platinum) are isomorphous and have unit cells of closely the same dimensions. The molecule is strictly planar in each case and the metal atom, if present, is situated at its centre. The formula usually assigned to phthalocyanine on chemical grounds is

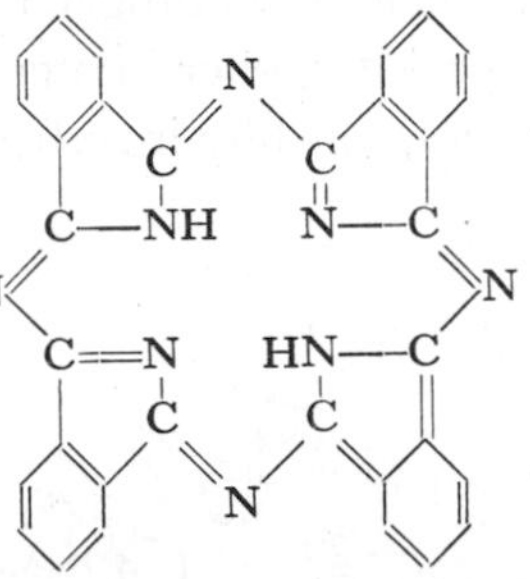

but as it stands this cannot be reconciled in detail with the form of the molecule observed in the crystal. The benzene rings all approximate very closely to regular hexagons, no distinction such as that implied by the formula being found between them, and in fact these rings are necessarily identical in pairs by virtue of the operation of the centre of symmetry which the molecule possesses. Moreover, the bonds within the inner system of sixteen carbon and nitrogen atoms form a very regular arrangement with a practically constant interatomic distance of 1·34 Å, indicating a state of complete single-bond–double-bond resonance in all the C–N bonds. The C–C bonds which connect this inner ring system to the benzene rings are also all equal, and of length 1·47 Å. This distance again indicates resonance, but with the bonds in question more nearly equivalent to single than to double bonds. All these facts are therefore consistent with the view that resonance takes place throughout the entire molecule and that it is to this resonance that its exceptional stability is to be attributed.

In the metal derivatives of phthalocyanine the metal atoms are co-ordinated by four atoms of nitrogen at the corners of a square, and this arrangement suggests that these atoms are bound by covalent dsp^2 hybrid bonds. Such an explanation is acceptable in the case of the transition metals, which readily form dsp^2 bonds, but cannot account for the existence of the beryllium compound. It is, however, notable in this connexion that this derivative is conspicuously less stable than the phthalocyanines of the other metals.

Hydrogen bonded structures

Resorcinol

14.36. The profound influence of hydrogen bonding in determining the way in which molecules pack together in a crystal is well illustrated by the structure of resorcinol, *m*-dihydroxybenzene, $C_6H_4(OH)_2$. This

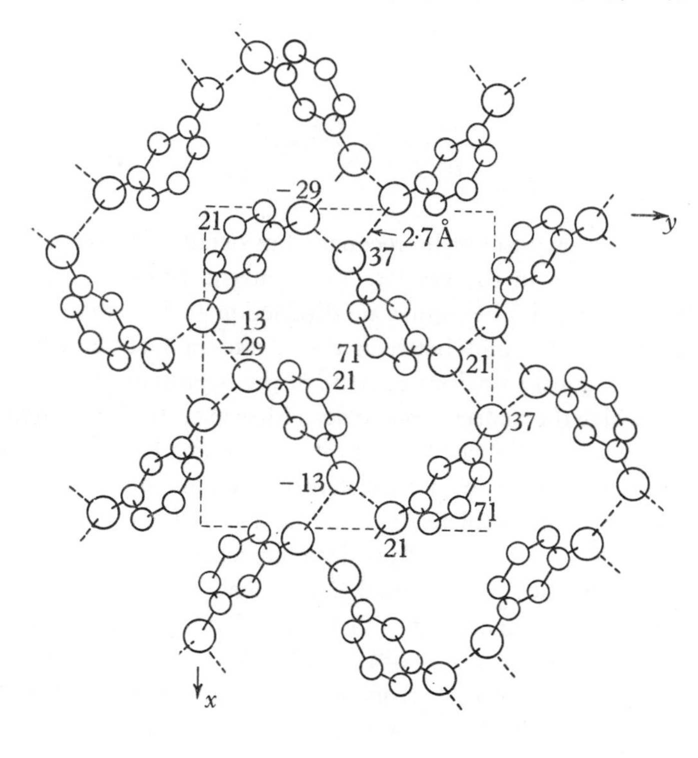

Fig. 14.26. Plan of the unit cell of the orthorhombic structure of α-resorcinol, *m*-$C_6H_4(OH)_2$, projected on a plane perpendicular to the *z* axis. The heights of some of the atoms are expressed in units of $c/100$ and hydrogen bonds between hydroxyl groups are represented by broken lines.

compound is dimorphous, and in the α form, stable at room temperature, the planar molecules are arranged as shown in fig. 14.26. Each hydroxyl group is bound to two others of two different molecules by hydrogen bonds of mean length 2·7 Å, and in this way the molecules are linked into infinite helices running through the structure in a direction perpendicular to the plane of the diagram. Each molecule, however, belongs to two such helices, so that the structure is strongly bonded both laterally and vertically. In spite of the close approach of the hydroxyl groups the carbon atoms of different molecules maintain their usual van der Waals separation of about 3·5 Å, and this gives rise to an exceptionally open structure, of specific gravity only 1·27, strikingly different from the relatively dense packing of flat molecules which obtains in structures where hydrogen bonds are not operative.

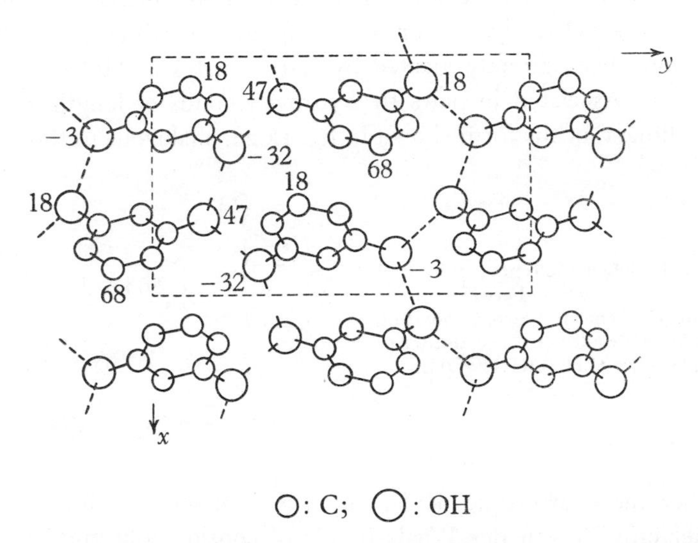

Fig. 14.27. Plan of the unit cell of the orthorhombic structure of β-resorcinol, m-$C_6H_4(OH)_2$, projected on a plane perpendicular to the z axis. The heights of some of the atoms are expressed in units of $c/100$ and hydrogen bonds between hydroxyl groups are represented by broken lines.

In the β form of resorcinol, stable above 74 °C, each hydroxyl group is again linked to two others of two different molecules by hydrogen bonds, but these bonds now form zigzag chains running parallel to the x axis of the crystal (fig. 14.27). Each molecule belongs to two such chains, so that the chains are linked into sheets parallel to the plane of the diagram and the structure as a whole is formed by the superposition of these sheets. In this way an array more compact than that in α resorcinol is formed and the specific gravity of 1·33 is correspondingly higher. It will be noted that the transition from the α to the β form is reconstructive (§9.14) in that a rearrangement of the hydrogen bonds is involved in the change, and it is interesting to observe in this connexion that the transition takes place only slowly at 74 °C. It is also of interest to note that in both the resorcinol structures the principle of maximum hydrogen bonding is satisfied: all the hydrogen atoms of the hydroxyl groups are involved in hydrogen bonds.

Salicylic acid

14.37. The structure of salicylic acid (*o*-hydroxybenzoic acid) $C_6H_4.OH.COOH$, has been determined with exceptional precision and is one of the relatively few organic structures in which the hydrogen atoms have been directly located by X-ray analysis. Molecules in the crystal are associated in pairs by hydrogen bonds of length 2·63 Å to form 'dimers' of the form shown in fig. 14.28, analogous to those found

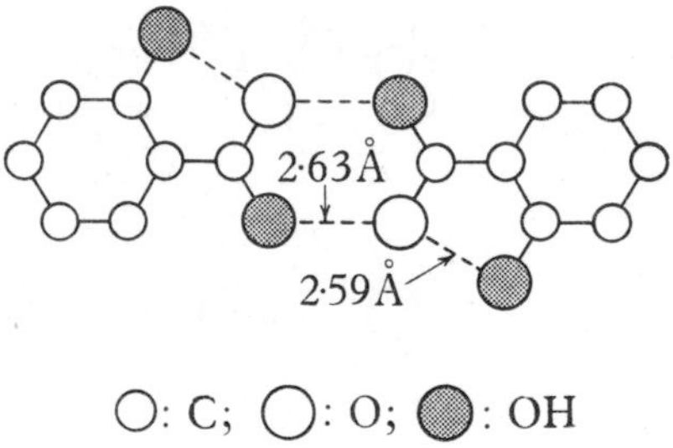

Fig. 14.28. Schematic representation of the hydrogen bonding between molecules and of the chelate hydrogen bonding within the molecules in the structure of salicylic acid, *o*-$C_6H_4.OH.COOH$.

in other monocarboxylic acids, and these dimers are linked to one another only by van der Waals bonds of considerably greater length. There is also, however, a strong intramolecular hydrogen bond, of length 2·59 Å, within each molecule between the hydroxyl group attached to the benzene ring and the oxygen atom of the carboxyl group, the first example we have encountered of such a *chelate* hydrogen bond. It has a profound influence on the chemical properties of salicylic acid; this acid is much stronger than the corresponding *m*- and *p*-hydroxybenzoic acids in which chelation cannot occur.

The dimensions of the salicylic acid molecule, shown in fig. 14.29, serve to illustrate the order of precision with which interatomic distances and interbond angles can be determined in crystal structure analysis. It will be noted that, as in other carboxylic acids, there is a significant difference between the lengths of the C–O and C–OH bond, and also that the positive location of the hydrogen atom of the carboxylic hydroxyl group at a distance of 0·91 Å from the oxygen atom confirms the unsymmetrical character of the hydrogen bond. It will also be seen that the benzene ring is not a perfectly regular hexagon, successive C–C bonds in the ring differing slightly but significantly in length. These and the other observed bond lengths in the molecule can be satisfactorily interpreted on the assumption that the molecular structure is a resonance between the configurations (I), (II) and (III) (fig. 14.30) to which (I) makes the largest contribution.

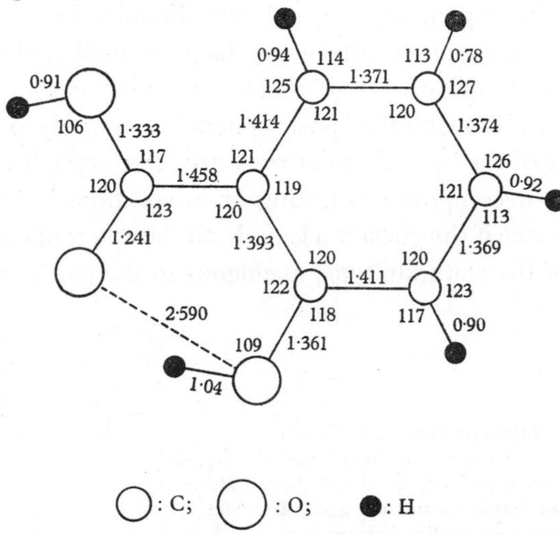

Fig. 14.29. Dimensions of the molecule of salicylic acid, $o\text{-}C_6H_4.OH.COOH$, in its crystal structure. Distances are given in Ångström units and angles in degrees.

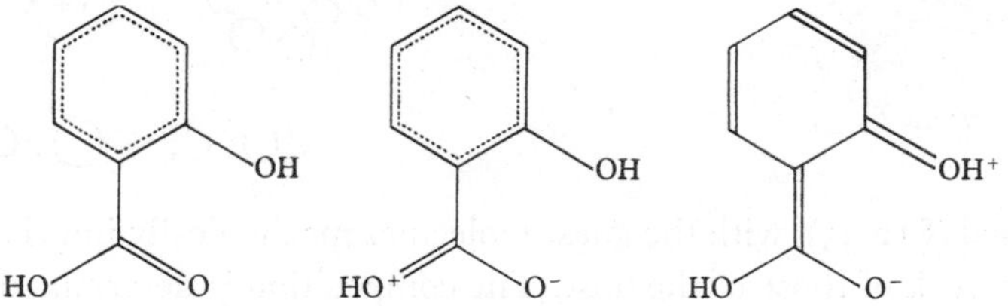

OH OH OH^+

HO O HO^+ O^- HO O^-

I II III

Fig. 14.30. Possible bond configurations in salicylic acid, $o\text{-}C_6H_4.OH.COOH$.

CLATHRATE COMPOUNDS

14.38. The compounds discussed above are but a few of the many hundreds of organic structures which have been analysed. Many more could have been described, but the examples chosen suffice to illustrate adequately the structural principles displayed by the majority of organic compounds. A few compounds, however, show new features, and we close with an account of one class of such compounds as an example of a field in which X-ray studies have proved particularly fruitful. The compounds in question exist only in the solid state and therefore do not readily lend themselves to chemical investigation.

14.39. It has long been known that phenol and quinol form crystalline compounds with argon, oxygen, sulphur dioxide, hydrogen chloride, methanol and many other substances having small molecules. These compounds are easily prepared, those containing gases, for example, being formed by allowing the 'host' (phenol or quinol) to crystallize in an atmosphere of the 'guest' under pressure. The crystals are stable and odourless, but decompose on heating or in solution, in the latter case with effervescence if the guest is a gas. Both these sets of compounds are found to be of the clathrate type, analogous to the gas hydrates already

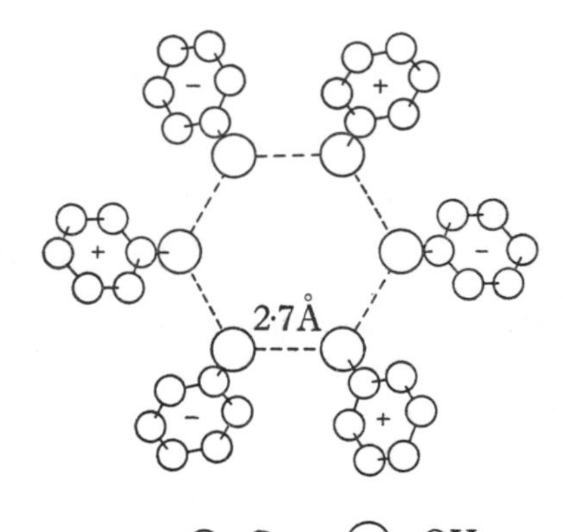

Fig. 14.31. Plan of the clusters of six phenol molecules in the structure of the phenol clathrates. The hydroxyl groups lie at the corners of a plane hexagon and the axes of the phenol molecules are inclined alternately above and below this plane. Hydrogen bonds between hydroxyl groups are represented by broken lines.

discussed (§ 12.35), with the guest molecules mechanically imprisoned in a framework of those of the host. The composition is determined by the number of interstices available in the framework, and only molecules which are small enough to be accommodated in these interstices can act as guests. The quinol compounds were the first to be studied by X-ray methods, but those of phenol are simpler to describe and will be discussed first.

In the *phenol clathrates* molecules of phenol (hydroxybenzene) are linked by hydrogen bonds, of length 2·7 Å, into six-membered clusters of the form shown in fig. 14.31. The hydroxyl groups of the six molecules lie at the corners of a plane hexagon, and the benzene rings are inclined alternately above and below this plane at an angle of about 45°; the molecules are also twisted about their long axes through approximately the same angle. These clusters are then arranged in pairs about the corners of the rhombohedral unit cell in the manner shown schematically in fig. 14.32, and each cell therefore contains twelve phenol molecules. Such an arrangement is not stable for phenol itself, which has a different structure, but is stabilized if the interstices *A*, at the centre of the cell, are occupied by guest molecules. There are, however, also smaller interstices *B*, between clusters of molecules, and these, too, can be occupied. Accordingly, there exist a number of clathrates of different

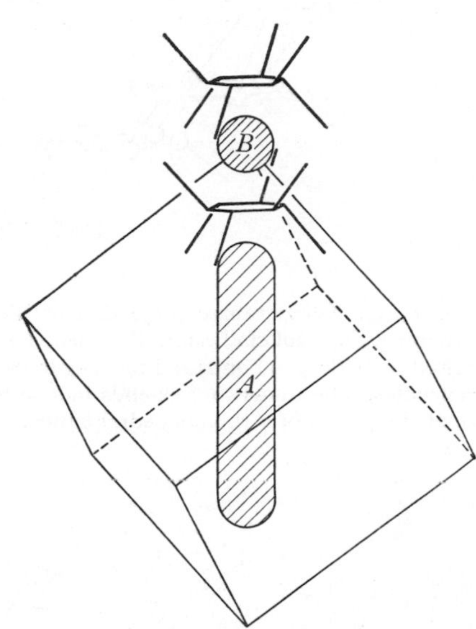

Fig. 14.32. Clinographic projection of the unit cell of the rhombohedral structure of the phenol clathrates. Clusters (fig. 14.31) of six phenol molecules are represented schematically, and these are disposed about the corners of the cell in the manner shown, for clarity, at only one corner. Guest molecules can occupy the interstices *A* at the centres of the cell and the interstices *B* at its corners.

compositions depending on the number of guest molecules which can be accommodated in the interstices *A* and on whether or not the interstices *B* are also populated. In $12C_6H_5OH.5HCl$ interstices *A* and *B* accommodate four and one molecules, respectively, of the guest; in $12C_6H_5OH.4SO_2$ only the *A* interstices are occupied; in $12C_6H_5OH.$-$2CS_2$ it is again the *A* interstices alone which are inhabited, but now there is room in them for only two molecules of carbon disulphide. In double clathrates, *A* and *B* interstices are occupied by different guests.

In the *quinol clathrates* the molecules of quinol (*p*-dihydroxybenzene) are again linked by hydrogen bonds between rings of six hydroxyl groups, but now such linkages take place at both ends of the molecule and bind the molecules into infinite three-dimensional frameworks of the form shown in fig. 14.33. This framework is very open, and in the crystal two such frameworks exist, completely interpenetrating each other but having no direct attachments, as shown schematically in fig. 14.34. (The arrangement has been likened to that of two ladders so constructed that the rungs of one pass through the gaps between those

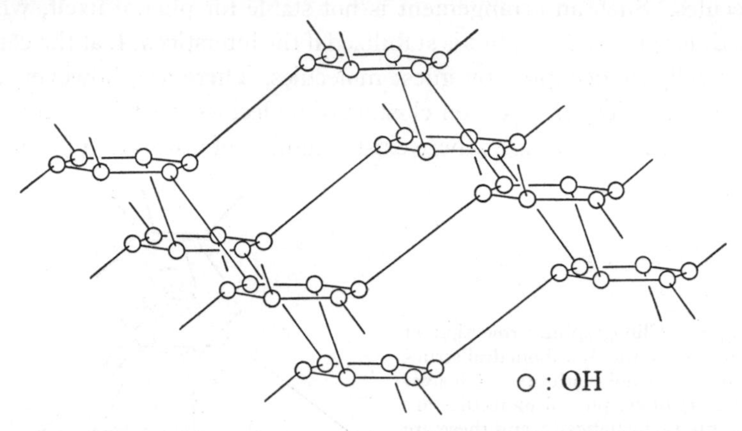

Fig. 14.33. Clinographic projection of the framework of quinol molecules in the structure of quinol clathrates. The plane hexagons represent rings of hydroxyl groups united by hydrogen bonds and the long lines represent the HO—OH axes of the quinol molecules. The framework extends indefinitely in three dimensions and the hexagons lie at the points of a rhombohedral lattice.

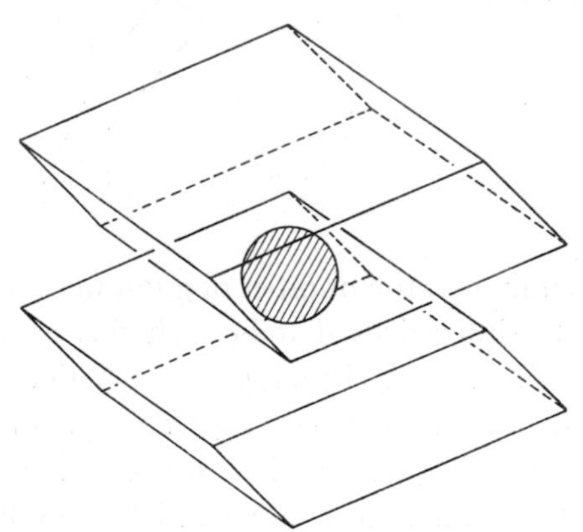

Fig. 14.34. Clinographic projection representing schematically the two interpenetrating frameworks in the structure of the quinol clathrates. The rhombohedra represent two independent frameworks, each of the form shown in fig. 14.33. Guest molecules occupy the interstices indicated between these frameworks. Both frameworks of course extend indefinitely but, for clarity, only one rhombohedron of each is shown.

of the other.) The guest component is then trapped in the interstices between the two frameworks, and the fact that this component may be an atom of an inert gas is evidence that the constraint is purely mechanical.

14.40. The formation of clathrates is yet another example of the profound influence of the hydrogen bond in determining the structure of a crystal. Clathrate compounds can exist only if the host component can form a coherent structure open enough to accept the guest molecules. Substances with small or symmetrically shaped molecules, especially if devoid of active groups, tend to form closely packed structures, but if the molecules are awkward in shape, and particularly if they possess groups which can give rise to hydrogen bonds, structures of lower density are formed. It will be noticed that it is just these latter characteristics which are displayed by the host molecules of the clathrate compounds described above.

CONCLUSION

14.41. We have now come to the end of our study of systematic crystal chemistry, and in reviewing the field as a whole we see what a large part of the whole realm of chemical science falls within its scope. In many directions the application of X-ray analysis is as yet very limited, but in almost every branch it has been sufficiently widely applied to indicate the broad outlines of the results to be expected and to establish beyond doubt its value as the most powerful tool yet available for the investigation of the solid state of matter. The results of X-ray analysis provide a welcome co-ordination between physical and chemical concepts, for on the one hand they afford a structural basis in terms of which observed physical properties may be explained, while on the other they furnish the chemist with a more rational scheme for the classification of his compounds, based on the physical nature of the forces responsible for their coherence. Indeed, a clear recognition of the different types of chemical binding and of their physical origin, together with an appreciation of the essentially geometrical basis of crystal architecture, are the most important general conclusions which have emerged from the systematic application of X-ray structure analysis.

On the purely theoretical side the results of crystal structure analysis have had no less important applications, and it is only necessary to refer back to our earlier chapters to appreciate that the recent developments of the quantum mechanics as applied to problems of chemical combination have very largely had their origin in the material made available by X-ray analysis. It is not unreasonable to expect that in the future no less than in the past continuous and rapid progress will attend the co-operation of the chemist, the theoretical physicist and the X-ray crystallographer.

APPENDIX 1
BIBLIOGRAPHY

Reference Works

DONNAY, J. D. H. (ed.). *Crystal Data.* Washington: American Crystallographic Association (Monograph 5). 2nd ed. 1963. A list of crystalline substances classified in terms of space group and cell dimensions.

GROTH, P. *Chemische Krystallographie.* Leipzig: Engelmann. 1906–19. Morphological and optical data on many crystalline substances.

Landolt–Börnstein, Zahlenwerte und Funktionen. Vol. 1, part 4. Berlin, Göttingen, Heidelberg: Springer. 1955. Cell dimensions, structures and physical properties of many crystalline substances.

PEARSON, W. B. *A Handbook of Lattice Spacings and Structures of Metals and Alloys.* London: Pergamon Press. 1958.

Structure Reports. Vols. 8– . Utrecht: Oosthoek (for the International Union of Crystallography). 1951– . A continuation of *Strukturbericht* in which the reports are even more comprehensive and detailed and in which they are supplemented by editorial comment. At the time of going to press Vols. 8–18, covering the years 1940–54, are available. Publication continues.

Strukturbericht. Vols. 1–7. Leipzig: Akademische Verlagsgesellschaft. 1931–43. A detailed record of *all* crystal-structure determinations published in the period 1913–39.

STRUNZ, H. *Mineralogische Tabellen.* Leipzig: Akademische Verlagsgesellschaft. 3rd ed. 1957. Cell dimensions and other data on minerals.

SUTTON, L. E. (ed.). *Tables of Interatomic Distances and Configurations in Molecules and Ions.* London: Chemical Society (Special Publication 11). 1958. A comprehensive list of interatomic distances and interbond angles determined by X-ray and other methods.

WYCKOFF, R. W. G. *The Structure of Crystals.* New York: Chemical Catalog Company. 2nd ed. 1931. A record of the more important crystal-structure determinations published up to 1930.

WYCKOFF, R. W. G. *The Structure of Crystals.* New York: Reinhold Publishing Corporation. 1935. A supplement to the preceding work covering the years 1930–34.

WYCKOFF, R. W. G. *Crystal Structures.* New York, London: Interscience Publishers. 1948–60. A record of the more important crystal-structure determinations published up to 1960. This and the two preceding works are distinguished particularly for their excellent illustrations.

History (Chapter 1)

BRAGG, W. L. *History of X-ray Analysis.* New York, London, Toronto: Longmans Green. 1943.

SPENCER, L. J. *Encyclopaedia Britannica*, vol. 7, p. 569. New York: Encyclopaedia Britannica Inc. 11th ed. 1929.

Lattice Theory (Chapters 2–6)

BORN, M. and HUANG, K. *Dynamical Theory of Crystal Lattices*. Oxford: Clarendon Press. 1954.

BROWN, G. I. *Modern Valency Theory*. London, New York, Toronto: Longmans Green. 1953.

BROWN, G. I. *Electronic Theories of Organic Chemistry*. London, New York, Toronto: Longmans Green. 1958.

CARTMELL, E. and FOWLES, G. W. A. *Valency and Molecular Structure*. London: Butterworths. 1956.

COTTRELL, A. H. *Theoretical Structural Metallurgy*. London: Arnold. 2nd ed. 1955.

COULSON, C. A. *Valence*. Oxford: Clarendon Press. 2nd ed. 1961.

HUME-ROTHERY, W. *Atomic Theory for Students of Metallurgy*. London: Institute of Metals (Monograph 3). 4th revised reprint. 1962.

JONES, H. *The Theory of Brillouin Zones and Electronic States in Crystals*. Amsterdam: North Holland. 1960.

KETELAAR, J. A. A. *Chemical Constitution*. Amsterdam, London, New York, Princeton: Elsevier. 2nd ed. 1958.

MOTT, N. F. and GURNEY, R. W. *Electronic Processes in Ionic Crystals*. Oxford: Clarendon Press. 2nd ed. 1948.

PALMER, W. G. *Valency*. Cambridge: University Press. 2nd ed. 1959.

PAULING, L. *The Nature of the Chemical Bond*. Ithaca: Cornell University Press; London: Oxford University Press. 3rd ed. 1960.

RICE, F. O. and TELLER, E. *The Structure of Matter*. New York: Wiley; London: Chapman and Hall. 1949.

SPEAKMAN, J. C. *An Introduction to the Electronic Theory of Valency*. London: Arnold. 3rd ed. 1955.

Systematic Crystal Chemistry (Chapters 7–14)

ARKEL, A. E. VAN. *Molecules and Crystals in Inorganic Chemistry*. London: Butterworths. 2nd ed. 1956.

BARNETT, E. DE B. and WILSON, C. L. *Inorganic Chemistry*. London: Longmans Green. 2nd ed. 1957.

BARRETT, C. S. *Structure of Metals*. New York, Toronto, London: McGraw-Hill. 2nd ed. 1952.

BIJVOET, J. M., KOLKMEYER, N. H. and MACGILLAVRY, C. H. *X-ray Analysis of Crystals*. London: Butterworths. 1951.

BRAGG, W. L. *Atomic Structure of Minerals*. London: Humphrey Milford. 1937.

BRANDENBERGER, E. and EPPRECHT, W. *Röntgenographische Chemie*. Basel, Stuttgart: Birkhäuser. 2nd ed. 1960.

BUNN, C. W. *Chemical Crystallography*. Oxford: Clarendon Press. 2nd ed. 1961.

COTTON, E. A. and WILKINSON, G. *Advanced Inorganic Chemistry*. New York, London: Wiley. 1962.

COTTRELL, A. H. *Dislocations and Plastic Flow in Crystals*. Oxford: Clarendon Press. 1953.

ELCOCK, E. W. *Order–Disorder Phenomena*. London: Methuen; New York: Wiley. 1956.

EMELÉUS, H. J. and ANDERSON, J. S. *Modern Aspects of Inorganic Chemistry*. London: Routledge and Kegan Paul. 3rd ed. 1960.

GILREATH, E. S. *Fundamental Concepts of Inorganic Chemistry*. New York, Toronto, London: McGraw-Hill. 1958.

HALLA, F. *Kristallchemie und Kristallphysik metallischer Werkstoffe*. Leipzig: Barth. 3rd ed. 1957.

HESLOP, R. B. and ROBINSON, P. L. *Inorganic Chemistry*. Amsterdam, London, New York, Princeton: Elsevier. 1960.

HUME-ROTHERY, W. *Elements of Structural Metallurgy*. London: Institute of Metals (Monograph 26). 1961.

HUME-ROTHERY, W. and RAYNOR, G. V. *The Structure of Metals and Alloys*. London: Institute of Metals (Monograph 1). 4th ed. 1962.

KITAIGORODSKII, A. I. *Organic Chemical Crystallography*. New York: Consultants Bureau. 1961.

KLEBER, W. *Kristallchemie*. Leipzig: Teubner. 1963.

LEWIS, J. and WILKINS, R. G. *Modern Co-ordination Chemistry*. New York, London: Interscience Publishers. 1960.

NYBURG, S. C. *X-ray Analysis of Organic Structures*. New York, London: Academic Press. 1961.

READ, W. T. *Dislocations in Crystals*. New York, London, Toronto: McGraw-Hill. 1953.

REES, A. L. G. *Chemistry of the Defect Solid State*. London: Methuen; New York: Wiley. 1954.

ROBERTSON, J. M. *Organic Crystals and Molecules*. Ithaca: Cornell University Press. 1953.

TAYLOR, A. *X-ray Metallography*. New York, London: Wiley. 1961.

WELLS, A. F. *Structural Inorganic Chemistry*. Oxford: Clarendon Press. 3rd ed. 1962.

WINKLER, H. G. F. *Struktur und Eigenschaften der Kristalle*. Berlin, Göttingen, Heidelberg: Springer. 2nd ed. 1955.

APPENDIX 2

THE REPRESENTATION OF CRYSTAL STRUCTURES

The unit cell

1. The structure of a crystal is characterized by the fact that it is formed by the indefinite repetition in three dimensions of the contents of a parallelopiped, termed the *unit cell*; if the contents of one unit cell is known, the structure of the whole crystal is given by stacking identical cells in parallel orientation in such a way that each corner is common to eight of these cells. (In an analogous way the pattern of a wallpaper is formed by the indefinite repetition in two dimensions of the design contained within a unit parallelogram.) Unit cells in crystals commonly have linear dimensions of the order of 10 Å, and contain a small whole number of formula units.

Choice of cell

2. Unit cells (all of the same volume) can be chosen for any crystal in an infinite number of different ways, but in practice it is convenient to select a unit cell whose shape conforms to the symmetry of the system to which the crystal belongs. Thus the edges of the cell, of lengths a, b and c, are taken parallel to the crystallographic axes x, y and z, and the angles between the pairs of edges bc, ca and ab are the interaxial angles α, β, and γ respectively. With edges chosen in this way the unit cells appropriate to the several systems are determined by the following relationships:

System	Relationship
Triclinic	$a \neq b \neq c;\ \alpha \neq \beta \neq \gamma$
Monoclinic	$a \neq b \neq c;\ \alpha = \gamma = 90^\circ,\ \beta \neq 90^\circ$
Orthorhombic	$a \neq b \neq c;\ \alpha = \beta = \gamma = 90^\circ$
Trigonal* / Hexagonal*	$a = b \neq c;\ \alpha = \beta = 90^\circ,\ \gamma = 120^\circ$
Tetragonal	$a = b \neq c;\ \alpha = \beta = \gamma = 90^\circ$
Cubic	$a = b = c;\ \alpha = \beta = \gamma = 90^\circ$

* It is sometimes more convenient to describe trigonal and hexagonal structures in terms of a rhombohedral cell with $a = b = c$, $\alpha = \beta = \gamma \neq 90^\circ$.

It follows from these conventions governing the choice of unit cell that the most convenient cell is not necessarily the smallest possible cell. Cubic structures, for example, are always described in terms of a unit cell which is a cube, even when, as is sometimes the case, a rhombohedral cell of smaller volume could be selected.

The clinographic projection

3. It is clear from what has been said above that in representing the structure of a crystal it is sufficient to represent the contents of a single unit cell, and this is the practice normally adopted. If the structure is simple the clearest representation is the *clinographic projection*, in which the structure is shown as viewed in a convenient direction from a point at infinity. Fig. 3.01 (p. 33) is such a projection of the unit cell of the structure of sodium chloride in which the centres of the sodium and chlorine ions are represented by conventional symbols. The black circles (say, sodium) occupy the centre of the cube and the mid-points of each of its twelve edges; the white circles (chlorine) occupy the corners and the mid-points of its six faces. It is easy from such a diagram to compute the quantity of matter in the unit cell. One sodium ion lies wholly within the cell and each of the other twelve is shared between four cells, giving $1+\frac{1}{4}\times 12 = 4$ ions. Each of the eight chlorine ions at the corners of the cell is shared between eight cells and each of the six at the centres of the faces is shared between two cells; the number of chlorine ions is therefore $\frac{1}{8}\times 8+\frac{1}{2}\times 6 = 4$. The unit cell thus contains four units of NaCl.

4. An important feature of the clinographic projection is that parallel lines in space are also parallel in projection; the receding edges of the cube in fig. 3.01, for example, are strictly parallel and do not converge towards infinity. For crystallographic purposes this is a valuable property because it enables directions to be readily appreciated and interrelated. In the structure of diamond (fig. 4.02, p. 62) the four bonds from any one of the carbon atoms will be found to be parallel to the four diagonals of the cubic unit cell and we may therefore at once conclude that they are directed from the centre to the corners of a regular tetrahedron.

5. Although it is always sufficient to represent only the contents of a single unit cell it is, nevertheless, often convenient to show also some of the atoms in adjacent cells. In the structure of sodium chloride it is easy to see (by considering the extended structure) that the co-ordination about all the sodium atoms is octahedral, but in that of molybdenum sulphide (fig. 8.14, p. 160) the co-ordination about the molybdenum atoms is far less obvious and is clearly appreciated only if some of the atoms in neighbouring cells are also represented. Similar considerations apply to some of the other structures represented in this book (e.g. figs. 8.02, p. 142; 8.15, p. 161; 8.17, p. 163; etc.).

6. It is desirable in the representation of a crystal structure to distinguish between interatomic bonds and lines (such as cell edges) which have no structural but only a geometrical significance. In this book, wherever practicable, interatomic bonds have been represented by full lines and cell edges by broken lines. In some structures cell edges and interatomic bonds coincide (as in sodium chloride), but in others, e.g. diamond, this is not so.

Packing diagrams

7. A clinographic projection showing atomic centres gives a clear representation of the co-ordination about the several atoms but it does not represent the relative sizes of these atoms or the manner in which they pack together. These aspects of the structure can be emphasized by drawing the atoms to scale, as in fig. 3.06 (p. 40), but in this case it will be seen that the co-ordination is far less clearly revealed. It is, however, generally of more importance to emphasize co-ordination rather than packing, and for this reason atomic-centre projections are usually to be preferred.

Atomic co-ordinates

8. Crystal structures may be conveniently described by quoting the co-ordinates of the several atoms in the unit cell, the co-ordinates being referred to the crystallographic axes and being expressed as fractions of the translations a, b and c parallel to these axes. The sodium chloride structures of fig. 3.01 would then be represented thus:

$$\text{Na:}\quad \tfrac{1}{2}, 0, 0;\quad 0, \tfrac{1}{2}, 0;\quad 0, 0, \tfrac{1}{2};\quad \tfrac{1}{2}, \tfrac{1}{2}, \tfrac{1}{2}$$
$$\text{Cl:}\quad 0, 0, 0;\quad 0, \tfrac{1}{2}, \tfrac{1}{2};\quad \tfrac{1}{2}, 0, \tfrac{1}{2};\quad \tfrac{1}{2}, \tfrac{1}{2}, 0$$

It will be noted that in this representation the number of ions of each kind is equal to the number within the volume of the cell; it is not necessary to quote all the ions shown in the illustration since an ion at, say, $(\tfrac{1}{2}, 0, 0)$ is necessarily repeated also at $(\tfrac{1}{2}, 1, 0)$, $(\tfrac{1}{2}, 0, 1)$ and $(\tfrac{1}{2}, 1, 1)$.

Atomic co-ordinates are not, of course, always simple fractions of the cell translations. In the tetragonal structure of rutile (fig. 8.05, p. 148), for example, the co-ordinates of the atoms are

$$\text{Ti:}\quad 0, 0, 0;\quad \tfrac{1}{2}, \tfrac{1}{2}, \tfrac{1}{2},$$
$$\text{O:}\quad \pm(u, u, 0);\quad \pm(\tfrac{1}{2}-u, \tfrac{1}{2}+u, \tfrac{1}{2}),$$

where $u = 0{\cdot}30$, while in the hexagonal structure of selenium (fig. 7.03, p. 124) the co-ordinates are

$$\text{Se:}\quad 0, -u, 0;\quad -u, 0, \tfrac{1}{3};\quad u, u, \tfrac{2}{3},$$

where $u = 0{\cdot}22$. In this latter example it must be remembered that the co-ordinates are expressed in terms of translations parallel to axes which, in this instance, are not orthogonal. In many structures some or all of the atoms occupy general positions (u, v, w), so the description of a structure containing n atoms in the unit cell may require the specification of as many as $3n$ positional parameters.

Plans

9. The clinographic projection shows clearly the atomic arrangement in a simple structure, but in structures of more complexity it is likely to be difficult to interpret. In such structures the atomic arrangement is usually revealed

more clearly in a plan of the unit cell in which the atoms are shown as projected on to a plane perpendicular to one of the crystallographic axes or to some other convenient direction. In such a projection two of the co-ordinates of each atom are explicit; the third, representing the height above the plane of projection, is conveniently indicated as a fraction of the cell translation in the direction in question. Clearly, the length of this translation, relative to those in the plane of projection, must be known if the structural arrangement is to be fully appreciated.

Fig. 8.13 (p. 159) is a representation of the carbon dioxide structure in both clinographic projection and plan. For a structure of this degree of complexity either presentation is acceptable, but for more complex structures the latter is to be preferred. Fig. 14.03 (p. 359), for example, reveals very clearly the atomic arrangement in the structure of tetraphenylmethane but a clinographic projection of the same structure would be well nigh incomprehensible. Sometimes two plans, projected in different directions, are desirable (fig. 14.22, p. 385), and occasionally a section through a structure may give the clearest representation of some feature which it is desired to emphasize (fig. 14.05*b*, p. 362).

INDEX

Substances not mentioned specifically may be traced under general references, e.g. ABX_3 *compounds, Elements, Halides, etc.*